CAD/CAM AND MIS IN JAPAN

Computer Applications in Japanese Industry

CAD/CAM AND MIS IN JAPAN

Computer Applications in Japanese Industry

Edited by

NOBURU TAKAGI

Union of Japanese Scientists and Engineers

and

JOHN BRANCH

Prosys Technology Ltd, Cambridge, England

1986

ACADEMIC PRESS

Harcourt Brace Jovanovich, Publishers

London Orlando San Diego New York Austin
Boston Sydney Tokyo Toronto

ACADEMIC PRESS INC. (LONDON) LTD.
24–28 Oval Road
London

U.S. Edition published by
ACADEMIC PRESS INC.
Orlando, Florida 32887

British Library Cataloguing in Publication Data

CAD/CAM and MIS in Japan: computer
applications in Japanese industry.
1. Industry—Japan—Data processing
I. Takagi, Noboru II. Yajima, Keiji
III. Branch, John
338′.06 HD45.2

ISBN 0-12-682580-7

Filmset by Eta Services (Typesetters) Ltd., Beccles, Suffolk
Printed in Great Britain by St Edmundsbury Press Ltd.,
Bury St Edmunds, Suffolk.

List of Contributors

According to company policy, a name of a contributor is sometimes omitted. Therefore the person from whom the reader may obtain further information is indicated parenthetically.

1 Toshihiko Hisada, Executive Director and Director of Institute of Construction Technology, Kajima Corporation, 1-2-7 Motoakasaka, Minato-ku, Tokyo, 107 Japan

2 Akira Masumoto, Elevator Engineering Department, Mitsubishi Electric Corporation, 2-2-3 Marunouchi, Chiyoda-ku, Tokyo, 100 Japan

3 Sazuku Tanaka, Masatoshi Yanase, Kanjiro Takahashi, Yoshihiro Ishida, Hitoshi Kurioka, Shuichi Nakayama, Information Processing Centre, Kajima Corporation, 1-2-7 Motoakasaka, Minato-ku, Tokyo, 107 Japan

4 Katsumi Takizawa, Institute of Design Research, Chief Researcher, Shimizu Construction Co Ltd, Sakai-Building 2-2-12 Kyobashi, Chuo-ku, Tokyo, 107 Japan

5 (Ryotaro Kanari, Office of Energy Conservation Engineering, Takenaka Komuten Co Ltd, 8-21-1 Ginza, Chuo-ku, Tokyo, 104 Japan)

6 (Michihiro Kohguchi, Industry Marketing, Distribution and Mass-media Department, IBM, Japan Ltd, 3-2-12 Roppongi, Minato-ku, Tokyo, 106 Japan)

7 Sachie Moriya, Distribution Control Centre, Tatuo Yamazaki, Plant Construction Division, Tokyo Gas Corporation, 1-5-20 Kaigan, Minato-ku, Tokyo, 105 Japan

8 Tamotsu Minakawa, Senior Consultant Manager, Power System Planning and Operation Department, Tohoku Electric Power Co, 3-7-1 Ichiban-cho, Sendai, 980 Japan

9 Toshiaki Tabuchi, Senior Staff, Technical Information Office, Ashigara Research Laboratories, Fuji Photo Film Co Ltd, 210 Nakanuma Minami-Ashigara-shi, Kanagawa, 250-01 Japan

10 Toshio Sunada, Manager, Management Systems Development Department, Systems Engineering Group, Fujitsu, Ltd, 1015 Kami-Kodanaka, Nakahara-ku, Kawasaki, 211 Japan

11 Masao Itoh, Executive Counsellor, General Manager, Computer System Division, Koji Yamada, Senior Manager, Computer System Division, Tomoshige Miya, Manager, Computer System Division, Nippon Steel Corporation, 2-6-3 Otemachi, Chiyoda-ku, Tokyo, 100 Japan

12 Tozo Hikichi, General Manager, Information Systems Administration Department, Hitachi Ltd, 4-6 Kanda-surugadai, Chiyoda-ku, Tokyo, 101 Japan

13 (Tadao Ito, System Centre Manager, Toray Industries, Inc. Mitsui No. 5 Annex-building, 2-6 Nihonbashi-Muromachi, Chuo-ku Tokyo, 103 Japan)

14 (Mitsuo Nakamura, Management Development Manager, Okamura Corporation, Tenri Building, 19th Floor, 1-4-1 Kitasaiwai, Nishi-ku, Yokohama, 220 Japan)

15 Koji Itakura, General Manager, System Development Department, Kobe Steel Ltd, 1 Wakinohama-cho, Chuo-ku, Kobe, 651 Japan

16 (Hirota Tamura, Koji Fujino, Hideo Uehara, Technical Department, Oita Works, Showa Denko K. K., 2 Nakanosu, Oita, 870-01 Japan)

17 Shinichi Itch, Associate Senior Vice President and Director, Nippon Electric Co Ltd, 5-33-1 Shiba, Chiyoda-ku, Tokyo, 108 Japan

18 (Minoru Ogura, Engineering Administration and Factory Information Systems Department, Fuchu Works, Toshiba Corporation, 1 Fuchu Toshiba-cho, Fuchu-shi, Tokyo, 183 Japan)

19 Toshio Doshida, Manager, Small Rotating Machine Department, Mie Works, Kazunobu Maruyama, Manager, Development Department, Heavy Apparatus Standard Products Division, Toshiba Corporation, 2121 Nao, Asahi-cho, Mie-gun, Mie, 512 Japan

20 Takeshi Nojiri, Executive Director, Yoshitsugu Odagiri, General Manager, Mikawa-Mizuho Plant, Toshiaki Ishihara, Manager, Production Engineering Department, Mikawa-Mizuho Plant, Minoruta Camera Co, 3-22-1 Honohara-cho, Toyokawa, Aichi, 442 Japan

21 Norimasa Kamii, Director, Plant Engineering Division, Engineering Business Operation Bureau, Ryoichi Mohri, General Manager, Equipment Department, Muroran Works, Kohji Inazaki, Manager, Electrical and Process Automation Section, Muroran Works, Osamu Ishihara, Manager, Plant Engineering and Technology Bureau, Nippon Steel Corporation, 2-6-3 Otemachi, Chiyoda-ku, Tokyo, 100 Japan

22 Takehiko Toyoda, General Manager, Katsuhiro Saito, Staff Manager, Eiichi Kawamura, Assistant Staff Manager, System Planning Department, Kashima Steel Works, Sumitomo Metal Industries, Ltd, Hikari, Kashima-cho, Kashima, Ibaraki, 314 Japan, Hiroyuki Tokuyama, Assistant General Manager, Control & OR Technology Department, Sumitomo Metal Industries Ltd, 1-3 Nishinagasu-Hondori, Amagasaki-shi, 660 Japan

23 Ikuo Nishioka, Takuji Kurimoto, Hisao Nishida, Central Research Laboratory, Sharp Corporation, Ichinomoto-cho, Tenri-shi, Nara, 632 Japan, Isao Shirakawa, Hiroshi Ozaki, Department of Electronic Engineering, Osaka University, Suita, Osaka, 565 Japan

24 (Makoto Oshima, General Manager of System Department, Kawasaki Steel Corporation, 2-2-3 Uchisaiwai-cho, Chiyoda-ku, Tokyo, 100 Japan)

25 Yukihide Hattori, Yoshihiko Ikeda, Kyuma Haga, Information Systems Department, Nippon Kokan KK, 1-1-2 Marunouchi, Chiyoda-ku, Tokyo, 100 Japan

26 Shuichi Hasegawa, Director, Engineering Division, Yoshihisa Hanamura, Assistant Manager, Engineering Division, Kenichi Kyoda, Engineering Division, Yokogawa Bridge Works, Ltd, 4-4-44 Shibaura, Minato-ku, Tokyo, 108 Japan

27 Noboru Tanaka, Tatsuo Mii, Tamano Laboratory, Tsuneo Yoshiki, Ship Design Department, Mitsuo Ito, Systems Headquarters, Takasaki Iwaki, Akihiko Maeda, Tamano Laboratory, Tadashi Ishii, System Department, Mitsui Engineering & Shipbuilding Co Ltd, 5-6-4 Tsukiji, Chuo-ku, Tokyo, 104 Japan

Foreword

Barely a century has passed since the Emperor Meiji encouraged overt contact between Japan and the Western world, although even under the Tokugawa Bakufu and Japan's long period of isolation, there was still interest in the foreigner and his skills.

As early as the beginning of this century, Japan gave notice of its emerging industrial power at Port Arthur, and it was some forty years later that a similar pre-emptive strike at Pearl Harbour precipitated the entry of the United States into the War in the Pacific.

Since the end of the Second World War, Japan has first of all challenged, and then surpassed, the Western world in many of their traditional areas of manufacture such as ship building, and the production of cars, motor cycles, cameras and watches.

Furthermore, Japan has also secured a significant percentage of the free world market in calculators, television sets and hi-fi equipment.

All this has been achieved by a nation possessing no natural oil resources. What, then, is the secret of this success? The answer lies partly in the single-minded application of Japanese management and worker alike to the efficient use of invested capital, to increased productivity, and to the pursuit of quality and customer satisfaction. These themes run like a connecting thread through all the diverse papers in this book.

Their belief in harmony and loyalty—which can trace its origins to 6th century Confucianism—may be an important factor in Japan's success, and one which contrasts sadly with the divisive attitudes of unions and management in our own country over the same period. But, perhaps equally significant, are the Japanese traits of pragmatism and flexibility which—with the exception of the Dark Ages—recur time and again throughout their long history from Shotoku Taishi, Nobunaga and Meiji up to the present day.

Strangely, my own deepest impression of present-day Japan was one of Victorian manners, attitudes and ethics—both within the family and within the business world. And yet, when one remembers that the Victorian age marked the zenith of British manufacture and achievement, perhaps the juxtaposition is not entirely without significance. Could it be that in our desire to dissociate ourselves so completely from these solid values we have laid the foundations for our fall from pre-eminence?

It is true that innovations still flow from the Western world, but it is in Japan that they find their most diligent and fertile application. Perhaps in reading the essays which form the subject of this book we shall discern a message for ourselves.

John Branch
Cambridge

Preface

It was in January 1982 when the publishers proposed a book on the Ishikawa Prize lectures.

In July 1982, JUSE approved the publication, and in January 1983 the editorial board was organized. Letters requesting manuscripts were despatched to the prizewinners in August, and the final manuscript was received in April 1984.

Much editorial effort was inevitable to co-ordinate the text and illustrations of the twenty seven chapters. Now that this work is finished, we wish to express our thanks to all the authors, and to the publisher's editors. Considering the international nature of the editorial work, we feel that it would have been difficult to shorten the time to publication appreciably.

This is the first publication on the Ishikawa Prize lectures—which reflect the efforts to modernise Japanese industrial management—and currently there is no book in Japanese on this theme.

The chapters have been grouped into the following categories: (1) Building and architecture, (2) Services, (3) Industrial management, (4) Industrial production, (5) Transport and construction. An historical description of the development of Japanese industry in terms of this classification, written by the economist Mr Akira Kakuta, is given in the Introduction.

Finally, we would like to thank the many people who co-operated in this work, especially Mr Toshihiro Arai of JUSE, Mr Jun'ichiro Minegawa and Dr Conrad Guettler of Academic Press.

Noboru Takagi
Keiji Yajima
Tokyo

Introduction to the Ishikawa Prize

The Union of Japanese Scientists and Engineers (JUSE) established the Ishikawa Prize in 1970 in commemoration of Mr Ichiro Ishikawa who had contributed much to the development of Japanese industry. This prize aims to promote the modernization of industrial management through the adoption of the most up-to-date technology.

JUSE is a non-profit-making foundation, established in 1946 and authorized by the Science and Technology Agency. The purpose of this Institution is to contribute to the development of industry by incorporating new scientific technologies into industrial activities. The soft technologies such as statistical and mathematical methodologies have been dealt with; their application especially to quality control practice has been a major field of endeavour. More recently the emphasis has been upon the application of computer aided design and manufacturing (CAD/CAM) techniques to industrial problems.

Mr Ishikawa (1885–1970) was born in Tokyo, and graduated in the chemical engineering department of Tokyo University in 1906. After the master course, he obtained a position as associate professor in 1911.

In 1915 Kanto Soda (Kanto Sansou, in Japanese)—which was the first chemical company in Japan, and in which his father had been working as a managing director—requested his participation as an engineer, and he resigned his academic position.

He became a managing director of Dainihon Fertilizer (Dainihon Zinzo Hiryo) Co, and in 1937 moved to become a managing director of Nissan Chemical Industries (Nissan Kagaku Kogyo) Co. During the period of the Second World War he worked as a president of the Chemical Industries Association from 1942.

After the war, he became a president of Economic Organization (Keizai Dantai Rengo-Kai), and also managed JUSE as president. In 1956 he became chairman of the boards of Nissan Chemical Industries, and in 1956 he was appointed a committee member of the Atomic Energy Committee of Japan.

He worked in the most difficult era for the reconstruction of Japanese industry, and throughout his life he typified the rationalism common to superior engineers.

In tribute to Mr Ishikawa, Dr W. E. Deming wrote: "It was through Mr. Ishikawa that I had the priviledge of addressing a gathering of 45 executives at the Industry Club in Tokyo on 25 July 1950, and again on 19 August at Hakone, with further meetings in 1952, and afterwards. It was in these meetings, thanks to Mr Ishikawa, that top management learned something about their own responsibilities to ensure success".

This book is the anthology of the Ishikawa Prize lectures delivered between 1970 and 1982, and it reflects the history and development of Japanese industry throughout these years.

K. Yajima
JUSE, Tokyo

Contents

Introduction

CAD/CAM in Japan: An Introduction to Japanese Industry

Akira Kakuta, An Economist

1 Innovation and Structural Change in Japanese Industry

Through the period of rapid growth in the latter half of the 1950s, and the period of reduced growth rate following the oil crisis, the structure of Japanese industry changed fundamentally.

In the first period industry began to change from its strong orientation towards primary industry, to the secondary and tertiary industries. The second period was characterized by increasing emphasis on tertiary industries, combined with a low level growth in secondary industries.

Table 1 shows the change in the number of employees in the three kinds of industries, the growth of the numbers in tertiary industry being due to those in the wholesale, retail sales and services.

There have also been many changes in the major industries. During recon-

CAD/CAM and MIS in Japan
ISBN 0-12-682580-7

Table 1 Structure of employment in industry (%)

Year[a]	Primary	Secondary	Tertiary[b]
1950	48.3	21.9	29.7
1960	32.6	29.2	38.2
1965	24.6	32.0	43.3
1970	19.4	34.0	46.6
1975	13.9	34.1	51.7
1980	10.9	33.5	55.3

[a] As of 1 October.

[b] The tertiary sector includes transportation, communication, electricity, gas and water supply, and public services.

The total for each year is less than 100%. The residual represents "other areas".

Source: National Census.

struction of the Japanese economy, the natural fibre, coal mining and chemical fertilizer industries had priority. But, from the latter half of the 1950s to the early 1960s, there was a rapid growth in the steel, petrochemical, synthetic fibre, and shipbuilding industries, all of which became heavy industries. There was also growth in the assembly industry, centring on home electric products, and the latter half of the 1960s saw the beginning of rapid expansion in the automobile industry.

The oil crisis marked a turning point. Whilst the material industries such as steel, and petrochemicals, and the heavy industries such as shipbuilding were forced to keep their activities at a low level, the electronics industry, with its high growth rate, was an exception. In fact, from 1973 to 1981, industries such as office machines, calculators and instrumentation, electronics, computer, medicine, automobiles and radio and television markedly exceeded the average production growth rate. On the other hand, the growth rate in the shipbuilding, metal products, steel, paper and pulp, and chemical industries was below average. Following the growth in the tertiary industries, the number of specialists, managers, and salesmen rather than workers engaged directly in production, has increased. The number of non-manual workers has also been growing in manufacturing industry, and these trends mark a switch-over to the service industries, and a softening of the economy (Table 2).

Computers have played a prominent role in accelerating these changes. The real use of electronic computers in industry started in the 1960s, and since that time their use has become widespread and both their efficiency and their numbers have increased. The number of computers used in Japan since 1972 are shown in Table 3, and the numbers in each industry are indicated in Table 4.

Table 2 Structure of occupation (1000 persons, %)

	Primary[a]	Secondary[b]	Tertiary[c]	Total
1960	14 172 (32.4)	10 150 (23.2)	19 385 (44.3)	43 719 (100)
1980	5996 (10.8)	14 189 (25.5)	35 362 (63.5)	55 665 (100)

[a] Primary occupation includes agriculture and forest workers.
[b] Secondary includes mining workers, skilled and industrial process workers, and manual workers.
[c] Tertiary includes all others.
Source: Economic Planning Agency, *Economic Year Book*, 1983.

The Ishikawa Prize aims to promote the modernization of industrial management, and the prizes have been awarded to those companies which have achieved outstanding results in the application of computers in the fields of management and production systems.

In the introduction, a general guide is given to the background of those branches of industry which have been given awards in chronological order, from 1970 onwards. Some which relate to management control have been omitted, as these are independent of any specific industrial application. The examples included are the total on-line system by Toray Industries, and the total management information system by Okamura Industries.

2 Construction and Land Development

From the Second World War up until the end of the period of rapid economic growth, Japanese construction industry has experienced extraordinarily rapid development. Its work has been concerned with restoration, mountain and river surveys, road construction according to the national land development plan, dam building, the assembly and construction of industrial factories and residential houses. The industry embraces a very wide area and, because of the great demand, the amount of construction work grew rapidly. Over the same period, technological innovation took the form of the introduction of new construction materials, and the use of large-scale construction machines and new construction methods.

The construction industry, like other industries, is now entering a different phase. Table 5 shows investment in construction, and indicates that investment in private enterprise in particular stagnated after the year 1974. However, investment in public construction, which had been developed positively as a means of combating recession, showed a growth rate of approximately 9% per year over the period 1973 to 1979. From 1979, the growth rate of public construction slowed down due to curbs on annual expenditure. At the

Table 3 Number of computer installations by size

		1972	1975	1976	1977	1978	1979	1980	1981	1982
No of installations	Large	1164	1725	2053	2232	2399	2691	2953	3222	3500
	Medium	3769	5441	5955	6276	6746	7364	8153	9522	11 130
	Small	3790	7337	9327	11 618	14 160	17 033	20 529	26 381	32 565
	Mini	4086	15 592	17 970	20 595	24 827	31 856	40 473	49 098	59 149
	Total	12 809	30 095	35 305	40 719	48 132	58 944	72 108	88 223	106 344
Value (billion yen)	Large	633	1105	1333	1519	1683	1929	2163	2441	2678
	Medium	391	588	627	663	717	778	839	946	1095
	Small	87	157	188	224	268	319	382	487	594
	Mini	25	97	110	127	153	192	240	290	349
	Total	1136	1946	2258	2533	2821	3218	3624	4165	4716

Source: MITI (Ministry of International Trade & Industry), Survey of computer installations and replacement. As of March 30 each year.

Table 4 Number of Computers in Industry

	Number	Value (million yen)
Construction	2416	61 335
Food	3590	82 529
Fibre	2205	49 039
Publication and printing	994	41 644
Chemical and petroleum	3815	178 459
Steel	1081	132 967
Metal	1843	59 752
Machine	2207	86 485
Electric machine	4730	536 842
Transportation machine	1774	201 374
Precision machine	919	47 227
Other manufacturing	3762	76 737
Wholesale and retail	47 786	728 708
Finance	6404	718 159
Securities	305	68 119
Insurance	661	151 825
Transportation, communication and news	3460	128 359
Electricity and gas	464	72 086
Service	8834	406 050
Hospitals and educational institutions	2789	188 428
Governmental offices	3502	619 810
Total (including others)	110 388	4 837 364

Source: As Table 3.
As of 30 June 1982.

same time, private construction still had a low level of growth, so the total growth rate in the period 1979–1982 was only 2% per year.

Although the growth rate of public construction slowed down from 1979, the slowing down of private construction was even greater. The proportion of public construction, which made up 35% of the total in 1972, has recently increased to 40%. This low level of growth in private construction has arisen from a slow-down in the rate of investment in equipment. Moreover, there has been an increase in those types of investment which do not greatly affect the expansion of construction, one example of which is the creation of new electronics firms.

In addition, the construction of residential houses, which made up 30% of the industry, has remained at a low level for a long time. The number of dwellings started in 1973 was 1.905 million, but after that year construction slowed down rapidly: 1.316 million in 1974, 1.356 million in 1975. Although the level rose to 1.5 million in the period from 1976 to 1978, it again declined

Table 5 Investment in Construction (100 million yen)

Fiscal year	Public enterprise	Private enterprise	Total
1970	49 090	97 251	146 341
1971	62 069	104 699	166 768
1972	76 080	138 546	214 626
1973	86 707	199 966	286 673
1974	104 386	189 558	293 944
1975	118 421	197 820	316 241
1976	122 135	219 830	341 965
1977	150 513	237 473	387 986
1978	175 812	251 048	426 860
1979	187 693	291 525	479 218
1980	196 171	298 582	494 753
1981	202 926	299 272	502 198
1982	203 424	306 450	509 874

Source: Ministry of Construction.
Figures for 1982 are estimates.

to a depressed level, with 1.152 million in 1981, and 1.146 million in 1982. The cause of this depression lies in the fact that the number of dwellings was high, which together with the rising price of land and constructional materials, had the effect of producing an upturn in the price of houses. Meeting this depression, the construction industry concentrated its efforts on obtaining overseas projects, the leadership of which has been taken over by major enterprises. The amount of orders received increased very rapidly, from 170 700 million yen in 1973 to 921 500 million yen in 1982.

The situation is the same in the case of land development. Table 6 shows that value in the supply of residential land was attained in 1972; thereafter, the supply decreased very rapidly.

In Japan large-scale land development has been carried out under the control of central government and public authorities, but from about 1960 large construction and real estate companies began to act as powerful developers.

The development of Tama Area (constructed by Tokyu business group) and Izumi new town (constructed by Mitsubishi Estate) has attracted attention. This was followed by a plan for the development of the Japanese Islands by the Government, and the speculative acquisition of land accelerated. However following the oil crisis of 1973, land development receded as a result of the depression; the subsequent decrease in investment from 1973 reflects this situation.

In the areas of construction technology and of land development systems, Ishikawa prizes were awarded to Kajima Construction Co. for computer-aided construction technology in 1971 and for its environmental impact

Table 6 Supply of residential land (hectares)

Fiscal year	Private supply	Public supply	Total
1970	15 600	5700	21 300
1971	17 100	5700	22 800
1972	17 900	5500	23 400
1973	18 300	4500	22 800
1974	14 700	4800	19 700
1975	11 700	4600	16 300
1976	11 000	4000	15 000
1977	9800	3700	13 500
1978	8000	3500	11 500
1979	7600	3700	11 300
1980	7000	3200	10 200
1981	7000	3200	10 200

Source: Ministry of Construction.
Supply total in newly developed and existing town areas. The areas shown are gross (i.e. including roads and amenity areas).

assessment system of sunlight in 1977. In 1980 Shimizu Construction Co Ltd received an award for its land development design system.

3 The Petrochemical Industry

The Japanese petrochemical industry began production in 1958–59, following design of the first stage of a plant complex. Therefore, even though its period of operation has only spanned a quarter of a century, its development rate has been very high. For example, in the case of ethylene plant construction, the initial maximum production was 24 000 tonnes a year. Later, 60 000-tonne plants appeared in 1961, 100 000-tonne plants in 1964, and a 120 000-tonne facility in 1967. Following this trend, the capacity of plants has increased even further since 1969.

Following economies, production costs have decreased rapidly, and the diversification of products and conversion to petrochemical production methods has progressed rapidly. As a result, total petrochemical production in 1965 attained 321 billion yen, and in 1970 it reached 1178 billion yen, which represented 21% of all chemical industry production. In 1981 petrochemical production reached 4730 billion yen, and its share of overall chemical production was more than 50%.

The factors contributing to the high growth rate of the petrochemical industry were: (a) raw naphtha was available at low price and in large quantities; (2) a demand developed for a number of new products based on petro-

Table 7 Trends in ethylene imports and exports (in 1000 metric tonnes)

Calendar year	Production	Export(A)	Import(B)	A − B
1971	3537	677	13	664
1972	3851	783	22	761
1973	4171	616	80	536
1974	4176	621	132	489
1975	3399	730	3	727
1976	3803	672	42	630
1977	3979	800	34	766
1978	4387	793	112	681
1979	4784	588	202	386
1980	4175	466	197	269
1981	3655	446	275	171
1982	3590	402	374	28
1983	3688	337	345	−8

All figures converted to ethylene.
Figures for 1983 cover from January to November.
Source: MITI.

chemicals, such as synthetic resins, synthetic fibres and synthetic rubbers; (c) the cost of production rapidly decreased with expansion in the scale of operation.

The oil crises in 1973 and 1979 triggered a new era in the history of the petrochemical industry. The reasons for this transformation were, primarily, the appearance of a decline in demand, due to the stabilization of growth in the domestic economy and, secondly, a decline in the exports and an increase in the imports of general products. Moreover, because the US petrochemical industry is based on indigenous resources of oil and gas, Japanese industry, following the sudden rise in the price of oil, was at a disadvantage.

The decline in exports and the increase in imports, as shown in Table 7, indicates that in 1983 converted ethylene imports exceeded exports. Although ethylene production recovered in 1978–1979, it declined again after 1980. From 1981 onwards production of ethylene has been less than 4 million tonnes, and the total deficit in the petrochemical industry amounted to 120–130 billion yen in 1981 to 1983.

Industry has concentrated its endeavours on the adjustment of supply and demand, which became imperative in 1982–1983 with the organization of an anti-depression cartel. In 1984 the situation seemed to have improved a little, but equipment in this industry is still over supplied. Although the industry has a production ability of 6.3 million tonnes, actual production has been limited to 4 million tonnes. The industry has also been taking other countermeasures, such as proceeding with a plan to reduce the capacity for vinyl

Table 8 Trends in steel production (in 1000 metric tonnes)

Fiscal year	Raw steel	Standard steel for hot mill press	Special steel for hot mill steel
1970	92 406	67 269	7365
1971	88 441	66 264	6834
1972	102 972	79 772	7 652
1973	120 017	93 496	9281
1974	114 035	87 267	9002
1975	101 613	78 096	8121
1976	108 326	84 222	10 259
1977	100 646	78 194	10 617
1978	105 059	82 333	11 771
1979	113 010	90 654	12 570
1980	107 386	84 906	13 094
1981	103 029	80 446	13 747
1982	96 299	76 350	13 168
1983	100 200	79 491	13 952

Source: MITI.

chloride by 30% and building a joint sales corporation for ethylene and vinyl products.

The petrochemical industry operates complex plants. Consequently, its prime importance is a balanced operation covering raw materials and semi-finished and finished products. In solving this problem Showa Denko K.K. was awarded an Ishikawa prize for its information processing system for an ethylene plant.

4 The Steel Industry

Through the latter half of the 1950s and 1960s the Japanese steel industry enjoyed rapid growth both quantitively and qualitatively. Large companies built continuous iron-works on the coast at competitive cost, and introduced the latest equipment, such as large-scale blast furnaces, Linzer-Drusen converters, new rolling mills and continuous casting machines.

Crude steel production in 1960 was only 23.160 million metric tonnes but reached 41.296 million tonnes in 1965. In 1970 production rapidly increased to 92.406 million tonnes as shown in Table 8 and in 1973 it exceeded 100 million tonnes, reaching 120 million tonnes. The reasons for this rapid growth in steel production were the increase in domestic demand, corresponding to overall industrial growth, and the increase in standard steel exports. The latter was an important factor; exports of ordinary carbon steel were more than

15 million tonnes, in 1970, rising to a peak figure of 32.5 million tonnes in 1976, accounting for nearly 40% of overall production.

With the coming of the oil crises, the rapid growth of Japanese economy ended, domestic demand slowed down and exports decreased. Accordingly, the production of crude steel decreased continuously from 1980 to 1983.

In the domestic markets, there was a stable demand for steel in the automobile industry, but in the construction and shipbuilding industries production was low, which had a great impact on the steel industry. The decrease in exports was due to the slowdown of world economic growth, and the increase in steel exports by newly developing countries. The self-restriction of exports to the United States arose from trade friction between it and Japan, which led to a decline in exports, and an increase in steel imports. In 1981 imports were 1.34 million tonnes; in 1982 imports totalled 1.89 million tonnes, and in 1983 they came to 2.65 million tonnes. For the most part, the countries which exported to Japan were Korea, Taiwan and Brazil.

The rapid increase in the price of oil caused by an upsurge of imported crude oil, and the decline in production brought on by the stagnation of demand, both domestic and foreign, created financial difficulties for steel manufacturers. The steel corporations counteracted with the following measures, including lowering costs. These are detailed below.

(1) The introduction of a rationalized production process, and the consequent lowering of costs by more efficient use of equipment, energy saving and a reduction in material unit costs. Of these measures, the introduction and spread of the continuous casting process has been the most important. Continous casting production was only 21.2% of overall steel production in 1973, but rose rapidly to 46.2% in 1978, 52.0% in 1979, 59.5% in 1980, 70.7% in 1981 and 78.7% in 1982. This shows that the Japanese steel industry has attained a position in the world ranking alongside Sweden and Austria. The continuous casting method brought efficient fuel consumption: 50 000–56 000 kilo calories per tonne compared with 300 000–350 000 kilo calories for the previous non-continuous (moulding ingots and blooming) method.

Before the oil crisis, a lot of heavy oil was blown into blast furnaces to raise productivity. In 1972, an average of 54 kg of oil was used per 1 tonne of pig iron. After subsequent saving the quantity of blown oil fell to 45 kg in 1975, and to 1 kg in 1981. Oil blowing was finally discontinued in all furnaces in August 1982.

The consumption of all kinds of oil including liquefied petroleum gas (LPG) to produce 1 tonne of crude steel amounted to 128 l in 1973 and was reduced to 37 l in 1982.

Progress was also made in recycling the heat energy from coke ovens, blast furnaces and converters. The heat energy consumed in producing 1 tonne of

Table 9 Computer installations and their rental fee in steel companies

	Computer installations	Monthly rental fee (million yen)
1972	409	1197
1973	481	1401
1974	509	1469
1975	550	1551
1976	600	1540
1977	723	1769
1978	744	1832
1979	871	1802
1980	893	1904
1981	984	1995
1982	1112	2128

As of 1 January each year.
Source: Information Processing Committee, Japanese Association of the Steel Industry.

crude steel was 11% less in 1981 than 1973, and oil consumption was at a very low level in the major steel-producing nations of the world.

(2) An increase in the emphasis on high-quality special steel which has extra value, and a strengthening of non-steel production. With special steel there has been a shift to high-quality products, such as coating or surface treating sheet, high tensile strength steel, seamless steel tubes and structural alloy steel for construction. In 1973 special steel amounted to 9% of overall steel production, whereas in 1981 it came to almost 15%.

The expansion of non-steel production includes the development of new materials such as ceramics, amorphous alloys, carbon fibres and powder metallurgy, based on knowledge accumulated from metal refining, fireproof materials and material test technology.

(3) The introduction of new, intensive production methods to replace low productivity equipment, and the concentration of production to a few efficient iron works. The largest manufacturer, Nippon Steel Corporation, which has many iron works, sees the problem of intensification as one of its main priorities, and is promoting the establishment of a new organization with a reduced labour force able to operate at a time when total domestic production of crude steel is less than 100 million tonnes.

Rationalization procedures for production management involving the use of computers in the steel industry began in the latter half of the 1960s; these have played a major role in future developments. The adoption of computers in the steel industry is detailed in Table 9. They have been widely used in many areas such as labour control, payroll, and in business applications including financial analysis, accounting, order acceptance and sales, shipping and transportation control, work recording, scientific calculation and process control.

Many Ishikawa prizes have been awarded to the steel industry. In 1970, the Nippon Steel Corporation received the first prize for the Kimitsu steel plant information processing system. In 1973, the Japanese Association of the Steel Industry won the prize for rationalization of acceptance inspection of raw material, the content of which is omitted in this book. In 1979, Sumitomo Metal Industries Ltd was awarded the prize for its production control system for a large shape mill. In 1980, the Nippon Steel Corporation received its second prize for a special steel bar production system, and in 1981 the Kobe Steel Corporation won the award for data analysis in Kakogawa Steel Works.

5 The Shipbuilding, Steelframe and Bridge Industries

After the Second World War, the Japanese shipbuilding industry adopted western technology, and in a relatively short time Japan became one of the leading countries in this field. The development was supported by government policies, such as the scheduled shipbuilding plan which provided scope for the promotion of shipping. Cheap prices and short delivery times contributed to the obtaining of overseas orders, and in 1956 the total tonnage launched reached 1.74 million gross tonnes, which exceeded that of Britain, making Japanese shipbuilding first in world production.

The development of the industry was especially remarkable during the initial period of from 1954 to 1958, and subsequently from 1963 to 1964. The Japanese shipbuilding industry achieved nearly 40% of the world market, and maintained its share at between 47 and 50% from 1966 to 1976.

Technological progress in the Japanese shipbuilding industry centred on welding methods and the block building system. These had been established very rapidly from the latter half of the 1950s. Emphasis was later placed on automatic welding, the implementation of large-scale blocks and on the increase of building equipment, shipyards and docks.

The shipbuilding industry has two fundamental features: it is highly technologically-oriented, and an expensive labour force. The major advantages of Japanese industry lie in its technological expertise and relatively low labour costs. The latter is a very important contributing factor in Japan's dominant position in international competition. The shortening of delivery times, which was achieved by an ingenious arrangement of process steps, was instrumental in obtaining a large number of orders from abroad.

With the technical progress of the block-wise building system, and the introduction of diesel engine technology in the latter half of the 1950s, Japanese shipbuilding technology attained the highest level in the world by the early 1960s. Also, during the same period, the development of more economical ships advanced alongside the scaling-up of shipbuilding equipment.

In the middle of the 1960s there was further construction of large docks, the shipbuilding companies competitively building 300–700-tonne deadweight docks. One million-tonne docks were subsequently built in 1972, but these were completed during the oil crisis, following which there was a decline in the demand for tankers; these docks therefore became a heavy burden for the companies.

From the latter half of the 1960s to the mid-1970s the Japanese shipbuilding industry was in a position to build half of the world's shipping but, on the other hand, the industry experienced big changes in management. The industry went through a period of so-called "busy work without profit" in the mid 1960s, during which it secured a large number of orders that turned out to be unprofitable. In 1971, when the dollar shock brought enormous losses due to the dollar exchange rate, the industry barely found its way to having a tanker boom. However, in 1972–1973, the shipbuilding industry enjoyed a worldwide boom, and Japan strove to obtain the maximum number of orders, which brought good profits.

After the first oil shock in 1973, there was a worldwide excess of shipping brought about by the world recession and subsequent decline in demand. From 1978, the total tonnage of launched ships in the Japanese shipbuilding industry decreased drastically. With a large number of previously accepted orders being cancelled, the industry entered a long-term depression. The figures indicate that the total number of orders accepted in 1978 decreased to 3.218 million tonnes, which is less than 10% of the figure for 1973, when order acceptance reached its peak (Table 10).

In order to recover from this difficult situation, the industry organized an anti-depression cartel, which regulated the operations of its members, initiated the phasing-out of outdated equipment, and every shipbuilder adopted severe counter-measures, including a reduction in the number of employees. The total amount of equipment phased out reached 3580 thousand CGRS (Compensated Gross Registered Tonnage) in 1981. After the latter half of 1978, the marine transportation market enjoyed good business, and the subsequent demand reflected the views of shipowners, both overseas and domestic, that prices would be raised because of the large-scale adjustments in the industry. There was also a market for medium and small-size tankers, so orders recovered temporarily from 1979.

But there was no change in the background excess of ships, and the number of orders in 1982 decreased greatly. In 1983 there were orders for bulk ships (which was a form of speculation), but there were still signs that

Table 10 Trends in domestic and export shipbuilding (1000 gross tonnes)

Fiscal year	Domestic	Export	Total
1970	4075	12 562	16 637
1971	7704	7223	14 927
1972	3651	17 772	21 423
1973	6084	27 706	33 790
1974	3659	5689	9348
1975	1605	6898	8504
1976	1598	6824	8422
1977	1252	3693	4946
1978	1277	1941	3218
1979	2877	6062	8939
1980	2847	6446	9293
1981	3063	5319	8383
1982	1714	2637	4351
1983	1900	9470	11 370

Figures for 1983 are estimates.
Source: Ministry of Transportation, Shipping Department.

there would be a low level of orders over the long term. Furthermore, the fact that newly developing countries, such as Korea, have been expanding their orders through low prices means that more competition has appeared for the Japanese shipbuilding industry.

In these severe circumstances the industry has been promoting the rationalization of management by constructing vessels which have additional value, such as liquefied natural gas carriers and highly automated ships. In addition, costs have been reduced by incorporating a computer-aided design system and by introducing industrial robots. There has also been a notable growth in engineering other than shipbuilding.

The use of computers in the Japanese shipbuilding industry started around 1962, when the design departments became centralized. In the mid-1960s computers were applied to production control. From the latter half of the 1960s they were being used for the cutting and bending of steel using numerical control machines in the engineering department, and later, in more complex systems, such as the group control of gas flame cutting. Computers have been embodied in high-grade integrated systems for design, processing and production control, and also in the field of process rationalization.

The award for the integrated-ship design and production system to Nippon Kokan, K.K. in 1975 was for an integrated system for the handling of figures in ship shell material, and the award for the general purpose finite element method program to Mitsui Engineering & Shipbuilding Co Ltd was for large-scale ship construction technology.

Table 11 Production of iron frames and bridges

Calendar year	Iron frame (1000 metric tonnes)	(Million yen)	Bridge (1000 metric tonnes)	(Million yen)
1970	1887	183 743	451	81 762
1971	1775	175 357	482	93 736
1972	2026	197 469	530	101 443
1973	2198	242 379	593	122 391
1974	1924	284 557	583	144 634
1975	1449	223 206	430	130 606
1976	1379	207 369	415	142 124
1977	1286	199 820	445	158 962
1978	1327	219 591	498	193 644
1979	1482	265 117	549	223 080
1980	1589	308 102	558	223 641
1981	1581	318 076	519	219 944
1982	1498	301 465	537	243 244
1983	1339	256 255	550	259 984

Source: MITI, Annual Statistical Report on Machines.

The steelframe and bridge industries have processes similar to the shipbuilding industry regarding steel cutting and assembly. Apart from specialized manufacture, the heavy industry companies, including shipbuilding, are all working in the same area. This is especially so in bridge production, which requires expertise both in design and construction.

The demand for steelframe depends on the trend of investment in private equipment such as building construction and factory erection, whilst the bridge-building industry relies mainly on public investment in roads and railroads. As shown in Table 11, the production of iron frames and bridges has changed in a fluctuating pattern. After 1970, which marked the end of the years of high growth, and especially after 1973 (the time of the first oil shock), the growth rate slowed down considerably. Both the iron-frame and bridge-building industries attained their peak production totals in 1972, after which, even though the total pofits increased, tonnage has not exceeded that of 1973.

To cope with the deceleration in private equipment investment and the slow increase in public investment, restricted by the financial situation, the iron-frame and bridge-building industry has been concentrating on obtaining overseas orders. Since the beginning of the 1970s, the proportion of overall overseas orders has gradually increased. In 1973, they amounted to less than 10% of the total. This percentage has increased steadily until in 1980 it stood at 30%, and it has recently reached nearly one third of total production.

In these industries, the large companies concentrated their efforts to the

Table 12 Production of office machines (billion yen)

Calendar year	Electronic desk calculators	Cash registers	Typewriters[a]
1970	131	17	14
1971	125	18	16
1972	118	18	15
1973	173	28	16
1974	180	36	16
1975	162	35	14
1976	185	59	23
1977	153	74	33
1978	182	72	33
1979	174	75	37
1980	202	87	52
1981	148	80	68
1982	142	74	57
1983	143	79	88

[a] Only roman character typewriters are included in the production figures for 1982 and 1983.

Source: MITI, Annual Statistical Report on Machines.

automation of design and production, and in 1979 Yokogawa Bridge Works, Ltd received an Ishikawa award for its work on the design and manufacture of bridges.

6 The Office Machine Industry

The office machine industry developed rapidly after the 1960s. Apart from electronic computers, electronic desk calculators and copying machines are two of the best known items of office machinery.

The electronic desk calculator was first industrialized in Japan, and the first calculator containing a transistor was produced in 1964, weighing 25 kg and costing 535 000 yen. A calculator equipped with an integrated circuit appeared in 1967, weighing only 4 kg. In 1970, calculators equipped with large-scale integration were produced, and three factors—their small size, a rapid drop in price, and their high performance—made personal purchases possible, which led to a rapid increase in production.

The total production of electronic desk calculators in 1965 was worth only 1800 million yen, reaching 54 700 million yen in 1969; with the appearance of LSI (Large Scale Integration) in 1970 it increased to 131 000 million yen. As shown in Table 12, production fluctuated, but had an overall increasing trend. The peak, attained in 1980, amounted to 200 000 million yen.

The electronic desk calculator is a uniquely Japanese commodity. Japan's

Table 13 Production of copying machines

Calendar year	Static electricity-type (1000 units)	(billion yen)	Diazo-type (1000 units)	(billion yen)	Total (1000 units)	(billion) yen)
1970	141	28.8	171	11.6	312	40.4
1971	156	35.4	147	10.2	302	45.6
1972	194	40.0	147	11.2	341	51.2
1973	318	65.3	181	15.1	498	80.4
1974	327	79.8	114	10.9	440	90.7
1975	306	99.5	95	10.1	402	109.6
1976	482	156.1	101	11.4	583	167.5
1977	622	196.3	117	14.2	740	210.5
1978	682	217.8	114	14.2	796	232.0
1979	799	244.9	102	14.4	901	259.3
1980	1034	323.6	90	13.2	1124	336.8
1981	1327	408.7	77	11.9	1404	420.6
1982	—	—	—	—	1420	415.7
1983	—	—	—	—	1924	491.9

Source: MITI, Annual Statistical Report on Machines.

share of the world market was 70–80% from the beginning, and the quantity of exports grew rapidly. In 1969 exports amounted to 50%, but recently more than 90% of production has been for export.

Another major product, the copying machine, enjoyed a rapid growth in production from the mid-1950s, and production grew very fast at the beginning of the 1960s. In this period the main type of copying machines were diazo-types. But, since the latter half of the 1960s, the growth rate of electronic copying machines has been striking. In 1965, the proportion of electronic copying machines to other copiers was only 24%, but in 1966 it rose to 40%, in 1967 to 52%, in 1968 to 62%, and in 1970 it increased to 71% (Table 13). In other words, electronic machines predominate.

There are two kinds of electronic copying machines, namely the electro fax (EF) and the xerography (or PPC—plain paper copier). From the 1970s onwards, the number of PPC electronic copying machines grew rapidly and the EF-types gradually diminished. Production of the PPC type electronic machine was started in 1962 by Fuji-Xerox Co, which is a joint enterprise of Rank Xerox Ltd, England, and Fuji Photo Film Co. From that time, PPC-type machines spread widely. Since 1970, when the basic patent on the Fuji-Xerox PPC expired, many companies including camera and electronics firms entered this area, and the growth rate increased considerably. As a result, the total production of the PPC-type was 66 700 million yen compared with a figure for the EF-type of 32 900 million yen in 1975. In 1977 this changed to

167 900 compared with 28 500 million yen, and 404 700 compared with 4000 million yen in 1981. These figures clearly indicate the decline of EF-type machines.

As the total production grew, the total exports of copying machines also increased rapidly. The sum total of exports was 12 000 million yen in 1970, 124 400 in 1977, 222 900 in 1980, and 319 500 in 1982, a figure representing 80% of total production.

The 1980s brought a new development in the office machine industry: the introduction of office computers and word-processors, which have been heralded as the standard bearers of office automation. In the office machine industry, an award was made to Sanki Industries Co Ltd and Minolta Camera Co Ltd in 1982 for their electronic copying machine production system. The rapid growth of electronic copying machines brought about a total production figure for copying machines in excess of the production of electronic desk calculators in 1977, and they took up the biggest share of business machines in production.

7 The Electronics Industry

Since the latter half of the 1950s the Japanese electronics industry has advanced as the major producer of domestic electronic appliances. In 1955 the Sony Corporation successfully developed a transistor radio, and in about 1957 it entered into proper factory production. In 1955 the production of black-and-white TV sets was launched in Japan. These two products enjoyed rapid growth as the leading domestic electronic appliances until about 1963. From that time the production of tape-recorders grew quickly, followed, from 1965, by colour TV. Throughout the 1970s these two products, together with the stereo record-player, occupied the leading place in domestic appliances. In the first half of the 1970s colour TV sets held the biggest share in the production of electric appliances, but in the latter half of the 1970s colours TV sets saturated the market, and their production growth rate lowered, and the production of tape-recorders and stereo record-players increased rapidly, compared with the production of colour TV sets.

From 1975, the video tape recorder (VTR) entered into wide-scale production; the figure in 1981 was more than nineteen times higher than that of 1976. In 1980, the production of colour TVs amounted to 711 000 million yen, but the sum for VTRs was 1 086 800; far in excess of the figure for colour TVs. In that year figures for tape-recorders and stereo record-players were 918 000 and 543 000 million yen, respectively.

The growth in the production of domestic electronic appliances depends not only on domestic demand, but on exports. Recently the proportion of exported colour TVs reached 50%, the figure for stereo record-players nearly

Table 14 Production of electronic products (billion yen)

Calendar year	Radio, TV, VTR, etc.	Communication devices	Computer and related devices
1970	1467	413	270
1971	1376	427	311
1972	1538	508	421
1973	1676	609	472
1974	1736	625	589
1975	1560	632	541
1976	2184	699	619
1977	2254	758	719
1978	2197	863	910
1979	2191	924	1124
1980	2814	1060	1293
1981	3516	1254	1468
1982	3406	1295	1733
1983	3733	1475	1957

Source: MITI, Annual Statistical Report on Machines.

80%, and tape-recorders 90%. Moreover, overseas production by Japanese corporations is growing.

The growth of domestic electronic appliances rests largely on solid state products. Consequently, very small products can be made by applying the transistor and integrated circuit (IC). The transistor was first adopted in the radio, and subsequently in tape-recorders and TV. The field of application became much wider when it was used for colour TV in 1970 (all types of colour TV were totally transistorized). The domestic production of ICs advanced in the latter half of the 1960s, and from about 1970 the IC began to be used for colour TVs.

The production of not only domestic, but industrial, electronic appliances increased from the latter half of the 1960s, and there was a special demand for electronic computers and communication equipment (see Table 14).

The trends in the production of electronic devices are shown in Table 15; special growth is noticeable in IC devices. Although the Japanese electronics industry experienced marked growth in domestic electronic appliances, from the latter half of the 1960s industrial electronics also grew in importance and in the 1970s that tendency increased. As is shown in Table 14, the total production of communications devices and computers was about 50% that of radio, TV and VTR production in the 1970s, but recently the figure has risen to around 80%, which means they have now reached the same production levels.

Statistics for the production of printed circuit (PC) boards for 1965 did not

Table 15 Production of electronics parts (billion yen)

Calendar year	General parts	Printed circuits	Electronic tubes	Semi-conductor elements	Integrated circuits	Total (including others)
1975	603	44	209	159	118	1137
1976	930	63	300	257	197	1742
1977	930	72	298	268	208	1761
1978	971	81	290	251	281	1847
1979	1090	98	297	254	383	2074
1980	1361	116	363	294	570	2641
1981	1681	142	435	378	689	3274
1982	1855	160	397	360	835	3490
1983	2189	195	453	421	1140	4246

Source: MITI, Annual Statistical Report on Machines.

appear in those of the electronics industry, production in that year being estimated to be under 1 billion yen. However, in 1976 production reached 63 billion yen and, in 1980, 116 billion yen, as shown in Table 15. There has also been a recent growth in PC boards for industrial appliances such as electronic computers and office automation machines. Complicated circuits and complex integration are required for these high performance machines; consequently, computer-aided design (CAD) techniques are necessary to produce the complex wiring patterns required. Because of the high quality equipment, the major electronics manufacturers include the board in intermediate products for sale.

The Sharp Corporation was given an Ishikawa prize in 1977 for a mini-computerized automatic layout system for two-layer printed wiring boards.

8 The Electric Power Industry

The growth of electricity consumption, even in the early 1970s, was high at around 10%, but after the first oil crisis in 1973 industrial demand declined and the growth rate decreased to around 5%. The second oil crisis emphasized this tendency later.

The relative proportions of three kinds of electric power production are shown in Table 16. In 1982 hydraulic power production made up 15%, thermal power generation 66% and atomic power generation 19%. The percentage of thermal power production is high, and although after the oil crisis the consumption of coal and liquefied natural gas (LNG) increased, crude and heavy oil were still the most important types of fuel.

Table 16 Trends in electric power production (million kWh)

Fiscal year	Hydraulic	Thermal	Atomic	Total
1970	73 637	229 370	4 581	307 588
1975	79 276	309 648	25 102	414 026
1978	68 941	369 118	58 956	497 014
1979	78 577	373 703	69 344	521 626
1980	85 146	346 895	82 009	514 050
1981	83 737	352 176	87 231	523 144
1982	77 342	343 299	101 299	522 465

Source: Association of Electric Power Production.

Breakdown of the demand structure shows that lighting takes 25% of all consumption, the remaining 75% being used for power, especially in industry.

After the oil crisis, the demand for electric power decreased, as a result of slow economic growth, the stagnation in overall industrial demand, and the progress made in energy saving both in industry and in the home. Also, as fuel prices increased substantially due to the foreign exchange rates which fluctuated on a large scale, and since it is a public commodity, the rise in the price of electricity was influenced by political considerations, which resulted in a fiscal imbalance in the electricity corporations.

In order to cope with these circumstances, the electricity industry is making efforts to decrease its dependency on thermal power production based on unstable oil prices. This will help to stabilize the cost of power production and increase the share of atomic power in electricity production. According to a recent report on the future demand for electric power, issued by a committee of the Electric Power Industry, atomic power production will reach 28% of total production in 1990, and 35 in 1995, to be compared with an estimated decrease in thermal power production from 39% in 1982 to 20% in 1990, and 14% in 1995.

The choice of a new location for an atomic power station is difficult, however, because of possible objections from the inhabitants of a particular district concerning safety and reliability. It takes 10–15 years of lead time for construction of an atomic plant because of legal regulations and for obtaining approval of design and construction documents. Moreover, the Japanese electricity industry, as a public utility, is governed by law. It is divided into nine districts where a monopoly of the power supply is guaranteed, but it is also obliged to operate impartial distribution, and the established price has to be accepted by the government.

Because of this situation, the electricity companies have to provide a long-range plan for their system networks and operations, in order to fulfil their function as a public utility properly to achieve efficient power construction and operation, while taking into account that it takes a long time to build a new power station. An Ishikawa prize was given to Tohoku Electric Power Co Inc for the dynamic performance of a power system which deals with the kind of problems outlined above.

9 The City Gas Industry

There are various uses of gas, domestic being the most important, especially for a hot water supply, Sales statistics for 1981 show that the sale of domestic gas made up 60%; 17% went on commercial use, 16% was for industrial use, and 7% was for hospitals and schools.

During the period of high economic growth there was a expansion in the demand for city gas. This was basically due to improvements in the standard of living, which brought a high rate of growth in the number of households built, resulting in further hot water supplies in kitchens. Over the period 1965–1973 the city gas demand showed an 11% growth rate each year. But after the first oil crisis, the growth rate slowed down, and from 1973 to 1981 it fell by one half, to 6%. This reduction was due to; (a) a reduction in the number of new households constructed; (b) limited expansion in the development of new domestic uses, including hot water and bath supply. There was also a cost-saving tendency among consumers because of the increase in the price of city gas, following the surge in the price of crude oil.

Regarding the raw materials for gas production, there was conversion from coal to oil in the 1960s, and in 1969 liquefied natural gas (LNG) began to appear as a raw material. Since then, the proportion of LNG used has increased rapidly, reaching more than 50% in 1979, as shown in Table 17.

In 1982 there were 248 city gas corporations throughout Japan generating 97 726 billion kilo calories per year, and supplying 17.432 million households. The Tokyo Gas Co Ltd is the largest city gas company. It has a dominant position, providing gas for 6.199 million households (36% of all households in Japan) and it supplies 35 355 billion kilo calories per year, which is nearly 38% of the total Japanese consumption. It provides gas for the highly populated districts in Tokyo and Yokohama which necessitates high quality planning both for gas supply and control, and for the import of raw materials. The Tokyo Gas Co received an Ishikawa award in 1975 for its production and control system on gas supply.

10 Finance

At present, Japanese finance is going through a period of great transforma-

Table 17 City gas raw materials (%)

	Fiscal year				
	1966	1971	1976	1978	1981
Coal	44.7	30.3	14.7	11.2	9.3
Crude oil	24.3	17.7	0.1	—	—
Naphtha	25.6	33.1	33.3	22.4	9.7
LNG	—	8.4	38.0	49.9	57.1
LPG	3.4	8.3	12.7	15.4	23.0
Other	2.0	2.2	1.2	1.1	0.9
Total	100.0	100.0	100.0	100.0	100.0
Production (billion kilo cal)	25 061	39 426	66 291	70 596	88 600

Source: Association of Japanese Gas Companies, *Handbook on Gas Supply Service*.

tion, which is sometimes called a financial revolution. To understand the background to this phenomenon one has to understand the rapid change in the money flow due to the overall decline in economic development following the oil crises (Table 18). Before this, incorporated companies constituted the greatest money losses. The financial losses of this sector and also those of the public sector were covered by surplus money in the private sector. But since 1974 the shortages in the incorporated sector rapidly disappeared, because of the decline of investment in corporations, while losses in the public sector increased rapidly. This was due to several factors: the public corporations

Table 18 Trends in the flow of funds (%)

	Fiscal year				
Dept	1971	1973	1975	1980	1981
1. Corporate business section	−55	−79	−34	−32	−29
2. Private section	92	79	95	82	100
3. Central Government, Public Corporation and Local Authorities sector	−20	−21	−66	−68	−66
4. Financial institutions	8	12	5	12	0
Overseas	−25	9	0	6	−5

Source: Nippon Bank, Flows of Funds Accounts.

Surplus and deficits in the flow of funds in each division are summarized.

Total amounts of surplus and deficit of departments in each year are equal to 100. The minus sign denotes a deficit.

Table 19 Trends in public securities (10 billion yen)

Fiscal year	Central government	Local government	Bond of public corporation	Total
1971	112	50	106	268
1972	186	62	121	369
1973	174	104	132	410
1974	212	147	161	520
1975	532	185	177	894
1976	711	209	221	1141
1977	985	195	250	1430
1978	1084	235	241	1560
1979	1370	254	284	1908
1980	1466	136	328	1930
1981	1177	117	347	1641

Source: Nippon Bank, Money Flow Table.

involving both central government and local authorities needed more investment, the income from taxes could not be increased, and expenditures such as social security had to be increased, which brought about a decrease in savings.

The growth of the money shortage in the public sector brought a rapid increase of bonds in public corporations, including government bonds (Table 19).

Due to these structural changes the banking system expanded with unrestricted interest rates and an increase in the business areas of the banks. On the administrative side of finance there has also been a change from protective administration to free competition. The signs of this in the Japanese financial system, involving the expansion of the areas of business of the overseas financial organizations, have already been mentioned.

The development of the electronic banking system is also a major factor in future bank management. The use of electronic computers in banking started in 1959. At that time, only the batch processes or off-line use could be applied in the branches, and data had to be processed at the headquarters. In 1965, however, the first on-line system was introduced. One major object of this system was to process ordinary deposits on-line, but business calculations in various areas also came within the scope of the on-line system. In 1974 the bank business entered into a second era of on-line calculation. In this, the computer performed not only individual account calculations, but was also used for the calculation of deposits, exchanges, loans and foreign exchange items. During this period the installation of cash dispensers and automated teller machines advanced, and in 1968 an on-line money order settlement system was started in the local banks. In 1973 a national bank communication system was developed, which integrated about 20 000 banks—including city banks, mutual finance banks and credit associations—into a network system covering the nation.

Computerization of the banking business made great progress in the 1960s and 1970s. According to the statistics for the end of June 1982, 718 159 million yen's worth of computers were used in business (Table 4).

From June 1984 a bidirectional data communication service system connecting the banks with business enterprises was introduced in several banks, following the release of communication lines from the National Telephone and Telegram Inc. This heralded the dawn of a new era, the electronic banking age. In this area the Japan Development Bank and Fujitsu Ltd received an Ishikawa award in 1972 for its decision support system.

11 The Newspaper Business

Newspapers published by Japanese corporations have a very wide circulation, and the number of copies per household is, along with Sweden, the highest in the world. But because of this high circulation and the low rate of population increase the growth rate in the number of printed newspapers is very low (below 2% a year; see Table 20).

The number of newspaper corporations which are members of the Newspaper Association is 114. Among these there are eight large companies. The mean number of copies, from January to July 1983, can be listed as follows: Yomiuri 8.859 (in millions); Asahi, 7.551; Mainichi, 4.325; Sankei, 1.974; Nippon Keizai, 1.974; Chunichi, 1.968; Hokkaido, 1.054; Nishinippon,

Table 20 Trends in newspaper circulation

	(Thousands)	(per household)
1970	36 304	1.30
1975	40 513	1.26
1976	42 140	1.24
1977	43 114	1.25
1978	44 277	1.27
1979	45 854	1.30
1980	46 391	1.29
1981	47 256	1.30
1982	47 993	1.30

Morning and evening issues are counted as one.
Source: *Yearbook of Japanese Newspapers*

0.751. These eight large companies dominate the market and, with the addition of two more newspapers (which are subsidiaries of Chunichi), account for 60% of total number of copies in Japan.

The managers of newspapers are faced with a difficult situation; stagnation in the number of copies, severe competition from the TV and mass media for advertising income, and the difficulty of raising the price of newspapers in the face of strong resistance from customers.

From the 1950s the newspaper business had incorporated technological innovation in both editing and printing. This has involved the adoption of a fully automatic mono-type casting system (which replaced hand typesetting by automatic composition), the installation of a Chinese-character teletype system for the transmission and reception of manuscripts, the utilization of large-scale facsimiles, and high-speed rotary presses. Computerized type-setting systems, which combine a computer with an automatic type-setting machine, have particularly advanced since the end of the 1960s, and the adoption of a complete system by Asahi and Nippon Keizai attracted great attention. In connection with this, in 1974 the Asahi Shimbun Publishing Co, the Nippon Keizai Shimbun Inc, and IBM Japan Ltd received an Ishikawa award for a Japanese publishing system. In the initial stages of this system, photographs and line-drawings had to be stuck on separately, but following progress in hardware technology, a full-page photoprinting system was later adopted. Nippon Keizai switched to this system completely in 1978, and Asahi transferred to it in 1980.

12 The Database Service

With progress in the application of computers, the Japanese information processing industry is now in the development stage, but it has not yet fully matured. The sales of the industry in 1977 were 412 600 million yen, which grew to 911 900 yen in 1982. The latter figure can be broken down into the following proportions: software production and programming, 33%: business computation, 30%: card punching, 9%: database services, 6%. But compared with western countries, such as the USA, the database service lags far behind.

According to the *Database Ledger* issued by MITI, there were 456 databases in September 1982. Of these, a batch service was provided by 64 databases, and an on-line service was available for 392 databases. On-line database services can be classified into the following area: business use, 171; science and technology, 148; sociology, 34; politics, law and others, 39. Almost all on-line database systems have been imported from abroad, notably from the USA, and these make up 84% of the total.

Business on-line database services started from 1980, but these are still in their infancy.

Private databases have accumulated few data in science and technology. For public organizations, there are two special agencies: the Japan Information Centre of Science and Technology (JICST), and the Japan Patent Information Centre (JAPATIC).

The former deals with technical information, and the latter with patents and utility models. JICST was established in 1957, and started its on-line service in 1976. The total number of data items processed was 460 000 in the fiscal year 1982. JAPATIC deals with around 400 000 items on domestic patents a year. It also has patent information on the USA, and its on-line service started in 1979.

In large private companies it is necessary to both accumulate and have the ability to process scientific and technical information on a computer, for which an Ishikawa award was given to Fuji Photo Film Co in 1978 for its technology information processing system.

1

Computer-aided Construction Technology

Kajima Corporation, Tokyo, Japan

1.1 Introduction

Use of computers within Kajima Corporation began in 1962. Since that date the number of computer personnel has increased year by year until at present over 200 personnel including mathematicians, civil engineers and structural engineers are currently engaged in the development of programs. Beginning with a program for seismic analysis, numerous programs have been developed independently in the fields of construction technology. The introduction of an automatic drafting machine in 1969 stimulated program development at an even greater rate. These programs have already been utilized in the design, planning, progressing, estimation and cost control of construction works with a value of over 200 billion yen, and in such diverse applications as skyscraper buildings, shipbuilding docks, residential development, pre-stressed concrete bridges, industrial buildings and nuclear power plants. Indeed, computer utilization is now indispensable to all activities in the field of engineering construction.

1.2 Present Status of Computer Utilization

A construction project such a skyscraper building, huge shipbuilding dock or

CAD/CAM and MIS in Japan
ISBN 0-12-682580-7

large residential development requires a total system which integrates such diverse functions as investigation, planning, design, construction and control.

In our system we are making full use of computers as a means to this end. The following paragraphs give a general description of the status of our computer-aided construction technology.

1.2.1 Utilization in Design

The computer has made it possible to obtain exact solutions for the dynamic and static analysis of structures using design theory. Such solutions were impossible in the past using hand calculation, but computer application has facilitated the work of preparing several comparative designs for a single project. Not only was Kajima Corporation the first in Japan to develop a design method using computer technology as an analytic tool in the design of earthquake-resistant structures for skyscraper buildings based upon the flexible structure theory, and the first to use this method in the construction of a series of skyscrapers including the thirty-six-storey Kasumigaseki Building (completed in 1968), the World Trade Centre Building and the Keio Plaza Hotel; but it also developed techniques for the earthquake-proof design of such new structures as building docks, sea berths for mammoth tankers, long span pre-stressed concrete bridges, sunken tube tunnels and pressure vessels for nuclear reactors.

1.2.2 Utilization in Construction Planning

During the past few years, construction projects have increased in scale and complexity and the computer has become a powerful tool in construction planning. The computer system which the Corporation developed for the planning of residential development projects is a novel system which can rapidly determine the optimum construction schedule, and the most rational plans for earthmoving and heavy equipment utilization. A representative example in the field of construction planning for building projects is the development of a production control system for skyscraper building.

1.2.3 Utilization in Construction Control

The Program Evaluation and Review Technique (PERT) is a method for controlling the progress of construction work in accordance with a predetermined plan. A computerized PERT method is used extensively for schedule control and the rationalization of construction work in the Corporation. It has led to a reduction in labour requirements and costs through the preparation and execution of procurement plans for labour, equipment, materials and operating funds.

1.2.4 Utilization in Cost Estimation

The many diverse items in construction work and the wide differences in working conditions between different construction sites present difficulties in the use of computers in cost estimation. The Corporation has completed programs for a variety of construction projects, such as skyscraper buildings and tunnels, which have numerous points of similarity, and is endeavouring to widen further the scope of such applications.

1.2.5 Utilization in Cost Control

The checking of construction costs is a most important factor in private enterprise, and we have almost completed a computer system for accurately and quickly estimating these. Two detailed examples of such computer systems are described below.

1.3. Skyscraper Production Control System

Besides the obvious increase in total floor area and the number of labourers employed on a skyscraper project compared with a lower conventional type building, the fundamental construction methods used are also different, and this presents numerous problems for scheduling and planning the movement of labour and materials. In the construction of the Kasumigaseki Building, the Corporation established a new method of schedule planning and control based upon the sequential–repetitive method of construction described below. In addition, network analysis technique was applied to the scheduling and control of lifting. The development of these production control systems contributed greatly to rationalization of construction, leading to a shorter construction period and a lowering of construction costs on the skyscraper building projects which were subsequently awarded to Kajima Corporation.

The sequential–repetitive operation method was developed so that construction progresses smoothly and continuously from the lower floors to the upper floors while maintaining co-ordination between the various operations.

The prime factors in making possible this use of the sequential–repetitive method were the following two developments.

1.3.1 Application of the Network Technique

In the planning and control of the construction schedule, the PERT method was adopted from the schedule planning stage. In the past schedule control has been performed simply by finding the critical path by the PERT/TIME method. On a skyscraper building project, because of the increase in the peak volume of loads to be lifted, the construction schedule must take into account the total number of labourers involved and the time for lifting loads. Because of this, adjusting all the various elements involves considerable calculation. With computer processing, an economical daily construction schedule was achieved, involving the preparation of a detailed working schedule, then making revisions to adjust the peak times in the lifting plans to level out the loads on labour and materials.

Furthermore, in order to simplify revisions to the construction schedule in accordance with the changes which occur during construction, and to facilitate progress control, the starting point network was introduced. In this new concept, interlinks between the various construction operations are intentionally cut, and the starting date for an operation is designated arbitrarily. This has the advantage that, when progress control is then used at the project site, only the starting time for each operation and the number of days required needs to be controlled. Moreover, with this system it is possible for the construction schedule, the number of labourers and the material lifting time to show any necessary change automatically by the computer, thus facilitating schedule control.

1.3.2 Development of a Lifting System

With the increased height of skyscraper buildings, the number of labourers required and the quantity of building materials to be lifted increase dramatically. In addition, because the locations for setting up elevators and other lifting equipment are strictly limited, the efficiency of the lifting operation has a great influence on construction both through its direct effect on lifting time and also the efficiency of other operations.

To solve this problem, the almost infinite variety of shapes, sizes and weights of materials to be lifted were classified by type of packing and weight as suitable for large, medium and small lifting equipment. Suitable lifting equipment was then installed, and an effective system developed for its unified control through the use of the computer and the network method.

1.4 Design System for Residential Developments

A large-scale residential development is essentially an integrated urban construction project. Until a few years ago, a project of 30 to 50 ha in area was considered to represent a large project, the scope of which involved the preparation of residential lots. When, however, the scale increases to 300 ha or more, it is no longer possible to handle such projects in the same manner. Because a development of 300 ha or more will have a future population of approximately 30 000, the project must be tackled as a problem involving the construction of a whole new town. When a development reaches that scale, consideration must be given to such factors as its interrelationships with the neighbouring major city and the local society. Consequently, the project must be planned to consider all aspects of the layout problem, right from the initial stages of the basic conceptual planning. With a development of this scale, the manpower spent on planning would be enormous, with 3000 or more drawings to be prepared during the investigation and design stages alone. To process this large volume of work by hand, thirty planners, designers and other technical personnel would take one and a half years, even assuming that all work from conceptual planning to the final design progressed smoothly without any unforeseen problems which might also increase the cost. Moreover, regarding quality, there is a natural limit beyond which, in manual operation, it cannot be improved.

With a large-scale project, there is just not sufficient time to prepare several alternative plans and make comparative studies on them, and even though there may still be doubts about the quality of a plan, time again restricts any further study. In large-scale residential developments which because of difficulties in acquiring land have gradually been forced to move into hilly areas where restrictive conditions may exist, the difficulty of harmonizing the image envisioned in the master plan with purely technical factors such as earthwork quantities and structures may already be too great for the technical specialists to handle. In these circumstances, our automated system for the design of residential development projects is an integral function in their execution, both qualitatively and quantitatively.

Items processed by this system and the results obtained are shown in Figs 1.1–1.3.

The system performs the following operations:

—Automatically determines design elevations.

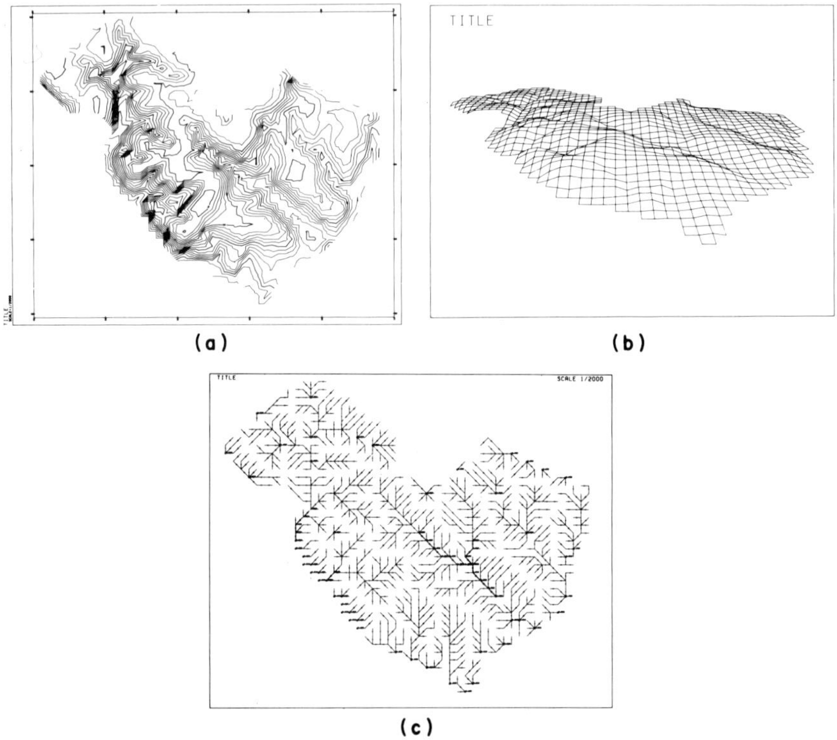

Fig. 1.1 Representations of land topology. (a) The topographical contour line map; (b) the perspective of land; (c) the stream line map.

—Calculates the quantities of cut and fill and determines the shortest haulage route taking into account gradients and points of detour.
—Makes these determinations in such a way that the amount of earth haulage at any given point of time is minimized.
—Automatically draws natural and design cross-sections.
—Automatically draws road plans.
—Prepares the drainage plan.
—Gives input of unit prices, estimates cost.

The following results have been achieved using this system:

—A reduction in manpower, and improvement in efficiency, as well as standardization and the assurance of accuracy.
—Accommodation of design changes during the preparation of alternative plans.

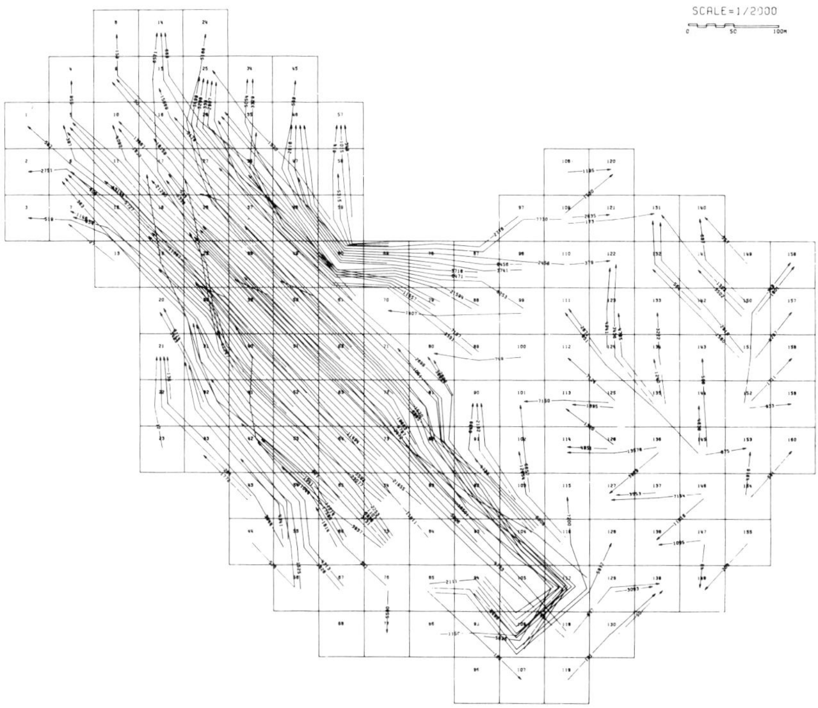

Fig. 1.2 The arrow diagram of earth movement.

—A reduction in both direct costs (design costs) and indirect costs (interest charges on land purchase).
—Thorough control of construction execution.

1.5 Conclusion

As can be seen from the above examples, the computer is being fully utilized by Kajima Corporation from the early planning stage of a project, and holds a central position in the operation of effective systems to answer the needs of various operational functions. In the preparation of these systems, the technical expertise of the various sectors of our organization is drawn together by the computer. A rigorous study is made of alternative plans, and an experimental follow-up performed from differing points of view, resulting in the preparation of the most appropriate system. With the perfection of numerous coherent system programs, the computer's capability to integrate

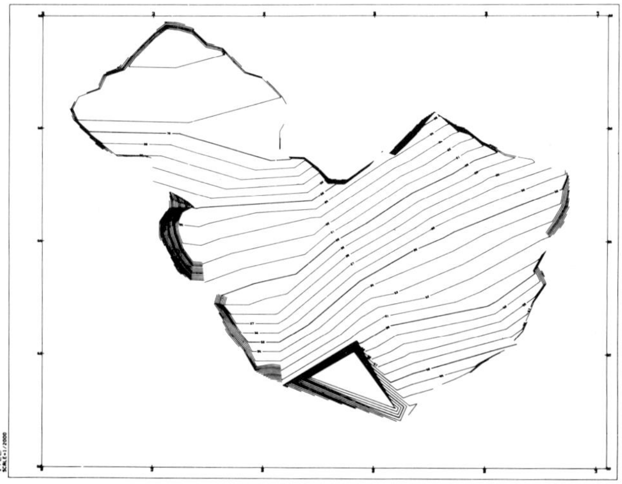

(a)

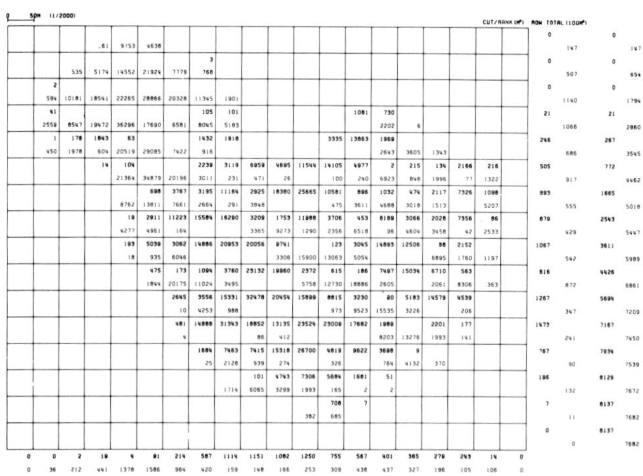

(b)

Fig. 1.3 (a) The finished shape of the ground. (b) Distribution map of soil volume (upper line shows cutting volume, lower line shows banking volume).

theoretical and experimental aspects has become the nucleus of construction technology promoted by the Kajima Corporation.
We intend in the future to:

1. Develop application programs which bring together the most advanced analytical techniques.
2. Develop programs which systematize all of the operations for a project from initial investigation to completion of construction.
3. Develop programs to secure efficiency of operation in accordance with the needs of all sectors of our organization.

Finally, we hope that by this contribution to the development of construction technology, Kajima Corporation may justify the honour of receiving this award.

2

Automatic Elevator Design and Draughting System

Akira Masumoto
Mitsubishi Electric Corporation, Tokyo, Japan

2.1 Introduction

With the tendency toward high city buildings in Japan, the demand for elevator equipment has increased rapidly since 1960. Above all, standardized units for small and medium structures and apartment houses have come to form 70% of all demand. For submission to the appropriate authorities, layout drawings of the elevator are indispensable for owners, architects and general contractors, as well as for installation work. This being so, the Elevator Engineering Department of Mitsubishi Electric Corporation started (in May 1971) to develop an automatic design and draughting system named the "MELDAC SYSTEM".

In August 1973, a system for the layout of standardized units was completed, and later it was extended to structural hole drawings, and to design drawings of elevator entrance and signal equipment.

In Japan, where ground space is so limited, buildings are constructed to use the ground most efficiently; hence there are many oddly shaped structures of the shaft and machine room. Furthermore, as earthquakes are a constant hazard, big pillars and beams often project into the shaft and machine room to supply the required resistance. These circumstances complicate the logic of

CAD/CAM and MIS in Japan
ISBN 0-12-682580-7

the layout of the elevator, and make the computer handling of the design very difficult.

2.2 Outline of System Development

The general development philosophy is as follows.

(1) Transfer design engineers to more creative work, therefore minimizing data input. The design should be automatic, making manual work unnecessary, and the drawings produced should be perfect without need of additions or amendments.
(2) The cost of the drawing should be as low as possible and certainly below the market price of hand drawings.
(3) To change the standard of drawings to suit the automatic draughting machine would result in unnecessary confusion. The conventional standard of drawings should therefore be adhered to in principle.
(4) The system may eventually extend to elevators and escalators made to order and embrace a total manufacturing system. Accordingly, the system should be as expandable and flexible as possible.

Table 2.1 shows the main points of research and study made preliminarily in the development of this system.

2.3 Outline of the Design and Draughting System

This system covers elevator layout with structural hole drawings, and the design of elevator entrance and signal equipment.

The system for elevator layout with structural hole drawings is for a standard elevator of one to three units.

The design work involves the most efficient layout of the elevator apparatus, which consists of cage, counterweight, separating beam, traction machine, beam, control panel, and governor, etc., for shafts and machine rooms of various shapes, taking into consideration the relevant codes and regulations and the efficiency of installation maintenance. The drawings resulting from the above are fully automated using a computer and automatic draughting machine.

The procedure for making drawings consists of the following steps.

(1) The elevator designer enters data concerning elevator specifications, and the dimensions of the building structure.

Table 2.1 Research items

Research Items	Remarks
(1) Social background of the elevator industry, and the future trends of demand for elevators.	—Forecast of the future trend of the Japanese economy and demand for elevators. —Forecast of the increase in volume of design and drawing work required and the corresponding capacity of handling.
(2) Design and drawing work,	—Analysis of design work. —Analysis of drawing work and its rationalization. —Research on the actual condition of drawings made by subcontractors.
(3) Standardizing the drawing method of elevators.	—Research on the standard of drawing based on JIS (Japan Industry Standard) and our own standard. —Research on the standard drawing method of elevators for automatic drawing.
(4) Capabilities of automatic drafting machines and study of the feasibility of the system.	—Investigation of the actual condition of automatic drafting machines in Japan and overseas. —Research and study on the technique of pattern handling by computers. —Study of the economic side and technical problems of automatic designing and drafting system.
(5) Classification of structural patterns for elevator equipment.	—Classification of structural patterns in connection with various shaped shaft and machine room. —Study of all kinds of symbols and hatching in plan.
(6) Elevator equipment layout and logic of laying out drawing elements in drawing sheet.	—Study of the elevator code and regulations related to elevator and study of working efficiency of maintenance and installation.
(7) Development of programs for pattern handling by computer.	—Study of the method of input of pattern information. —Study and development of universal pattern handling programs and symbol drawing programs. —Study of the control program for drafting machine using a minicomputer.

(2) Input data is transferred onto punched cards and input into the main computer.
(3) As well as the above input data, the following master file data (on DISK) input is entered as reference data for calculation, and supplies supplementary drawing information.

The master data involves unit specifications, fundamental dimensions for layout, elevator dimensions, size of drawing sheet, etc. The pattern master file is made in order to draw elevator equipment with variable dimensions. The master file for characters and words involves registered Chinese chatacters, Japanese kana, alphabet, numerals, words and sentences, etc.

(4) The following calculations are made in the main computer, and the information necessary for the drawings is output on magnetic tape and instructions and messages to the operator of the draughting machine are output to the teleprinter. These are:
   (a) check input data;
   (b) calculate the detailed dimensions of each section of the structure;
   (c) extract pattern for drawings and reference data for necessary apparatus from master file;
   (d) design the best layout of elevator cage, counterweight and beams in shaft;
   (e) design the best layout of elevator traction machine, machine beams, controller panels, receiving panel, governor, etc. in the machine room;
   (f) calculate the positional dimensions in order to layout the drawing elements on a sheet;
   (g) decide the size and number of sheets required;
   (h) calculate the information (per-select, per-up-down, moving co-ordinate).
(5) The magnetic tape into which the data is output from the main computer is sent to the draughting machine, and the drawings are made automatically.

These processes are summarized in the system flow chart (Fig. 2.1). The principal devices used in the automatic draughting system are shown in Table 2.2 and Fig. 2.2.

2.4 Program Output

The designing and drawing program generates data on a magnetic tape to ar-

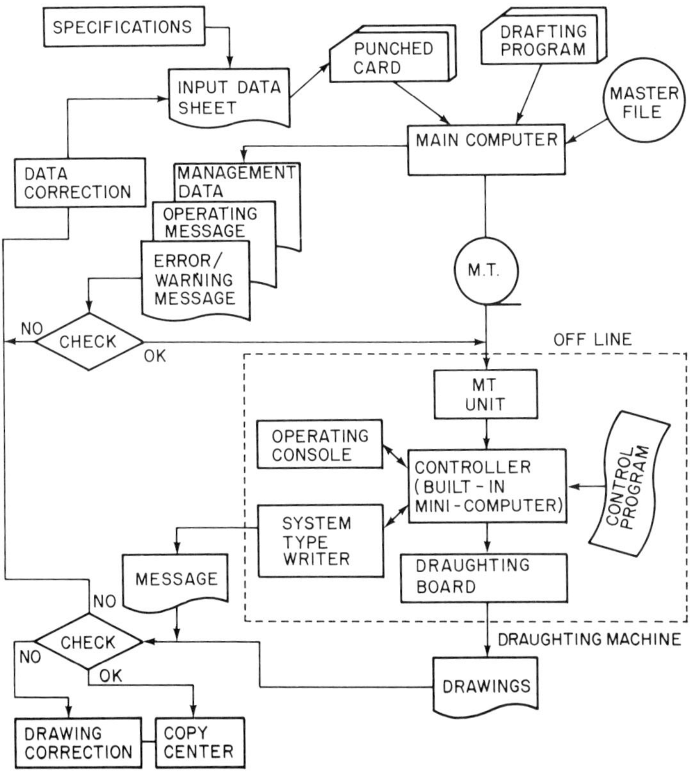

Fig. 2.1 Process flow chart.

range elevator equipment, to draw shaft and machine room layout, and produce longitudinal section drawings and elevator specifications.

The construction of the program is shown in Fig. 2.3, the fundamentals of the automatic draughting program in Table 2.3, and the general flow chart in Fig. 2.4.

Designs produced by these programs give not only the layout of the elevator equipment in the shaft and machine rooms, but also calculate positions and dimensions for the drawing and erase hidden lines. Here, we give a brief account of automatic design using these programs by explaining the automatic design for the elevator cage, counterweight and separating beam.

(1) The cage is first arranged in the shaft to see if the counterweight can be dropped in the back.
(2) If this is imposssible, dropping the counterweight on either side of the cage is examined as an alternative. The traction machine and governor

Table 2.2 Devices in automatic draughting machine

Name of device	Functions	Remarks
Mini-computer	Linear and circular interpolation, translation of drive command and operating command, speed control of machine etc., after reading in the draughting data.	MELCOM-70 (Mitsubishi Electric Corp., 16 kilo word)
Magnetic tape unit with interface	Data from the main computer on magnetic tape is read into the mini-computer by this unit.	9-track, 800 BPI
System typewriter	This unit is for print-out of operating messages and for manual control command.	Built-in punch tape reader/punch
Draughting board	Draughting is done on the surface of this board. The effective drawing area is 900 mm × 1500 mm. Automatic 4-pen selector. Maximum pen speed is 102 m/min.	Automatic draughting board TYPE S-915 (Mutoh Kogyo Co.)
Operating console	This console, which consists of various switches and lamps, is for operation of the draughting board.	
Interface for draughting board	Convert computer signals (M-70) into driving signals for the draughting board.	

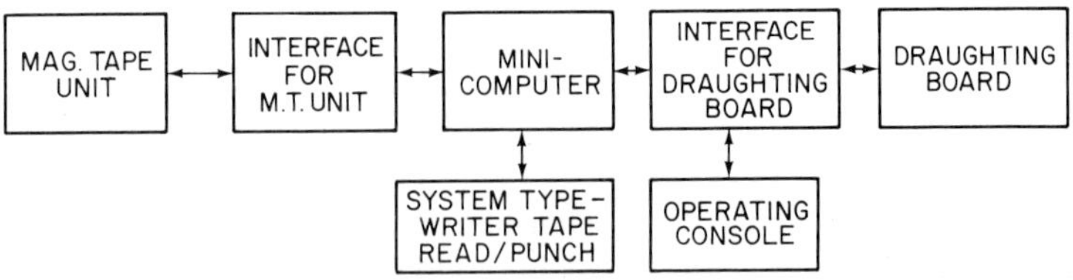

Fig. 2.2 Construction of draughting machine.

are placed in the machine room if there is enough space for their installation and maintenance. The positions of the counterweight, cage and separating beam are then finally determined. If there is not enough space for maintenance of the traction machine and governor, the positions of the cage and counterweight are modified.

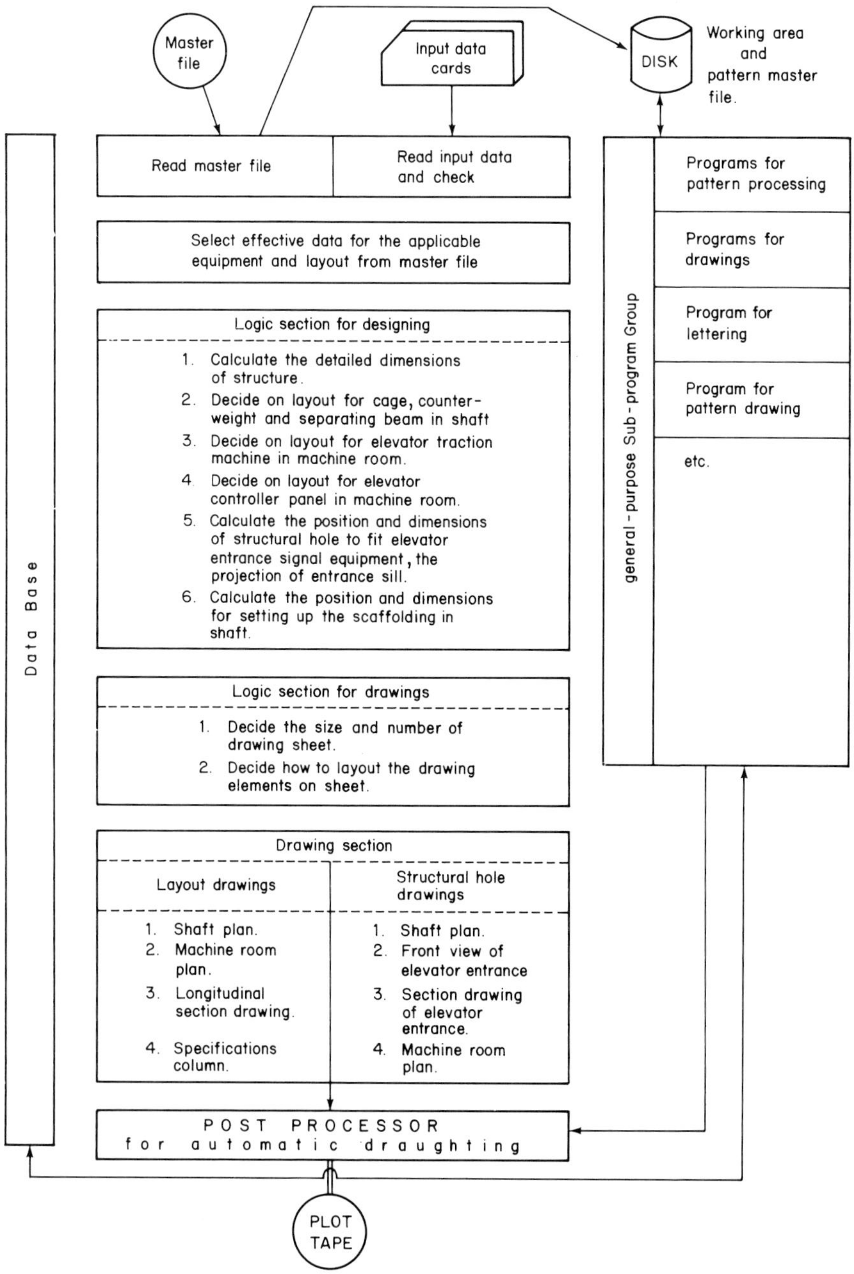

Fig. 2.3 Construction of the program.

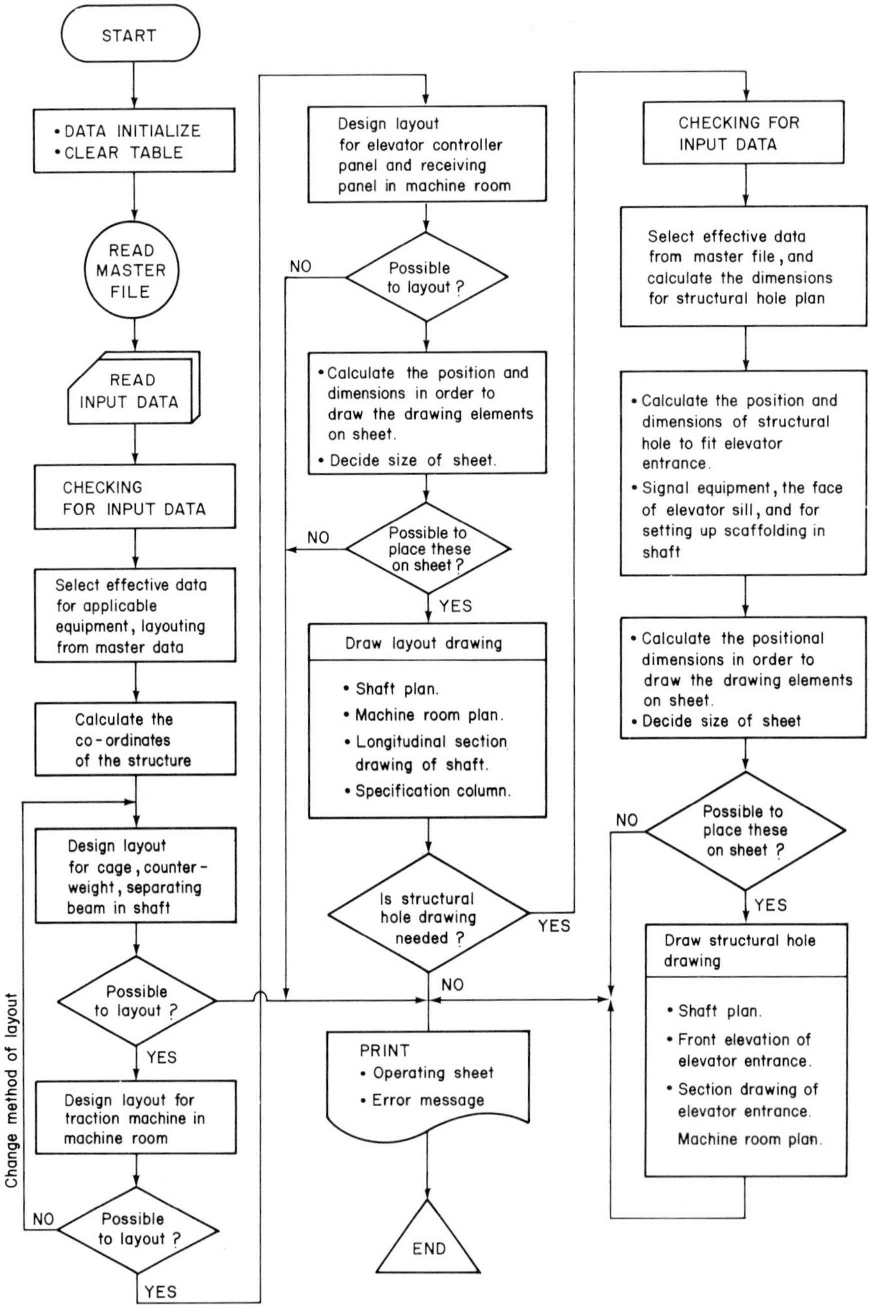

Fig. 2.4 General flow chart.

Table 2.3 The construction of the program

Program language	FORTRAN (IV)	
Number of subprograms	Elevator layout with structural hole drawing	200
	Design drawing of elevator entrance and signal equipment	110
Number of programming steps	Elevator layout with structural hole drawing	30 000
	Design drawing of elevator entrance and signal equipment	15 000

Design drawings of the signal equipment are available for 45 types of panel shapes and fifteen kinds of lettering. Design drawings of the elevator entrance are available for many types and combinations of buildings and cages with 12 basic shapes of the entrance jamb.

2.5 Merits of the Automatic Design and Draughting System

This is a revolutionary system which has the following advantages compared with conventional systems:

(1) It can handle building structures of unusual shape.
(2) Automatic design and draughting is perfected using little input data.
(3) Automatic design of the best arrangement is made within the restrictions imposed by codes and building structure, taking into account efficiency of installation and maintenance. The layout is designed so that the elevator cage, counterweight and separating beam are properly arranged in the shaft; the best arrangement of the traction machine in the machine room is met by considering the proper placement of equipment in the shaft; and the most suitable arrangement of the controller, receiving panel and governor achieved by taking into consideration the entrance, pillars and windows, etc.
(4) The type and size of drawing sheets, the placement of each element (scaled or not) and the position of each element are determined automatically. These drawings are produced by selecting the pattern from many combinations of drawing layout.
(5) High-quality drawings are produced by
   (a) erasing the hidden lines (overlapped lines);
   (b) avoiding overlapped placement of dimensions, pattern drawing and letters, etc.;

(c) using special drawing methods (for example, hatchings);
(d) placing dimensions and letters, etc. in the best position;
(e) employing dimensions which fit building structures of irregular shape.

(6) Alphabetical letters (capital and small), Chinese characters, Japanese kana, numerals and special characters can be reproduced freestyle and in various ways (vertical, horizontal, height and width of letter, space between letters). A maximum of 2500 letters, characters, numerals, etc., and 500 words and sentences are registered in the master file.
(7) The cost of drawing is lower than that produced manually.
(8) New means have been devised for ease of program maintenance, decrease in drawing cost, better program development and efficiency.

2.6 The Effects of Introducing the System

The following results have been brought about by the introduction of this system.

Labour saving.

A case which previously took a skilled draughtsman 3 or 4 hours now takes only about 15 minutes.

Quality improvement.

Designs and drawings of elevator equipment can be made as efficiently as by skilled designers and draughtsmen. Also the drawings produced are standardized.

Elimination of mistakes.

Since all information is checked by the computer, mis-entries, omissions and miscalculations are eliminated, and rechecking of the drawings is no longer necessary.

Decrease in cost.

The price of the drawing is not affected by increases in personnel cost. Long-term plans for draughting can thus be stabilized.

Shorter delivery time.

Previously it took 7–10 days to deliver drawings; now it takes only a couple of days. The Mitsubishi Electric Corporation is therefore in a position to cope with any rapid increase in demand.

3

Environmental Impact Assessment System for Sunlight

Kajima Corporation, Tokyo, Japan

Summary

This paper describes the environmental impact assessment system for sunlight developed by Kajima Corporation.

The system is used to analyse various environmental factors on sunlight—that is, shadow, radiation, brightness, glare, view and spaciousness, and at the same time to create and preserve the best conditions possible for human life.

The system consists of five subsystems which have the following functions.

(1) It draws shapes of shadows cast at various hours of the day, and draws distribution charts of equivalent shaded time or radiation at the actual building site.
(2) It determines more detailed information, such as starting and end time of shadow, and variation of radiation during a day at any selected points.
(3) Monthly or annual variation of shadow and radiation can also be determined on the basis of actual weather data.
(4) It calculates the brightness by skylight and the range of view which represents spaciousness to human eye is also determined.

CAD/CAM and MIS in Japan
ISBN 0-12-682580-7

(5) It calculates and draws the locus of reflected light by mirror-glasses attached to the surface of the structure.

The system can be regarded as a general purpose program not only for the general shape of any building but also for a spherical gas holder, an elevated bridge or undulating ground such as a mountain or a hill. All results are obtained in a graphic form so that quick judgement for environmental conditions can be made.

Three-dimensional geometric problem such as the calculation of shadow cast by building or undulating ground is generally believed to be very difficult to solve accurately and economically by digital computer. However, by introducing several new concepts, these difficulties can be completely resolved.

3.1 Introduction

Urban redevelopment, new town construction, the preparation of housing and plant sites, the installation of energy facilities such as nuclear reactors and LNG (liquefied natural gas) tanks, and the construction and development of facilities to dispose of household and industrial wastes are all creative work. However, their impact on the preservation of a favourable environment for man and nature needs to be considered. This is accepted as common sense today, regardless of any legal regulations which may be in force. Any new construction or development planning should therefore be preceded by a detailed investigation and analysis of the current natural and living environment of people. Since the middle of the 1960s in keeping with these social needs, the Kajima Corporation has centred its research and development efforts on the environmental impact of large-scale development.

Computerized impact assessment systems have been developed, covering the following fields:

(1) Changes in sea currents, tidal flows and water quality due to plant and industrial waste disposal facilities, and harbour construction.
(2) Spread of river/lake contamination due to nuclear reactor and general plant waste water discharge.
(3) Spread of air pollution due to emissions from very tall chimneys, process plants and expressways.
(4) Spread of noise and vibration from process plants, roads, railroads and work sites.
(5) Wind damage and radio interference due to the construction of high-rise buildings.

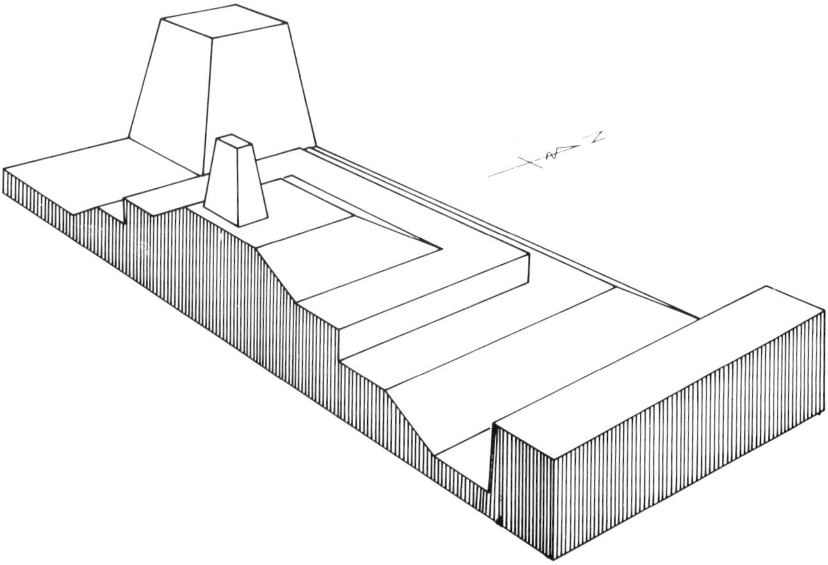

Fig. 3.1 Sunshade drawing.

(6) Sinking or displacement of foundations in association with the construction of tunnels, subways and downtown buildings.
(7) Changes in layout of land and rainwater courses associated with the preparation of residential sites, golf links, etc.

As a major engineering constructor, supported by superior technical capabilities in both design and implementation, these systems have already been adopted in many projects and have produced overall improvement in the environment.

The Environmental Impact Assessment System for Sunlight aims to analyse and assess the environmental conditions of the site and of peripheral areas in terms of sunlight. It is therefore a system for the preservation or creation of more favourable living conditions.

3.2 System Outline

The system not only calculates the extent of shadows cast by structures at various times of day, but is also designed to assess the impact of sunlight on life in terms of warmth, brightness, glare and aspect.

The system consists of a main program and five subsystems. The first three subsystems are used to investigate sunshine, shade and radiation parameters.

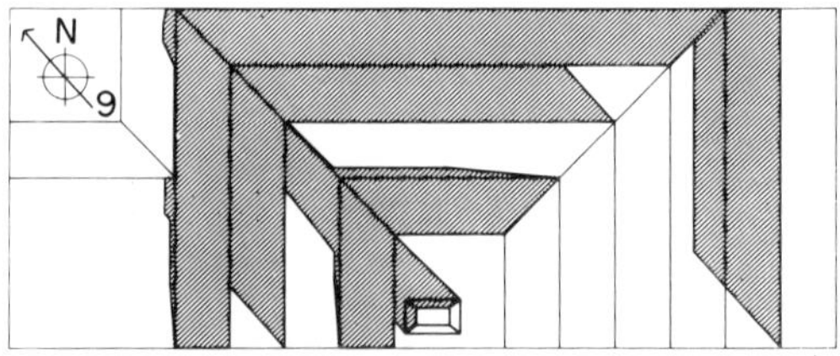

(9.00 am)

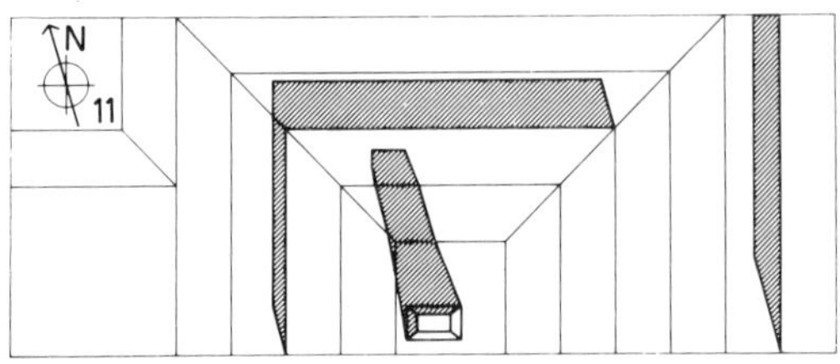

(11.00 am)

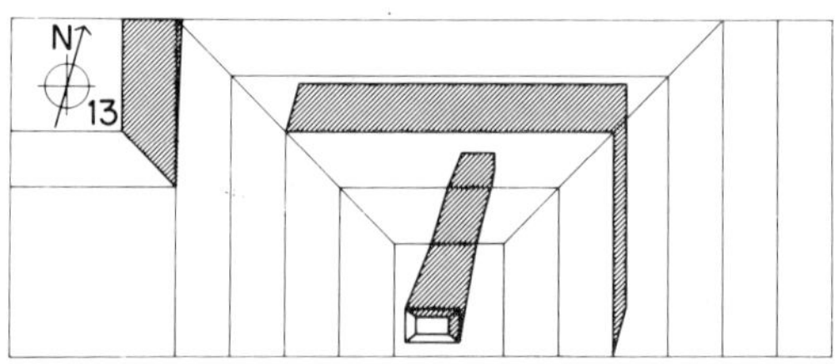

(13.00 pm)

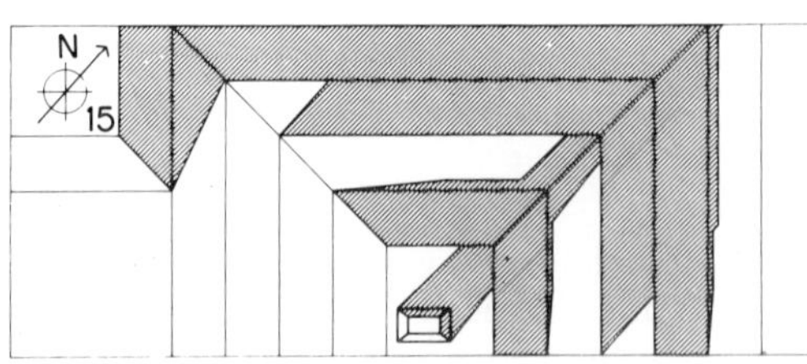

(15.00 pm)

Fig. 3.2 Impact of building and lay of ground considered.

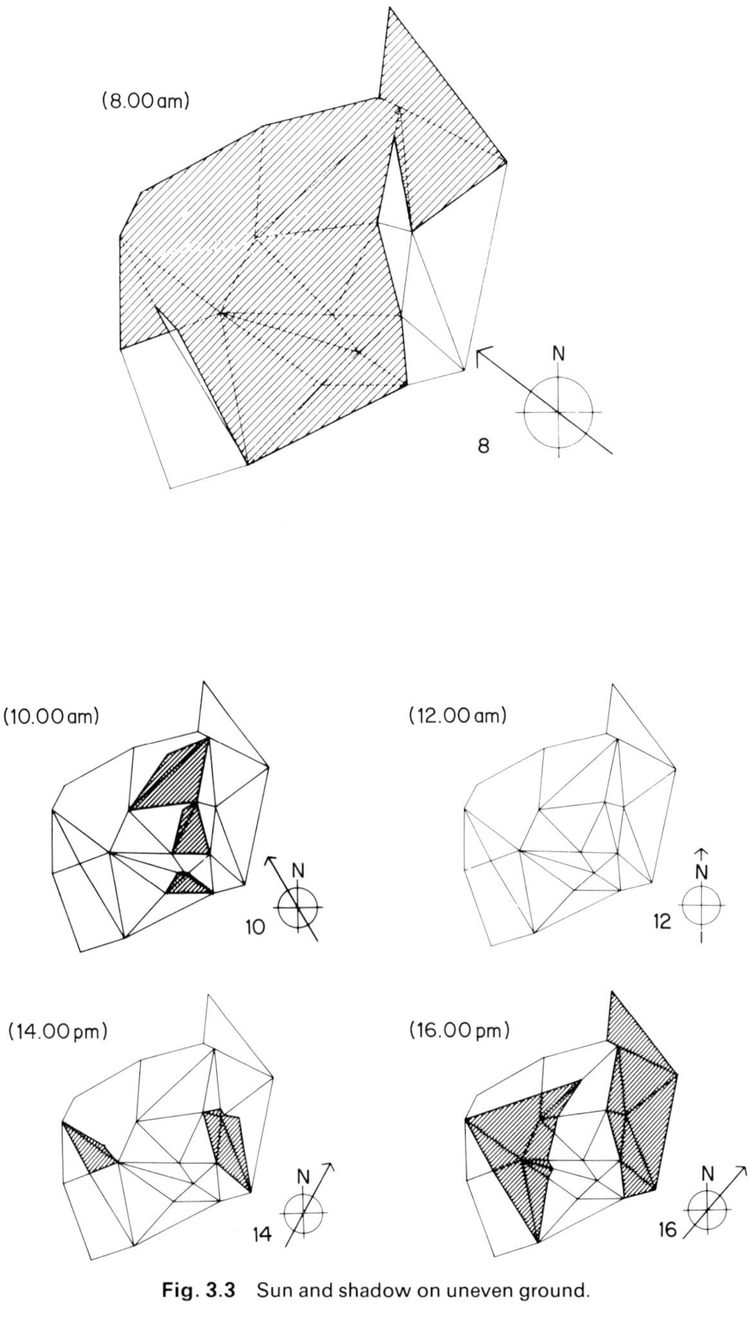

Fig. 3.3 Sun and shadow on uneven ground.

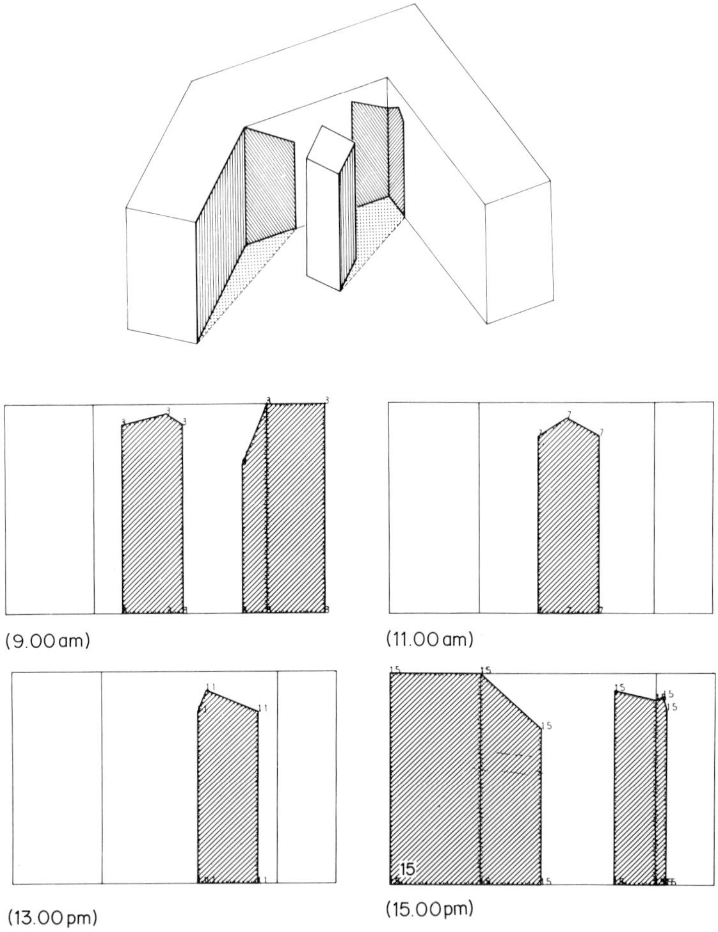

Fig. 3.4 Shadows cast on adjacent wall.

The fourth subsystem assesses lighting, spaciousness and view, whilst the last is used to study the effect of reflected light. The functions of the five sub-programs are as follows.

(1) The shape of shadows cast at various hours of the day are drawn; this produces a distribution chart of shaded and illuminated time for the building site (Figs 3.1–3.3).

(2) More detailed information, such as the start and end time of shadow is determined, plus the variation of illumination during the day at any selected point.

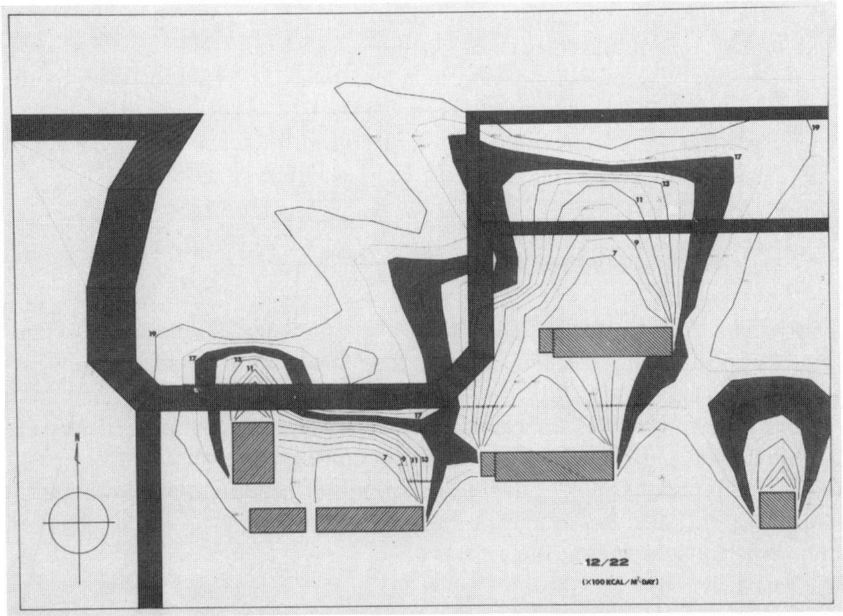

Fig. 3.5 Distribution of solar heat received on uneven ground allowing for impact of shade cast by buildings.

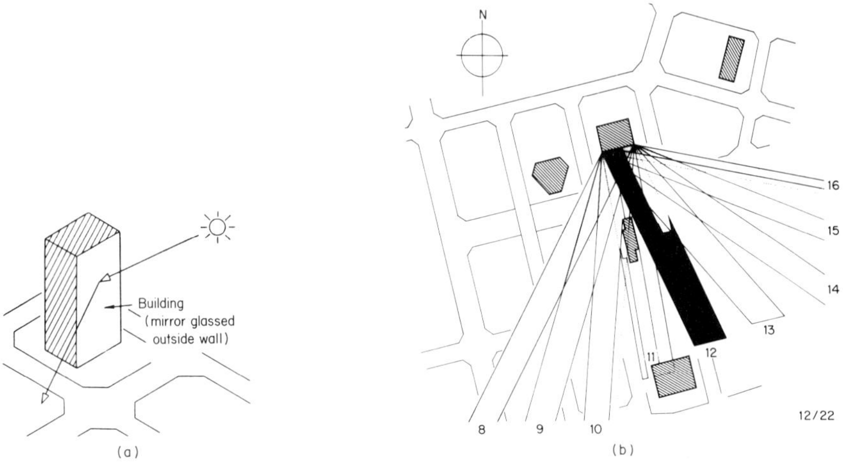

Fig. 3.6 (a) Reflections locus drawing. Assessment of sunlight reflected from outside wall. (b) Impact on adjacent buildings.

(3) Monthly or annual variation of shadow and radiation can be also determined from actual weather data (Figs 3.4 and 3.5).
(4) The intensity of illumination is calculated. The range of view to the human eye is also determined (Fig. 3.6a).
(5) The locus of reflected light by mirror-glasses attached to the surface of the structure is calculated and drawn (Fig. 3.6b).

3.3 System Applicability

The system is designed for use firstly in the planning and design stages of construction of buildings, reclaimed land, green areas and agricultural land. Its usefulness is also demonstrated in analysis of sunlight-related environmental factors, especially psychological factors such as aspect for which standards of quality have yet to be established. The radiation subsystem not only estimates the impact of any decline in the amount of heat by solar radiation but is also appropriate for other applications, for example in development of more efficient solar energy collectors.

4

Land Development Design System

Katsumi Takizawa, Shimizu Construction Co Ltd, Tokyo, Japan

4.1 Introduction

Two decades have passed since computers were introduced into the construction industries in Japan. In the early years they were used simply to process vast amounts of numerical data with precision and at a high speed. Subsequent software development has upgraded and diversified computer usage in civil engineering and building. Problems which used to be difficult to deal with because of their complexity (such as the analysis of internal soil and rock formation, or the dynamic stress analysis of building structures) were among the first to benefit. As a result, the level of structural design in civil engineering and architecture has improved markedly.

Initially, in the field of land planning and design, computer application lagged by at least five years. In the last decade, however, there has been a development of computer peripherals capable of rapidly handling graphic information. A number of programs which utilize computer graphics for visualizing topological information have been developed. As a result, we have seen a rapid development of computer application in land planning and design in the last few years.

4.2 Japan's Geographic Characteristics and Land Development

4.2.1 Topographic Characteristics of the Japanese Archipelago

The Japanese Archipelago, located at the eastern end of the Eurasian conti-

CAD/CAM and MIS in Japan
ISBN 0-12-682580-7

nent, is a group of islands stretching over 3000 km from north-east to south-west in an arch shape. It is composed of four major islands and about 4000 small islands, with a total land area of about 378 000 km^2.

"Habitable" land, which is usually interpreted as having a slope of eight degrees or less, accounts for only 31% of the total land area; consequently, Japan has one of the highest population densities in the world.

4.2.2 Historic Trends of Population and Land Use

When Japan opened its doors to international trade and began to adopt modern western industries at the end of the Edo period, population started to grow at a rapid rate. It reached 50 million by the end of the Meiji period (1868–1912) and 80 million by the end of the Second World War (1945). This rapid increase of population was accompanied by swift urbanization, and in particular a heavy concentration of population and industries in the Tokyo region.

As post-war industrial recovery was followed by a period of continuous and high rate of economic growth, population exceeded 100 million, creating serious social problems such as environmental pollution and regional disparities.

Recently, policy placing more importance on the growth of regional urban centres gained recognition. This shift in priorities in urban development led to our development of a computer-supported land development design system. Changes are now being made from previously uncontrolled conversion to urban use of agriculturally productive low-lying land, to the utilization of untouched hilly land.

4.2.3 Land Development after the Second World War

In the first decade after the Second World War, the urgent priority was to restore land which was in a state of ruin. There were problems demanding immediate attention such as reforesting and flood control, restoration of urban infrastructure, increase of food production, reconstruction of industries and so forth. No significant project in land development was undertaken during this period.

In the 1960s and early 1970s, Japan had achieved a continuous and rapid rate of high industrial growth, helped by low energy prices and free international trade. The main thrust of industrial growth was in the heavy chemical industry. In order to provide the necessary infrastructure for this industrial development, a number of large-scale construction projects were carried out in hydro-electric generation, water resource development, expressways and supertrain networks, deep-water port facilities, and industrial land development, all at the same time.

With industrial development, concentration of productive activities and population into large urban centres was intensified. This movement was responsible for the "land boom" in and around large cities, creating a large demand for residential housing, and public and recreational facilities.

These large-scale and speedy construction projects would not have been possible without the introduction of large construction machines and new civil engineering technology. This was revolutionally progress compared with pre-war technical standards. Successful as it was, it created a myth that through the development of technology nothing is impossible. No questions were raised about the wisdom of changing the natural environment in such a large scale. However, by the early 1970s this unabashed optimism generated by the unconditional belief in efficiency and productivity was about to be challenged.

In land development, as everywhere in the country, a civic movement has been organized to protest against further destruction of the natural environment. Simple application of regulatory measures or engineering solutions alone could not solve these disputes; a holistic attitude towards solving this problem by all those involved is necessary.

4.3 Concept of the Computer-Supported Land Development Design System

4.3.1 Social Background

Generally speaking, those engaged in engineering are target-oriented which up to the present has benefited the development of technology. However, to cope with the complexities of present society this is no longer sufficient; all the implications of what is proposed, technical, social, economic and cultural, must be examined.

At the initial stage of the development of a computer-supported land development design system, we therefore examined the background of the project in order to analyse its significance and role in the social context. What follows is the result of this examination.

4.3.2 Land Development Projects and Their Participants

Any land development projects are founded on a complex system of relations with their environment. To examine their true feasibility, we should study not only the prevailing conditions but also the social implications. This requires a careful survey and analysis of regional conditions, without which proposals may not be realized at all.

Traditionally, the role of each participant was clearly defined and it was possible to plan a project so long as each performed his defined role within his domain of responsibility. This is no longer the case because of the increasing complexity of our contemporary society. As everything is interrelated to (and interdependent upon) everything else, it is no longer possible to plan on a simple delegation of responsibilities. At each stage of the planning process, there is a need for the co-operation of all participating professionals to integrate and exchange their views across disciplinary boundaries.

4.3.3 Elements of the System

Our task is to find out how computers can be utilized in a planning process which is experiencing a shift in priorities. We have to understand what computers can do and what they cannot do if we want to use them for our benefit. Tasks which computers perform extremely well include massive data handling, high-speed scientific calculation and precise and fast graphic output. Possible objectives for the introduction of computers into the planning and design process are labour saving, improvement of design quality through fast data handling and optimization of design through mathematical programming and visualization.

In order to decide what jobs can be assigned to computers for benefit in land development design, it is necessary to analyse the work of engineers in practice. Its stages may be summarized as follows:

(a) examination of topography;
(b) on-site inspection and other surveys;
(c) analysis of the topographic characteristics of the site;
(d) formulation of goals and boundary conditions;
(e) sketches of the plan;
(f) presentation of tentative plans in terms of drawings;
(g) technical examinations of feasibility;
(h) evaluation of possible alternatives;
(i) preparation of working drawings.

In each of the above stages, division of labour between man and machine can be specified on the basis of their relative merits.

Computers are best utilized in those procedures which require little or no human judgement but which require fast data handling, or graphic presentation, or both. Corresponding to stages (c), (f), (g) and (i) above, we therefore selected the following four items as our development priorities:

1. Tools for land form recognition.
2. Tools for simulating alternative plans.
3. Tools for technical feasibility studies.
4. Automated draughting system.

To assist co-operation among experts from different backgrounds, efforts should be made to make computer outputs as legible as possible and to make best use of graphic presentation.

4.3.4 System Development Policy

Development of a large software system such as the one we envisage requires much effort and large investment. No matter how technically good the system, it has no economic merit if its level of precision is beyond that required, or if it is too difficult to use. This was avoided by adopting a policy of incremental improvement. Once the major structure of the system was decided, any improvements were immediately tested on the job, and worked upon. We believe it is important to structure a large system of this type from the start in such a way that each element can be readily withdrawn (or added to) without jeopardizing the effectiveness of the system.

Our experience in computer application in land development and earthwork design shows that a technically sound program may fail to achieve its objective because of a conceptual gap between programmers and users. In order to develop a system which is truly effective and user-friendly, it should be programmed by one who uses it as a planner. It is therefore desirable for a planner to acquire the necessary programming skills to develop a system which best fits his job, which is the practice of our system development policy.

4.4 Outlines of the System

4.4.1 Necessary Peripherals for Graphic Output

Many different types of peripheral devices have been developed for graphic output. They may be classified into one of two types: that which is directly connected to the central processing unit (CPU), and the other which is offline but communicates with the CPU via magnetic tapes.

On-line devices, which are usually connected directly to the CPU, include electrostatic plotters, computer output microfilming devices (COM) and graphic displays. Since these devices do not involve the mechanical movement of a "pen", plotting time is short. Graphic display is almost mandatory for an interactive system, allowing step-wise checking of graphic display.

Off-line devices usually use a mechanical plotting pen with a separate control unit equipped with a magnetic tape handler. Although precise drawing up to A-O size is possible, their draughting speed is generally slower than on-line devices.

Basic instructions such as the plotting of letters or lines are best prepared as subprograms which may be called by other main programs. A set of well-prepared subprograms allows relatively easy handling of complex graphic output.

4.4.2 Digitization of terrains

Terrain information is usually presented graphically in map form. Since digital computers handle digitalized information only, it is necessary to transform the mapped information into numerical data. We call this digitalized terrain information a "digital terrain model" (DTM).

The DTM expresses a terrain as a set of discrete points with x, y and z coordinates. There are several types of DTMs depending on the method of selecting data points, but any DTM must satisfy the following conditions:

(a) efficient collection of numerical data;
(b) ability to replicate terrain to the desired level of precision with a minimal number of data points;
(c) efficient calculation of the ground height at a given location.

Among various DTMs, the one which uses a square grid for data collection has the merit of easy sampling (relative to random spatial sampling, for example) and simplifies mathematical handling because the distance between data points is regular, allowing very fast calculation of the ground height at any given point. With a square grid of 10–20 m between sample points, the ground height can be estimated to within 30 cm plus or minus if the terrain is replicated by a cubic function.

We therefore decided to employ the square grid DTM for our system. Two different methods of data collection are possible at present: one which uses a digitizer to read contour maps and estimates the ground height at grid points from the data along contour lines, the other which manually reads directly from the map with a square grid overlay. Ground heights are interpolated by approximating the terrain with a cubic function from data at the nearest sixteen grid points.

Once the terrain information has been digitized and the method of interpolation of ground heights established, it is possible to analyse the given terrain by a computer.

4.4.3 Analysis of Terrain Characteristics

Very often not enough care is taken in the design of earth-work since it is not technically difficult to move a large volume of earth by construction machines. As a result, there have been few efforts in research and development in this field in comparison with those made for bridges and tunnels.

Without a thorough analysis of the terrain characteristics, however, it is almost impossible to develop design images for earth-work. To realize this, one need only to think about such fundamental problems as balancing the amount of banking and cutting, changes in drainage patterns and discharge, preservation of vegetation, or fitness of artificial land forms with the natural topography. The more complicated the terrain, the more detailed should be the analysis.

The following five items are the geometrical characteristics of the terrain which can be analysed on the basis of the DTM and presented quantitatively in graphic forms.

1. Gradient

The predominant gradient and its direction are calculated for each mesh from the ground heights at its four vertices. An arrow in each mesh shows the direction of the main slope and the number defines the gradient in degrees. A circular graph of the frequency of slope directions and two histograms (one showing ground heights and the other gradients) are then produced.

2. Drainage pattern

We can obtain the streamline of surface water for each mesh by selecting the steepest line connecting the centre of the mesh with the centres of the adjoining eight meshes. By cumulatively counting the number of meshes in the upstream, it is possible to calculate the total area of watershed.

3. Ridges and troughs

Meshes may be classified into ridges and troughs by checking the convexity of the curved surface of the cubic function approximated from the ground heights at the adjoining sixteen data points. Shaded areas with a letter 'R' indicate meshes with a ridge form (Fig. 4.1).

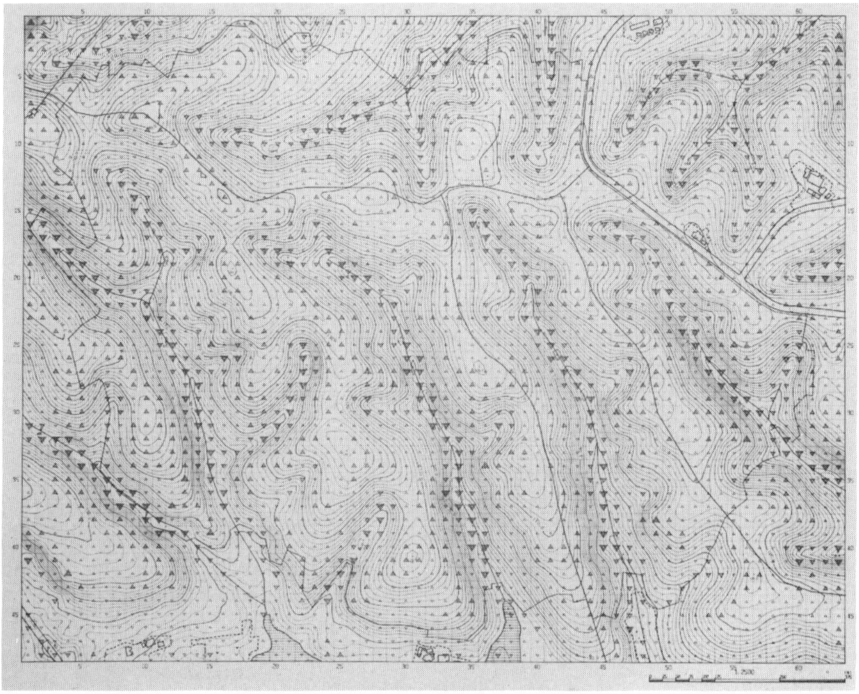

Fig. 4.1 Ridges and troughs.

4. Perspectives
Figure 4.2 is an example of a three-dimensional perspective made from the grid data. The picture was drawn by connecting the ground heights of cross-sections interpolated from the grid data perpendicular to the line of view.

5. Hours of sunshine and the amount of solar energy
It is possible to calculate hours of sunshine and the amount of solar energy on a given day of the year from the DTM. Input data are the height and orientation of the sun, the gradients and slope directions of meshes, and the ground heights of nearby high points which may cast shadows.

4.4.4 Simulation of Plan Proposals

Once the terrain characteristics are analysed, the next step is to develop images of plan proposals which would fit the stated goals and objectives and external conditions. Our practice is to develop several alternative plans and to simulate them on the DTM in order to examine and evaluate their implications. Results of the simulation are given in graphic form for easy comprehension.

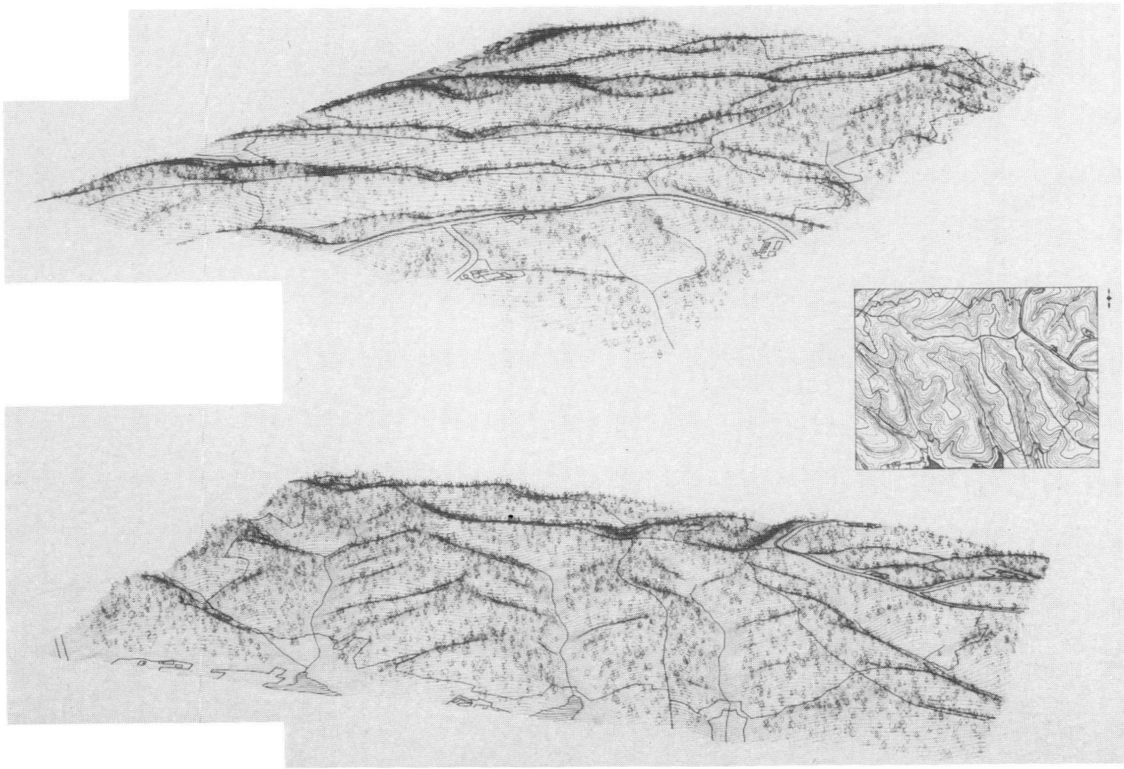

Fig. 4.2 Perspective of existing terrain.

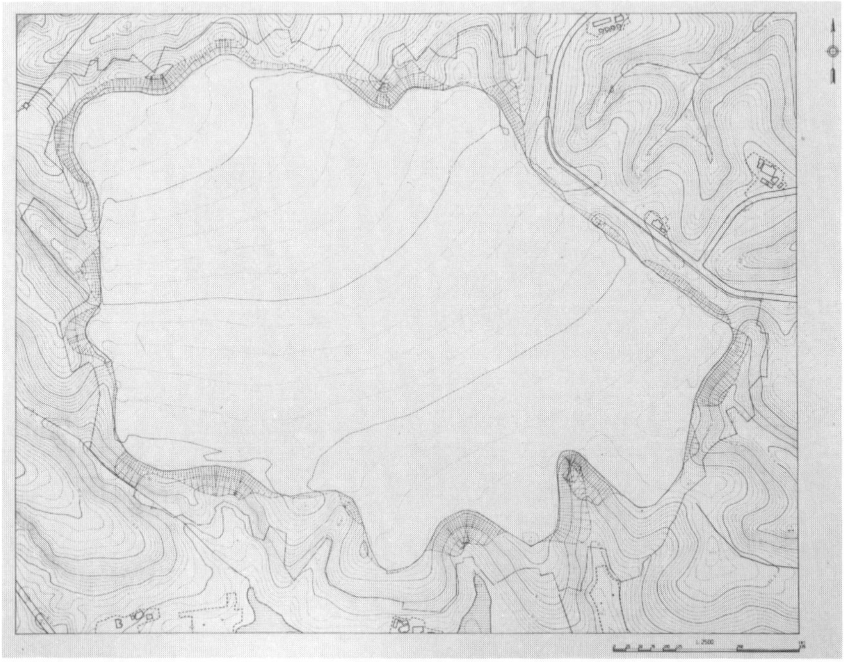

Fig. 4.3 Contour map of planned ground heights.

1. Automatic setting of planned ground heights

The purpose of earth-work is to obtain the land form which will meet project objectives. It is certainly desirable if this is to be achieved with the least change in the natural terrain. Once the planned ground heights are decided, one should try to make the earth-volume to be moved as small as possible, taking the balance between banking and cutting into consideration. Making excess gaps between the planned area and the surrounding area at the boundaries of the site should also be avoided.

The basic principle of the automatic setting of planned ground heights may be described as incremental smoothing by gradually "shaking" the proposed site. When mesh gradients reach the allowed maximum after a number of "shakings", the optimal planned ground heights which require the least earth-volume are obtained together with a balance between banking and cutting (Fig. 4.3).

2. Calculation of earth-volume

Earth-volume is calculated by taking the ground height of each data point as the average height of the corresponding mesh. The difference between the

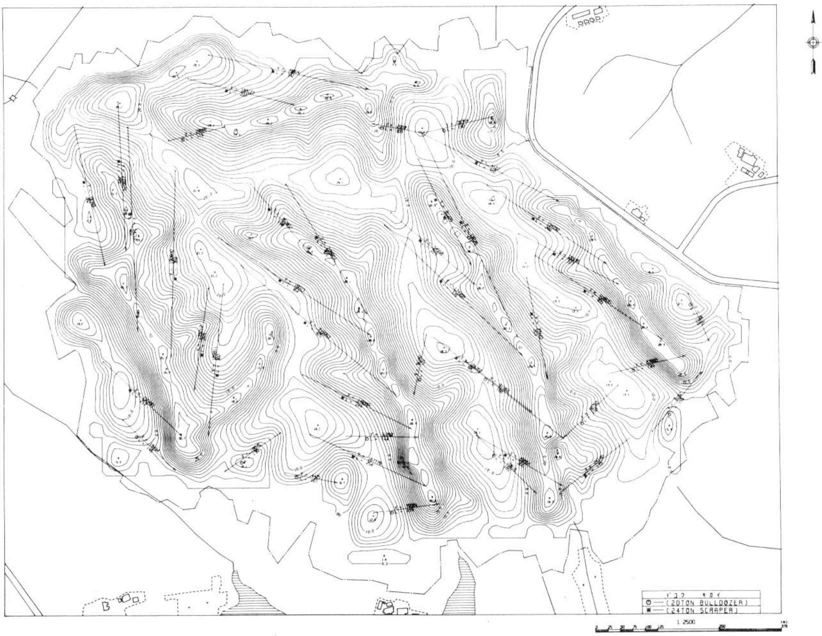

Fig. 4.4 Earth moving plan.

original height and the planned height is multiplied by the area of a mesh to obtain the volume of earth to be moved. This method is sometimes called "column method".

3. Earth-moving plan

Earth that has to be cut and banked is distributed widely over the site. The earth-moving plan specifies which earth has to be moved and which place has to be banked. The structure of the problem is similar to the standard "transportation problem" of linear programming. Places to be cut are the sources and places to be banked are the destinations, and the cost of moving earth is the transportation cost. Earth-volume is the weight to be loaded. The objective of the program is to minimize the sum of products of earth-volumes and distances, assuming that transport cost is proportional to the distance (Fig. 4.4). An approximate cost estimate can be made once the type of earth-moving machine is selected to determine the unit transport cost.

4.4.5 Technical Examination of Proposals

Parallel to the simulation, alternative proposals should be examined from various technical aspects. These are three examples of this procedure.

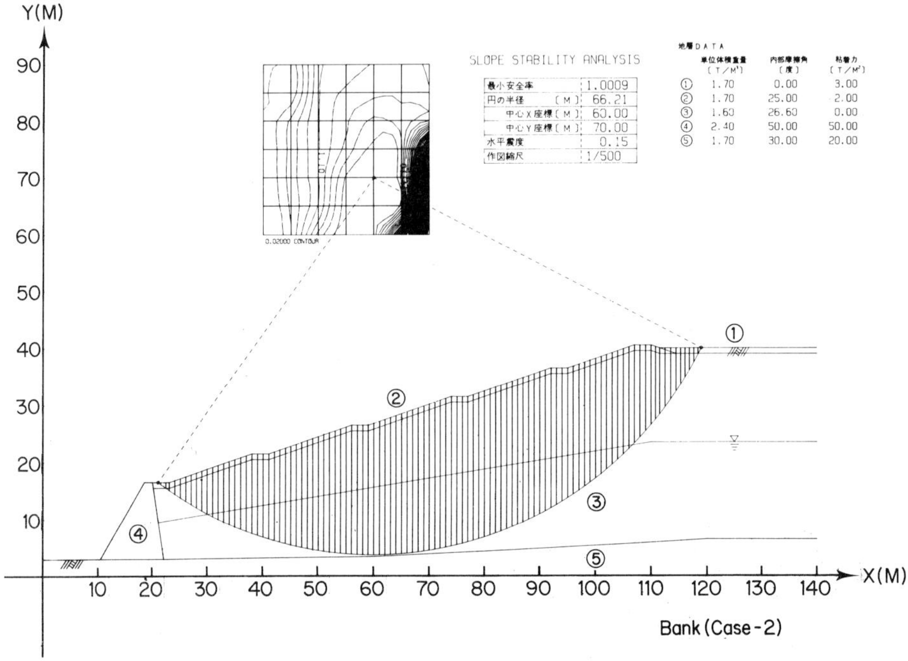

Fig. 4.5 Slope stability analysis.

1. Checking increased discharge

Where land development takes place, the treed area which can retain surface water for a longer period, usually declines and is replaced by building sites and pavements. The resulting increase of discharge places a heavy load upon the rivers downstream. Regulating ponds can temporarily retain the increased amount of discharge and help regulate the outflow.

It is possible to calculate the necessary volume of the regulating pond by calculating the amount of precipitation, inflow and outflow in time series.

2. Stability of slope faces

Man-made slope faces should not slide under their own weight and should remain stable against erosion and earthquakes. We must select the appropriate height and gradient, taking the soil characteristics of the site into consideration. Retaining walls may have to be constructed when sloping does not provide enough protection. Since there are a number of different types of retaining walls, one should select the most appropriate.

Figure 4.5 is an example of the output of the slope stability analysis.

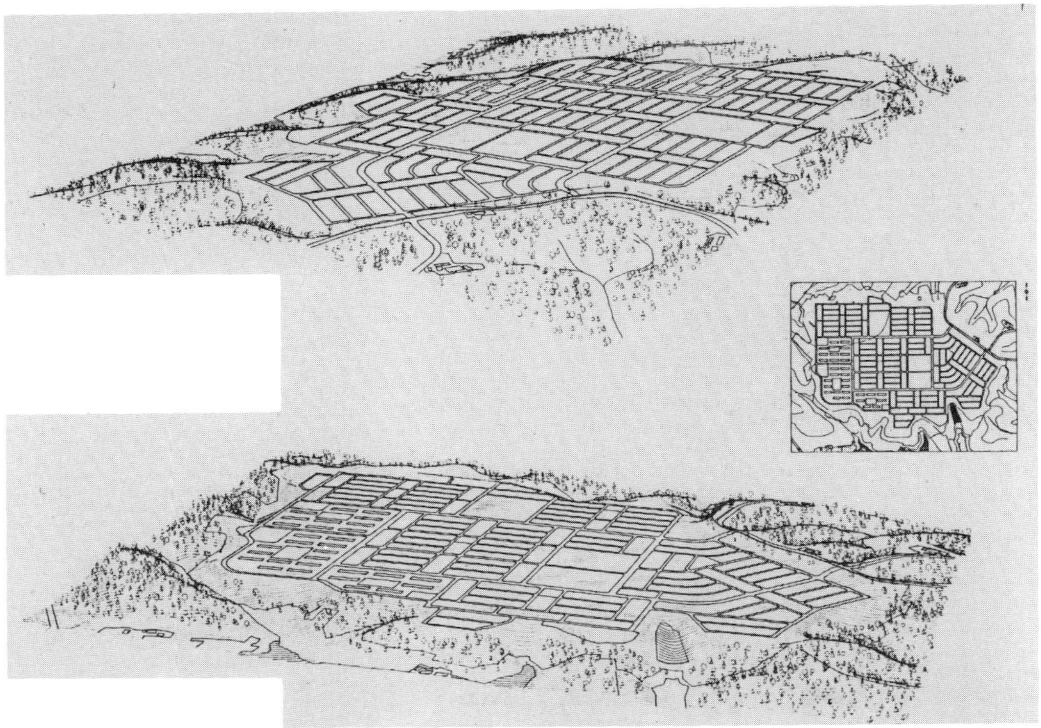

Fig. 4.6 Perspective of proposed plan.

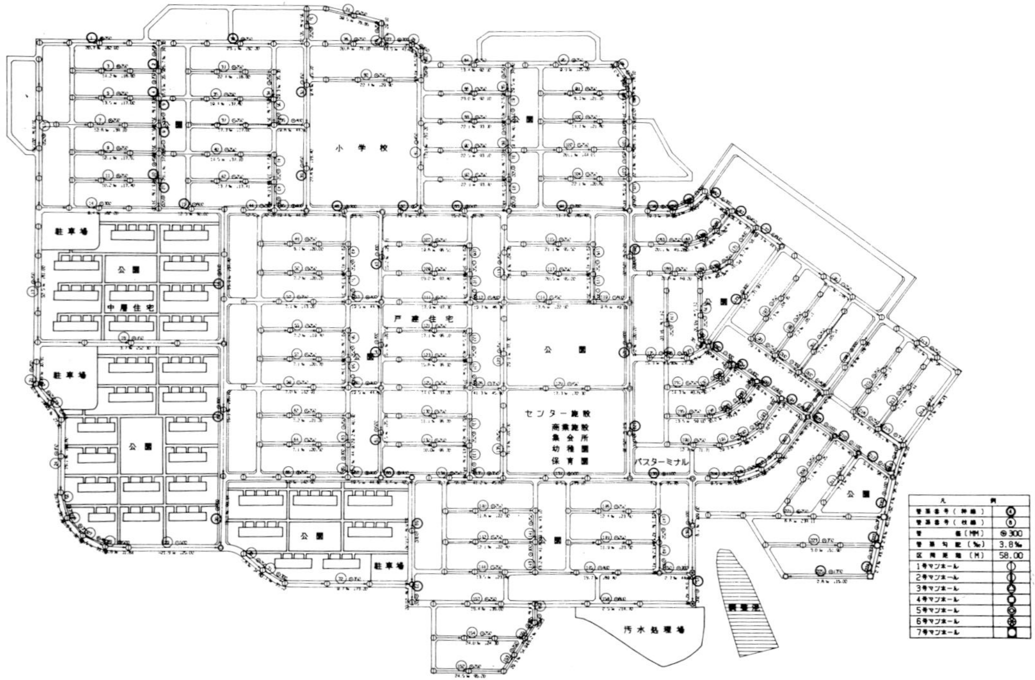

Fig. 4.7 Land development plan and rainwater sewer plan.

3. Landscapes
We can predict the topographic characteristics of a proposed land development by applying the same tool developed for the analysis of existing conditions. Figure 4.6 is an example of the perspective drawing of a proposed land development. By comparing Fig. 4.6 with Fig. 4.2, we can visually check changes in landscape caused by earth-work.

4.4.6 Automated Design and Draughting System

When the optimal alternative is finally chosen, the next step is to proceed with detailed design and prepare working drawings. In land development this procedure is relatively standardized, compared with other fields of civil engineering. Although many drawings have to be produced, their formats are few in number and much of the work is routine. This field is therefore suitable for the development of an automated system.

However, experience shows that complete automation of every aspect of design is not economically viable: provision should also be made to allow minor manual corrections to the output. It is also necessary to make the system interactive in such a way that a designer can change input data so that every stage of the design procedure can be monitored to reflect his design intentions, and which require his judgement.

Figure 4.7 shows a residential development plan produced by the automatic draughting system. Only the locations of centre lines of rights-of-way and their widths and turning radii need be specified to produce a residential development plan. In the case of the sewer system, pipe dimensions and gradients are automatically selected and their plans and cross-sections prepared.

Our design system for roads produces plans, cross-sections and longitudinal-sections and calculates the volumes of earth to be moved.

4.5 Concluding Comments

Computer applications in land development and earth design will certainly advance in the future, allowing planners and designers to prepare better land use plans. However, no matter how far a computer-supported design system is developed, we should never forget our desire to build an environment suitable for living, working and playing.

5

Energy Conservation in Building Design

Takenaka Komuten Co Ltd, Tokyo, Japan

5.1 Introduction

In recent years, the problem of resource conservation has been forced upon all aspects of industry. To cope with a lack of energy resources, it is essential to ensure a stable supply of petrochemical energy (for which Japan is highly dependent on overseas supplies), and to encourage energy conservation coupled with the development of new energy sources.

In the building field, energy conservation covers a wide range of functions. Air conditioning and heating, lighting, ventilation and hot water supply account for about 60% of the domestic use of energy, and this percentage is expected to increase in future.

In the building industry, reducing energy consumption as much as possible without harming such environmental aspects as comfort, safety and sanitation, fulfils a social responsibility with subsequent economic advantage.

To meet such social needs and promote energy conservation in building, Takenaka Construction has developed its Energy Conservation Building Design System. Here, correlation techniques and data integration are used to analyse the building with emphasis on energy conservation. Software such as load analysis programs are integrated with the energy-saving system.

The system is currently incorporated in the integrated planning-design-construction set-up of the company. The design and construction of energy-

CAD/CAM and MIS in Japan
ISBN 0-12-682580-7

efficient buildings achieve a high technical level and, depending on building designs, attain 20% to 50% energy savings, compared with past designs without system analysis.

5.2 Outline of the Development

5.2.1 Fundamentals of System Development

Before the development of this system in the company energy conservation in buildings had been done at the research laboratory and design department. This involved a piecemeal introduction of techniques, dependent upon the designer's own ideas. In order to systematize energy conservation equally throughout the company, the following fundamental policy decisions concerning the system were taken.

Flexibility

—Ability to cope with buildings of any type.
—A system where individual judgement may be applied in the course of design.
—Can readily cope with alterations in the design.
—Can be utilized at any phase of the design.

Versatility

—The software shall be designed to be versatile.
—Contents shall conform to legal standards.

Wide application

—Shall be usable as a tool for planners and designers.
—Usable in presentations to parties concerned.

Savings in manpower

—Readily determine design conditions.
—Select energy conservation techniques and calculate the effects quickly.

Shall be open to future development

5.2.2 Organization of Development

The formation of the software and hardware was carried out under the following guidelines.

(1) The needs of the respective departments.
(2) Organizational considerations relating to the speedy transfer of results.
(3) Smooth transfer from program development to application.
(4) Techniques required in the various specialized fields.
(5) Flexibility regarding time and order.
(6) Backup functions necessary inasmuch as this is a company-wide, long-term project, involving co-ordination of departments, policy and dealings with outside firms.

Whilst developing the system, we have made changes in the development organization and have reallocated members of the production staff.

Every year experts from the laboratory, building and mechanical engineers from all the design offices, and computer engineers meet to form a development working group. The executive director acts as the project leader. The Technical Administration Department of the Office of Production Management takes charge of the Project Promotion Secretariat and manages the company-wide development organization. Project co-ordination is achieved through a committee consisting of the heads of departments.

5.3 Outline of the System

This system is designed to achieve comprehensive energy conservation in buildings from the planning stage to completion. It is centred around the unsteady thermal load analysis program, which plays an important role in designing energy conservation buildings. The system comprises the "Energy conservation design manual", the "Building thermal load analysis program—THERMACS", "New technology of energy conservation", "Air conditioning system measuring manual" and the "Energy conservation building enforcement group". These techniques are systematically applied in the existing design flow. (Fig. 5.1).

5.3.1 Energy Conservation Building Design Manual

This manual consolidates the existing energy conservation techniques, rules and regulations regarding buildings and facilities. It has been compiled in

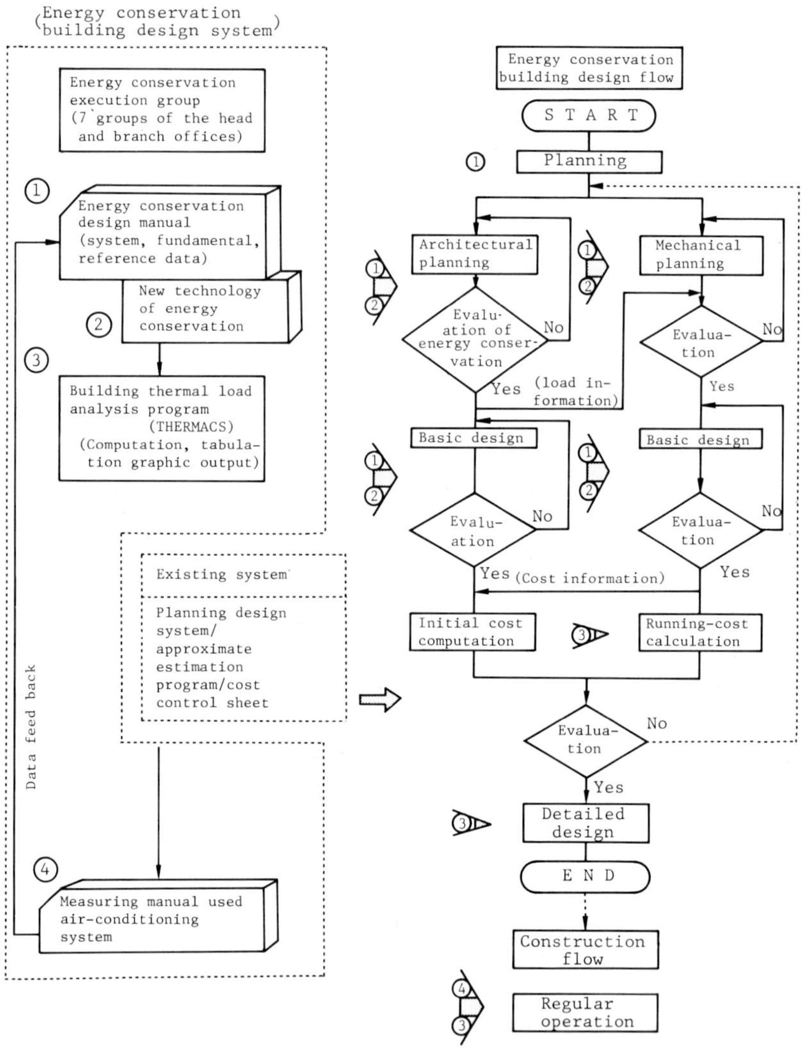

Fig. 5.1 Formation of the system and its involvement with the flow of design and construction.

Table 5.1 New technologies for energy conservation

Technology	Item	Effect
Apparatus of heat source	Reciprocal heat recovery, air thermal source heat pump (TACNES-HR)	High efficiency utilization of energy and use of surplus energy
	Heat balancer	Utilization of surplus energy
Energy transportation	Power recovery pump	Utilization of surplus energy
	Zoning of C/H water transportation	Optimum utilization of energy
Operation and maintenance of building	Building management system by computer (optimum control)	High efficiency utilization of energy
	Outdoor air load control system	Reduction of energy demands
	Insulation load control system	Reduction of energy demands and utilization of natural energy

two parts: the Fundamental Volume and a Reference Volume which acts as a check list for design.

The designer can systematically introduce energy-saving mechanisms from the planning stage, and can perform detailed studies of the various techniques. Preparation of the input data for a load analysis program, planning, building facilities and basic design can all be effectively performed.

5.3.2 New Techniques for Energy Conservation

The basic requirements for energy conservation in buildings are as follows.
(a) minimizing the energy required;
(b) utilizing energy efficiently;
(c) utilizing surplus energy effectively.

In compiling the foregoing design manual, the new techniques shown in Table 5.1 have been added to existing techniques, enabling more effective energy conservation design.

5.3.3 Building Thermal Load Analysis Program (THERMACS)

Meteorological, building and thermal load data are all necessary in the design of energy conservation for buildings. This requires considerable calculation, tabulation and analysis of results. In the past design was performed based on the designer's knowledge of steady thermal loads; it could not also cope with energy conservation. The present program was therefore developed as an essential element in energy conservation applied to building design.

Outline of the program

The original program which was based on the Program HASP/ACLD (developed and proposed by the Society of Heating, Airconditioning and Sanitary Engineers of Japan) and relying on the unsteady heat transmission calculation method, has since been widely extended (Fig. 5.2). Programs for tabulation, statistical treatment and graphical output have now been developed. In order to resolve past problems, these functions have been combined with development of a program to enable the study of energy conservation at any given design stage.

Control of the calculation routine is shown in Fig. 5.3, and the handling of input-output data is indicated in Table 5.2.

Effect of the program application

The application of the program at a given stage in the design work is shown in Table 5.3. The program can also be applied in the simulation of energy saving measures for existing buildings, and as a technique for the study of improvements.

5.3.4 Manual for Measuring Related to the Air Conditioning System

This manual was prepared as a back-up function for the building energy conservation design system. It was compiled as a common guide for personnel at the planning stage of building, and for those evaluating the energy conservation design after completion of the building (Fig. 5.4).

5.4 Effects of Application of the System

This system is extensively used in the design departments of the head and branch offices of the company, and to date about 40 energy conservation buildings have been designed (326 buildings in 1983). These extend from

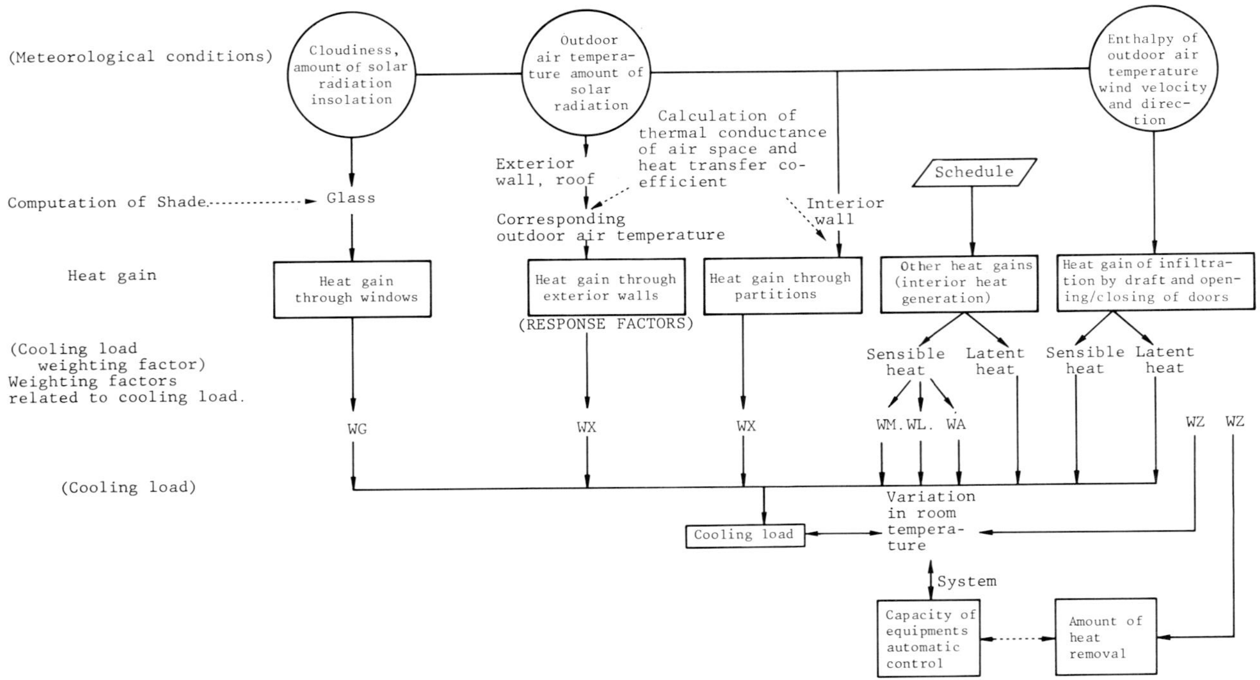

Fig. 5.2 Unsteady heat load calculation flow.

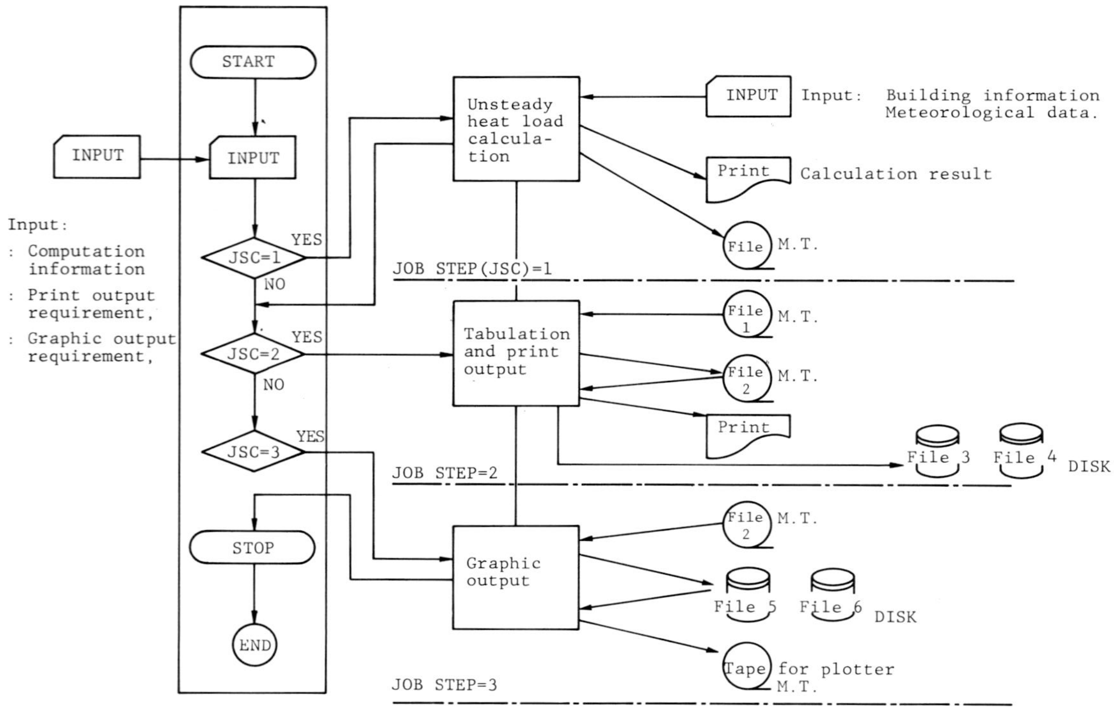

Fig. 5.3 Flow chart of computation.

JSC	Print output requirement (IPC)						Graphical output requirement (IGC)				
	1	2	3	4	5	6	1	2	3	4	5
1	●	●	●	●	—	—	—	—	—	—	—
2	●	●	●	●	●	●	—	—	—	—	—
3	●	●	●	●	●	●	●	●	●	●	●
Remarks	By time, by position, details Close to HASP/ ACLD/ 7010	Statistical values (histogram) Close to HASP/ ACLD/ 7301	By time, by element, important with ±	By time, by element, details, with ±	Daily integration, monthly integration, important with ±	Daily integration, monthly integration, details, with ±	Max./min. load variation including peak time; weekly	Max./min. weekly load variation including peak day (daily integration)	Load variation over any period on hourly basis	Load variation over any period on daily integral basis	Load variation over any period on monthly basis

Note: ⤻ mark indicates one or the other

Table 5.2 Code relations between calculation information (JSC), print output (IPC), and graphical output requirements (IGC).

Table 5.3 Effects of applying the program in designing

Design phase	Main works	Type of output	Effects of application
Planning	Deciding shape and the orientation of building, deciding arrangement of the buildings	2	Heat load of the various parts of the building on a monthly and annual basis and the totalled values can be grasped, enabling design from the overall viewpoint
Basic design	Design of exterior walls, windows and eaves Feasibility study of heat recovery and design of energy storage capacity Design of heat-source equipment capacities such as refrigerators and boilers	1 2 4	Optimum heat design from the heat load and formation of the cladding under set conditions can be performed (reduction in energy demands) Heat load is output simultaneously as gain (+) and loss (−). Therefore, comparing re-use of heat can be judged and energy storage capacity can be obtained from the usable amount of the heat (utilization of surplus energy) Actual heat capacity, to be deducted from the heat load, can be obtained and the heat source capacity be selected (high efficiency utilization of energy)
Detailed design	Study of design of secondary side apparatus capacities and equipment operation patterns	1 3 4 5	Apparatus capacities considering safety factors (ratio of risk can be determined and study of control method matching the anticipated heat load pattern can be made)

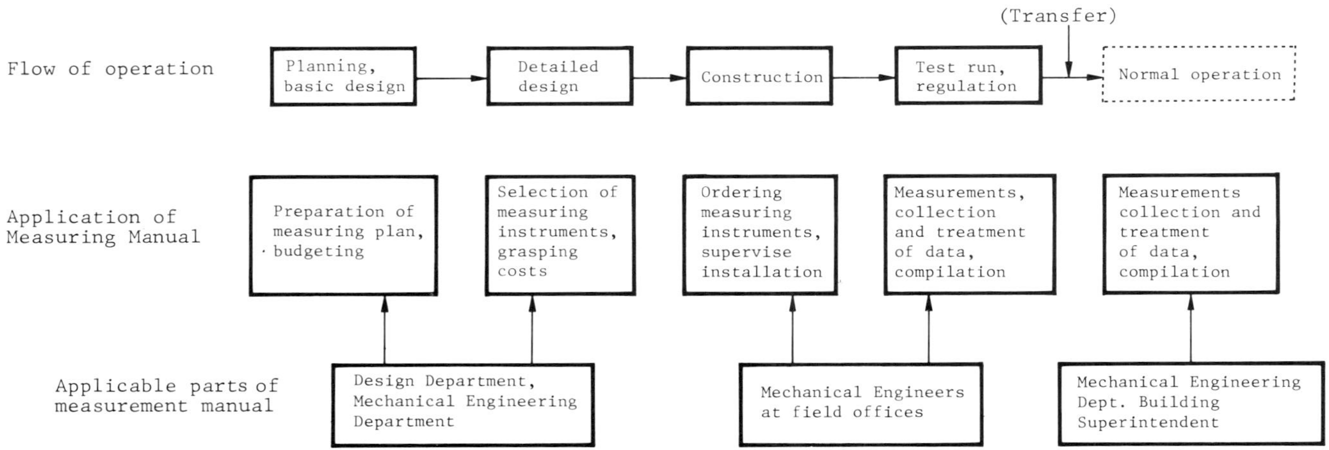

Fig. 5.4 Application of the measuring manual and relation of the various departments.

Hokkaido to Kyushu and Okinawa, and the energy saved in terms of power amounts to 32 560 000 kWh per annum (264 000 000 kWh in 1983).

In particular, by applying this system the Shibuya Building is expected to attain a 53% energy saving, but many other buildings have achieved even greater savings through its introduction.

Effective energy conservation design

At the planning stage, studies of energy conservation techniques can be carried out on building and facilities and maximum energy conservation can be designed when estimates are presented to the owner.

Improvement of quality of design

By standardizing a part of the total design operation and making possible unsteady thermal load analysis , designs of high accuracy can be achieved.

Effect of manpower saving

Through the selection of energy conservation techniques, thermal load analysis and facilities system studies, greater operation efficiency results, allowing the company to cope with any necessary changes in the design in a relatively short time.

5.5 Summary

A design system to cope efficiently with the social needs of energy conservation has been established with significant effect. This system is not designed for a single application, but for the wider propagation of energy conservation techniques. The details of this system should be applied to all buildings to be constructed, and feedback of collected data and improvement of the program over a long-term scale should be undertaken.

6

Japanese Publishing System

IBM Japan Ltd

6.1 Introduction

The Japanese Publishing System (hereinafter called JPS) was completed in November 1971, thereby realizing the long-cherished dream of the newspaper and the printing industries for a cold type system without lead type. JPS, a total system featuring its conversational mode of operation, is the first of its kind in the world in that it fully utilizes a computer system in the newspaper editing and production process. With JPS, all necessary activities such as input of news stories into the computer, proofreading, preparation of layouts and output of film on phototypesetting machines are handled on an on-line system.

The printing method which has been in use since Gutenberg's invention of alloy type in the mid-15th century is now out of date for the following reasons.

(1) The problem of noise and the possibility of contamination by lead can-

CAD/CAM and MIS in Japan
ISBN 0-12-682580-7

not be completely eliminated from the process of preparing original plates made of alloy type.

(2) Skilled workers are hard to find and the training is difficult, leading to a severe shortage of manpower.

(3) An ever speedier method of information processing is required to meet the needs of a rapidly changing society, and rationalization through mere mechanization (now seemingly approaching its limit) does not offer any fundamental solution.

JPS was developed to cope with the problems arising from the older printing method; it has been designed for use by both the newspaper and the printing industries.

6.2 Current Newspaper Processes

Most articles appearing in everyday newspapers are either collected by reporters or provided by news agencies. They are first channelled to respective contributing departments (politics, economics, sports, etc.), and after evaluating their importance, selected and reduced, converted into paper tapes and sent, if necessary, to the local head offices. Some news items are covered locally and sent to the head office from correspondents or branch offices, some of which are exchanged among the head offices. Paper tapes are used to obtain copies for easier transmission.

As can be seen from the figures, conversion into paper tapes allows concurrent handling of a series of operations such as transmission of news stories to remote locations, production of type by casters, and printing of articles on "monitor prints". Monitor prints are checked against the original manuscripts (proofreading), and any errors are marked on the monitor prints. They are then forwarded to the Editing Department, who then consider the amount of space which is available, discuss the use of particular articles, make appropriate "headlines", and specify boxes to be made. Layouts are made and reconstructed until just before printing.

Arranging and editing are the most important, complex and time-consuming activities in the whole newspaper production process. For example, in addition to the proofreading of monitor prints, the editing staff might decide on corrections which would necessitate resetting of the type. Also, with the Japanese language, there exists certain rules peculiar to the text composition which must be observed in setting lead type composition. This requires considerable skill and experience.

At the page-forming stage, photographs and advertisements are also put together and ink is applied to the rows of type bars to produce several copies for the final proofreading. Even at this stage proofreading is necessary to check for possible errors due to the resetting of types by hand.

Finally, a paper mat which is pressed on the rows of type bars forming a newspaper page is taken out and melted lead is poured into the mat to form a semi-cylindrical stereotype for the rotary press. The stereotypes cast in this manner are placed on the high-speed rotary presses which produce millions of copies of the newspaper.

6.3 Philosophy Behind JPS Development

The development of JPS began with a comprehensive survey and analysis of all the newspaper production processes. The philosophy behind the development can be summarized thus: "The conventional functions throughout the type printing processes should be retained as much as possible." This has been taken into consideration during the analysis of the conventional system. JPS, which has been designed with this in view, in (Fig. 6.1):

(a) an interactive system using display screen;
(b) a computer system in which all type printing functions are adopted;
(c) a general purpose system which may be introduced by any newspaper company.

JPS makes it possible to integrate various activities, from manuscript handling to preparation of original plates, into one whole system, without using alloy type, thus breaking new ground in printing processes, and solving the difficulties which conventional printing operations have by replacing manual work with an accurate, steady and highly productive computer.

JPS was developed in co-operation with both the Asahi Shimbun and the Nihon Keizai Shimbun. They added the user-supplied subsystems (user programs, system pattern files, etc.) to the system, and completed NELSON (New Editing & Layout System of Newspapers) and ANNECS (Automated NIKKEI Newspaper Editing & Composing System), respectively. They are now working on improvements and further developments of the systems, to ease the operation still further. These two systems are being used in part of their newspaper production.

6.4 Application of JPS

The process of producing a newspaper breaks down roughly into four operations:

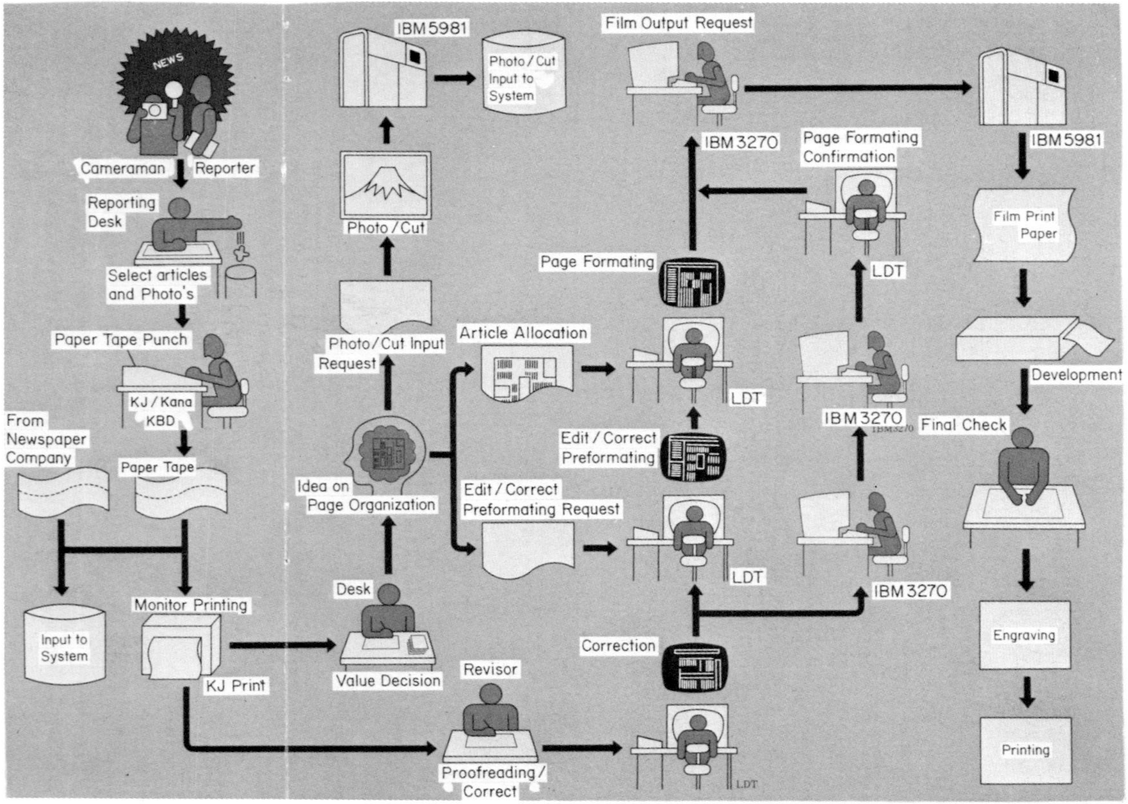

Fig. 6.1 JPS Job Flow.

Fig. 6.2 Proofreading on an IBM 5937 layout display terminal.

(1) Evaluating news stories and giving instructions concerning their handling.
(2) Correcting errors by proofreading.
(3) Laying out and composing pages according to a set of rules.
(4) Preparing high quality plates.

Although, for smooth handling, some human intervention is necessary for each of these operations a considerable part can be transferred to a computer system.

In newspaper production the following operations can be undertaken with a computer system:

(a) Modifying/adding to a data stream of news stories.
(b) Arranging elements of different sizes such as characters, illustrations and photographs within a certain region according to specified rules.
(c) Outputting final results on to original plates.

JPS has the necessary software to process these three operations, the instructional input from operator, and the various information devices such

Fig. 6.3 IBM 5981-D01 scanner/plotter.

as IBM 5937 Layout Display Terminal (Fig. 6.2), IBM 3270 Information Display System and IBM 5981 Scanner/Plotter (Fig. 6.3).

6.5 Input

6.5.1 Text

News stories from various contributing departments are first selected, then, with the specifications of date, page and edition, they are punched on to paper tapes. Not only the text of news stories, but also the necessary information such as the size/font of characters to be used, are also punched. If certain characters within a piece of news need to be changed, e.g. from Mincho (primer-face type) to bold-face type, instructions to that effect are punched.

Data specifying composition method can also be input, which is especially useful when using "patterns" (described later in the "Page-forming" and the "Programs" sections) to lay out a page all at once. In this case, however, appropriate layout specification information must be input beforehand by means of paper tapes.

Punched paper tapes are placed on an IBM 1017 Paper Tape Reader, and when the "READY" status is entered, the information on the tapes is automatically read into the system and stored on an external storage device.

The proofreading operation is performed by means of an IBM 5937 Layout Display Terminal (LDT). Since each LDT can be used for both proofreading and layout operations, the operator can select either before commencing.

When the identifier (ID) of a particular news story is keyed in on a LDT in the proofread mode, the news story together with the accompanying information is displayed on the screen in Kanji (Chinese) and Kana (phonetic) characters. The operator reads the story and makes the necessary deletions, insertions and replacement, using Program Function Keyboard (PFK), and Kanji-Kana Character Keyboard (KKB).

The operator indicates the error with a cursor and gives a command from PFK. This command is analysed by the Command Analyser routine of JPS, and the appropriate program performs the necessary modifications to the text. When the modifications are carried out, the images on the screen are changed accordingly, and the proofreading operation is completed.

6.5.2 Illustrations

Photographs and illustrations are input to the system by IBM 5981-D01 Scanner/Plotter. The photographs and illustrations set on the drum surface of IBM 5981 are optically scanned and digitized to store in the external storage device (Fig. 6.4).

IBM 5981 has two types of the scanning functions which can be selected by a switch; black/white scanning for monochrome line drawings/illustrations, and grey shade scanning for half-tone photographs. For monochrome drafts, the scanned density is compared with the predefined value and if the density is greater than the value, the scanned spot is considered black. If on the other hand the density is smaller than the value, it is considered white. For photographs, the density is classified to 256 levels of meshpoints, which will be converted to 51 levels when it is output on to film/printing paper.

It is also possible to adjust the tone of the photographs at the density level conversion. With this function, the final output has a uniform tone although a draft's tone is either dark or light.

For photographs, there is usually a difference in size between the scanned

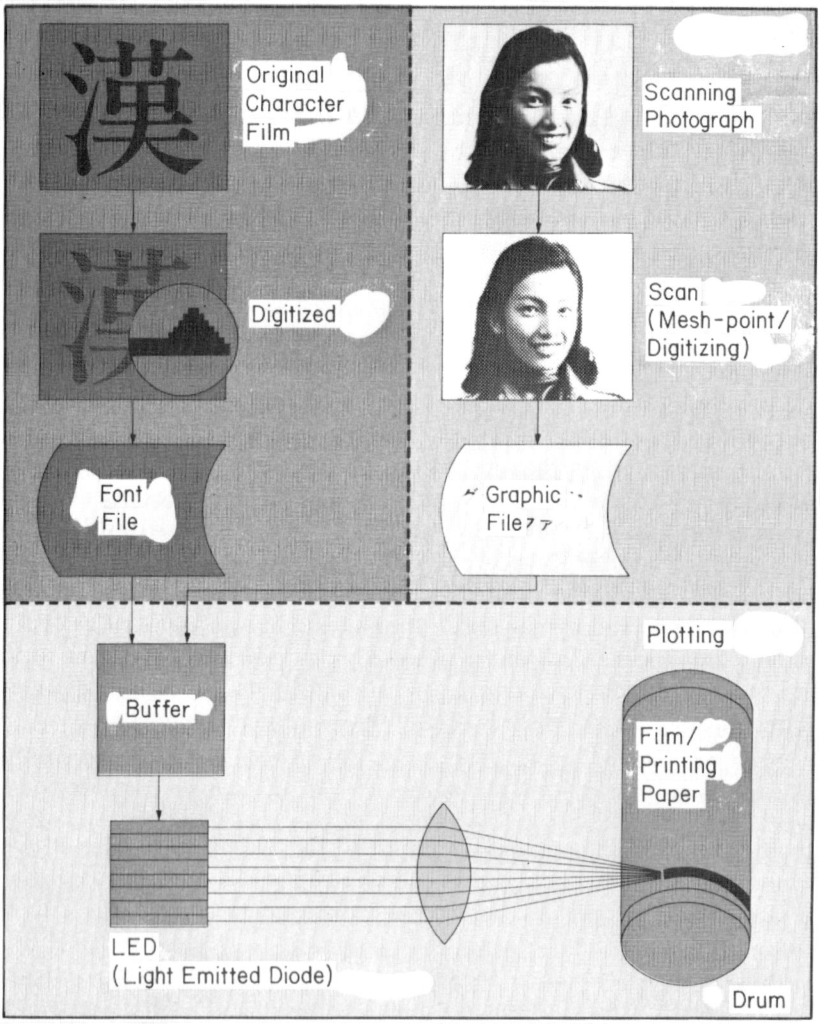

Fig. 6.4 IBM 5981 scanner/plotter function diagram. IBM 5981 scanner/plotter is used for both scanning of photographs/illustrations and plotting full newspaper page size films and printing papers.

draft and the final output in the newspaper. This size adjustment is made by the operator who specifies the scanning by either enlargement or reduction.

6.6 Page-Forming

After news data have been proofread, the layout is made for the newspaper page. The LDT or 3270 terminals to be used in the layout operation must first be placed in layout mode. This operation follows the page layout draft prepared at an earlier stage in the process.

The layout operation can be carried out in two ways. Either a newspaper page is laid out step by step, by entering a sequence of single-function commands, or it is laid out completely in one operation by calling the layout command stream which was previously read from cards and stored in the system pattern file, and the specifying necessary parameters such as dimensions and IDs given to news data. The second layout method is especially useful when laying out headlines and boxes with fairly fixed format. The use of "patterns" eliminates careless input mistakes and simplifies the operation. The proofread news data are run into specified regions which are formed at layout time. Running of the news data is accompanied by various operations such as taboo handling, space reservation for ruby type and multiplier conversion. After the data are run, the results are returned to the operator, so that he can confirm if the right action was taken.

If the text of a particular news story is too short or too long for the specified region, he displays the story and decides whether he should enlarge or reduce it to fit the space, or whether he should change the size of the space by repeating the layout operation. Photographs and illustrations are positioned by using IDs specified at their scanning time.

It is also possible to trim a photograph by operator instructions when the size of the photograph already stored in the file has to be changed at layout. When favourable layout results are obtained and confirmed on a LDT or 3270 terminal, the operator gives instructions to have them stored on an external storage device (system composing file). Once stored, the layout results can be recalled any time when they are needed by using the IDs given them at the time of storage. The results of pre-forming are always retrieved at the page-forming time and positioned as assigned.

Also within the system, any spaces of a newspaper page are filled with character images of a certain size, or photographs, illustrations and headlines, all according to specified rules and given instructions. This is essentially the same as the operation carried out manually on the makeup table, where the workers lay out a newspaper page. However, the layout operation by a computer system is superior in that mistakes due to carelessness (leaving the types out, placing them upside down, etc.) are prevented, and proofreading after pre-forming and page-forming is simplified.

6.7 Output

In the conventional newspaper production process, the newspaper page is completed when the page-forming operation is finished on the makeup table. With the JPS process further operations such as outputting film/printing paper from 5981–D01 Scanner/Plotter, and developing of the whole negative are required before the page is completely finished. The only data available concerning the composed and laid out pages at this stage are the size information of characters, photographs and illustrations, the ID information, and the positioning information. Images of characters themselves are not yet included as data. When the operator enters an output command on a 3270 or LDT key-boards, the completed layouts (page-forming, pre-forming) are retrieved. They are first rearranged in the order of output from the 5981. Then, the necessary images are taken out of the character image file (font file) and pictorial information file (graphic resource file) and are output to a film or a printing paper, the size of which is a full page. The output can be selected as positive/negative or ordinary/reversing. For negative output, the developed film is sent directly to the engraving process.

6.8 Programs

The software of JPS consists of programs to be used both on-line (Japanese Publishing System/On-Line System, herein-after abbreviated to JPS/OLS) and off-line. The total number of steps included in these programs amounts to approximately 170 000 (Fig. 6.5).

JPS/OLS consists of some 800 program modules and approximately 120 000 steps. The most prominent feature of JPS/OLS is its flexibility. An IBM user can easily customize it and process complex applications by specifying, at the system generation time, more than 100 parameters to determine the method of taboo handling and the characters concerned, the format of commands to be entered from LDTs and 3270s, the basics which constitute the page layout and the organization of data files. Other features of JPS/OLS include the "pattern" and the user program interface. The pattern is a stream of single-function layout commands which are read from cards and stored in a file (system pattern file). Patterns can be called at the layout time by a single command and, given a few parameters, can compose boxes and headlines of fairly fixed formats all at once. The use of patterns eliminates careless command input mistakes, and simplifies the operation. The user program interface is available for incorporating user programs into JPS/OLS. Since the method of processing input/output of news stories differs among users, it may be necessary for a particular user to develop his own programs for processing input/out in the most convenient way. With the help of interface, the user programs can operate in the same way as other JPS programs.

The following programs have been developed to constitute the off-line portion of JPS:

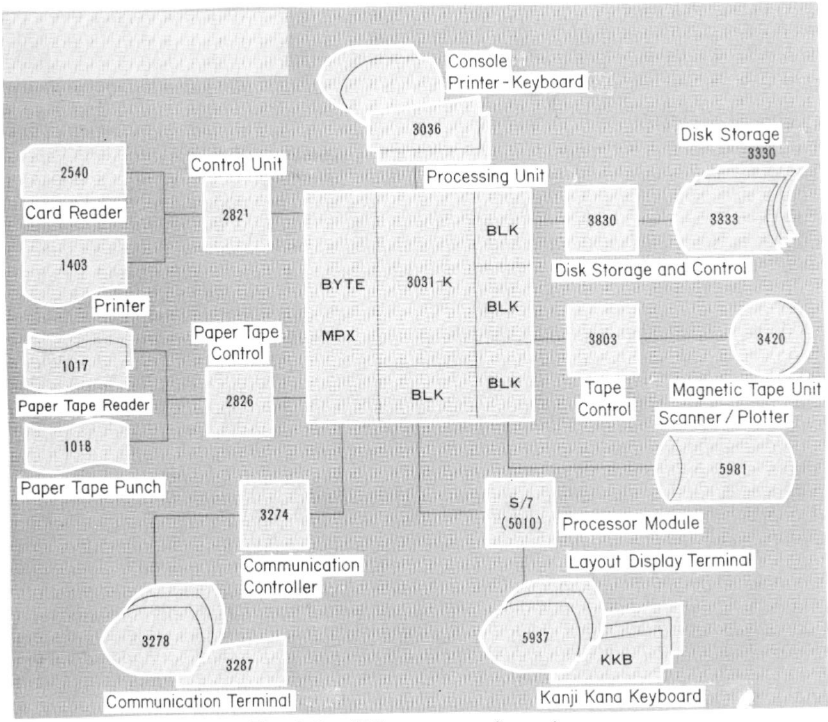

Fig. 6.5 JPS system configuration.

(a) Font Creation and Maintenance Program for IBM 5981 Scanner/Plotter.
(b) Font Library Maintenance Program.
(c) Format Conversion Program.
(d) Proof Print Program.
(e) JPS/OLS System Generation Support Program.
(f) Utility Scan Program.
(g) Utility Plot Program.
(h) Symbol Storage Utility Program.

The System Control Program (SCP) controls other application subtasks, allocates core storage, manages JPS/OLS data files and controls application programs.

The Terminal Subsystem handles some 120 commands that can be entered from LDT and 3270 terminals, controls the display screens, and performs proofreading operations.

The Input/Output Subsystem handles the processing of news tapes (paper tapes as well as magnetic tapes) which form the basic data.

The Composition/Layout Subsystem performs the main tasks of page organization such as positioning of headlines, running of news stories and framing boxes. It also processes the 36 composition commands which are directly included in the news articles.

The Scanner/Plotter Subsystem inputs photographs and illustrations to the system by scanning them, and plots them on to the film of the completed page and pre-formed page; the characters, photographs, and illustrations are all output on to a full page sized film and printing paper simultaneously.

There are also Data Management which handles the passing of data between subsystems, File Management which writes data into files as well as reading data from them, and Font Utilities for LDT.

6.9 Consequences of JPS

The conversion to a Cold Type System (CTS) raises problems of organizational structure, system operations and conversion.

It is essential, therefore, to plan a phased implementation of CTS with adequate time for study and thorough preparation/training in advance. For example, CTS can initially be applied to a limited number of the pages and then gradually expanded to full conversion. Expansion of system configuration should also be planned in parallel, taking into account:

(a) number of pages to be produced by CTS;
(b) contents of the pages to be produced by CTS;
(c) production time allowance;
(d) operational procedure;
(e) data volume to be processed

Only when the above elements are fully investigated, can a decision be made on: required system processing capability and performance and number of units. Since JPS is designed as a complete system, it is possible to reconfigure it to fit any required environment. Figure 6.5 shows a typical configuration.

6.10 JPS Current Position

In March 1978, the Tokyo main office of Nihon Keizai Shimbun switched its operation completely from the hot-type system to ANNECS, the first fully computerized cold-system for newspaper production in Japan.

In September 1980, NELSON, the most advanced JPS system, produced 150 pages a day at the new Tokyo main office of the Asahi Shimbun.

NELSON, combined with several mini-computers, carried out the following functions in addition to those of the original JPSs:

(a) producing heading-cuts ornamented with various patterns;
(b) scanning advertisements by flat bed scanners, which are docked with textual materials in the system;
(c) printing full page proof copies after the page is made up;
(d) transmitting the digitalized image data of the pages by digital facsimiles to the remote plants.

Two local newspaper companies, with circulations of about 3 million and 1 million have successfully made trial runs of their JPS systems.

As the JPS is gaining ground in the newspaper industry, IBM Japan is now developing an advanced JPS, called JPS-N, based on the NELSON of the Asahi Shimbun, to apply it to newspaper companies.

7

Control System for City Gas Production and Supply

Sachie Moriya, Distribution Control Centre, Tokyo Gas Co Ltd, and Tatuo Yamazaki, Plant Construction Division, Tokyo Gas Co Ltd, Tokyo, Japan

Summary

TGCS is a centralized system for the control and operation of gas production and supply using computers.

It enables plants and holder stations to be operated under optimum and safe condition at all times. This paper describes the computer hierarchy used, which consists of large computers (IBM 370/155), medium computers (IBM 1800), mini computers (IBM S/7), telememer/telecontrol equipments, and groups of plant computers including the DDC (Direct Digital Control) computers. Planning, co-ordination and control of production and supply are achieved through man–machine communication with the aid of graphic displays, character displays, and announcement devices.

7.1 Introduction

At the Tokyo Gas Company, city gas is manufactured from various raw materials at five plants on the coast, transferred to one point and then distributed through the pipeline network to about 5 million customers located in the metropolitan region. In the production and supply system for city gas, gas demand varies considerably with the season and time of day. Therefore to operate the system effectively and safely, it is essential to control the gas

CAD/CAM and MIS in Japan
ISBN 0-12-682580-7

production at the plant site, the input and output of the gas holders, and the pressures at principal points in the supply network. Account must be taken of the projected demand based upon such data as the rates of consumption in the whole supply area, gas pressures, storage in holders and the working condition and the production capacities of gas plants.

Application of a centralized control system satisfies all these requirements.

7.2 Outline of the Production-Supply Control System (TGCS)

The configuration of the TGCS software and hardware are shown in Figs 7.1 and 7.2. The production and supply control system, which has a multi-level structure, has the following features:

(1) The computing capacity to process rapidly complex calculations such as large-scale LP and network analysis problems.
(2) Interactive communication with the computer at high speed for information retrieval, trial and error calculations, diagrammatic representation of the network condition (especially DOCS: the daily operation control system, which corresponds to the co-ordinating level), and functional ability to discriminate at high density. The gas production control system (GPCS) and automatic supply network control system (TACS) are also interactive systems which are easy to operate, providing announcement devices, CRT and others.
(3) Commands from the upper to the lower levels are sent by data transmission. Control commands computed by the DOCS are transmitted to the GPCS and TACS of the lower level systems. The DDC in individual plants also automatically controls the start up, shut down and operating rate of the plants. Fourteen holder stations and about twenty pressure control stations in the supply network control gas storage and pressure automatically.
(4) The reliability of the whole system has been improved by appropriate separation of the hardware hierarchy. This facilitates back-up as each provides an effective and pertinent back-up function when required.
(5) The necessary data for information retrieval, computation and tabulation are collected from the production and supply networks and are stored in large capacity files for future access.

7.2.1 TGCS Development

The on-line realtime system using TOSBAC-3400 as the main computer was

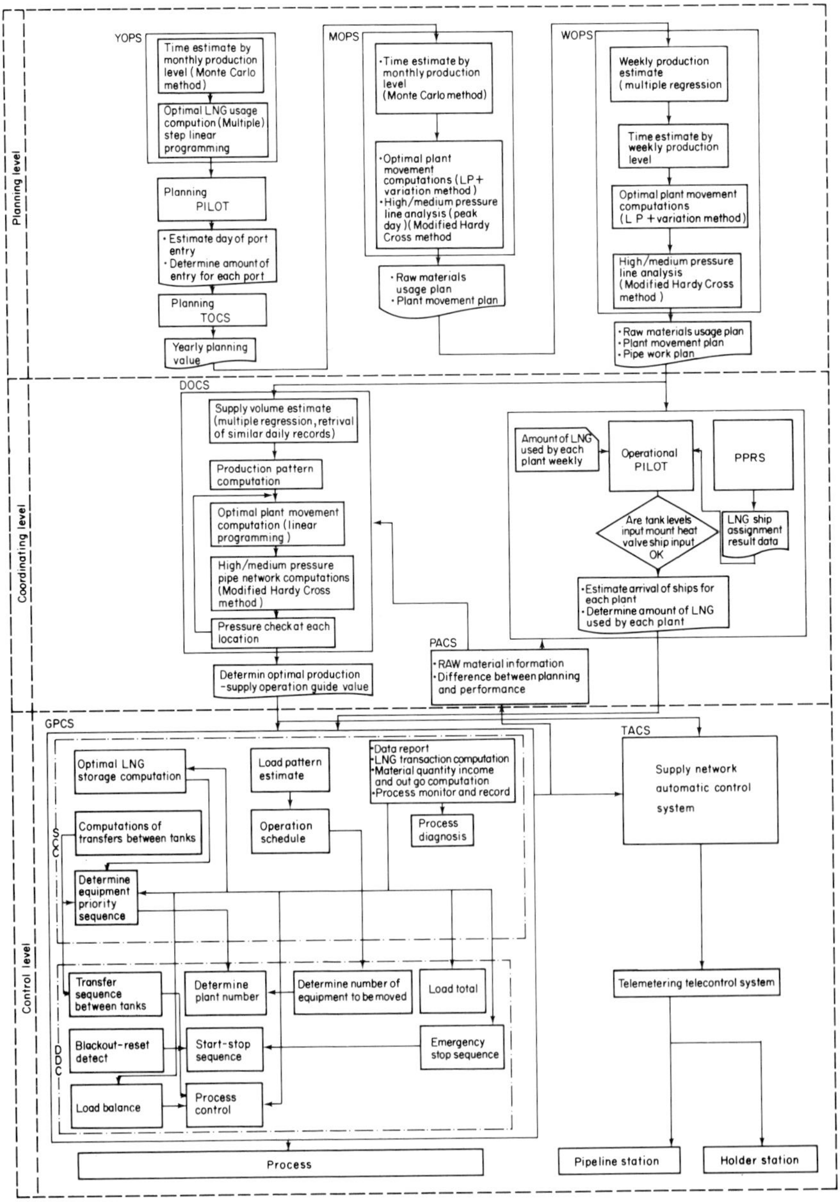

Fig. 7.1 TGCS Software system flow diagram.

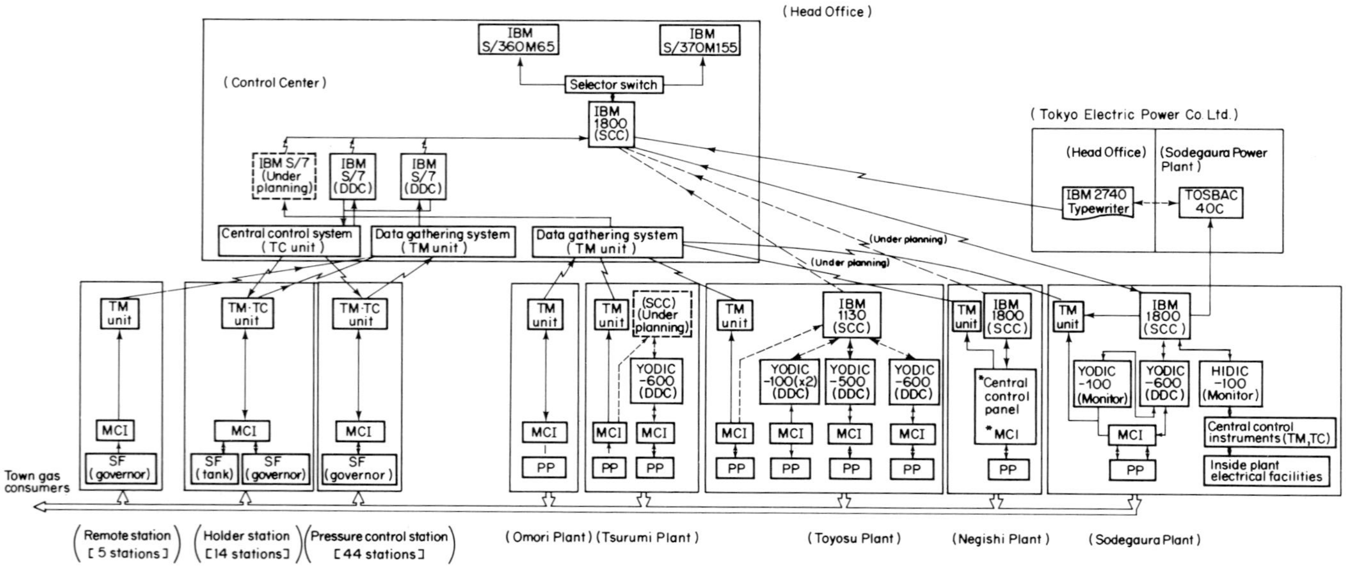

Fig. 7.2 Production control system hardware configuration general design. MCI = measuring and control instruments; SF = supply facility; PP = production plant.

established in 1965. It was called GCS (Gas Computing Supervisor) and provided production-supply monitoring, demand forecast, optimal control commands for the supply networks, automatic listing and various off-line computations.

Later, the production-supply networks were expanded and became more complicated following the introduction of LNG (liquefied natural gas) and construction of the Sodegaura Works and the Chiba High Pressure Gas Trunk Line.

From limited amounts of raw materials, it is very difficult with a rapidly shifting daily demand and seasonal fluctuations to maintain hourly optimum production and supply whilst at the same time ensuring an effective long-range program. Therefore, to operate supplies it became necessary to integrate the control system to include raw materials.

TGCS was developed to satisfy the realtime operation of the above dynamic adaptive control requirements, and is currently satisfying most expectations.

The systematic compilation of production-supply network realtime data together with the updating and efficient information retrieval facilities of the system form the basis for:

(a) Long and short range programs for optimal intake and ultilization of raw materials.
(b) On-line and realtime optimal control of the production-supply networks.
(c) Supply network station telecontrol.
(d) On-line realtime commands to the production networks.
(e) High grade monitoring of the operation of the production-supply networks. In particular, the correct hierarchical structure and the interactive communication system facilitate easy and reliable operation of the large-scale system.

7.2.2 Outline of the production and supply networks

As shown in Fig. 7.3, there are at present five works around the periphery of Tokyo: Sodegaura, Toyosu, Omori, Tsurumi (including Suehiro) and Negishi. The gas produced at these plants is fed into the supply network and sent to the consumers.

There are four types of supply network classified according to gas pressure: high, medium A, medium B, and low pressure. General household consumers are supplied from the low pressure networks. The different pressure tubes are connected through a pressure control facility and these pressure tubes are

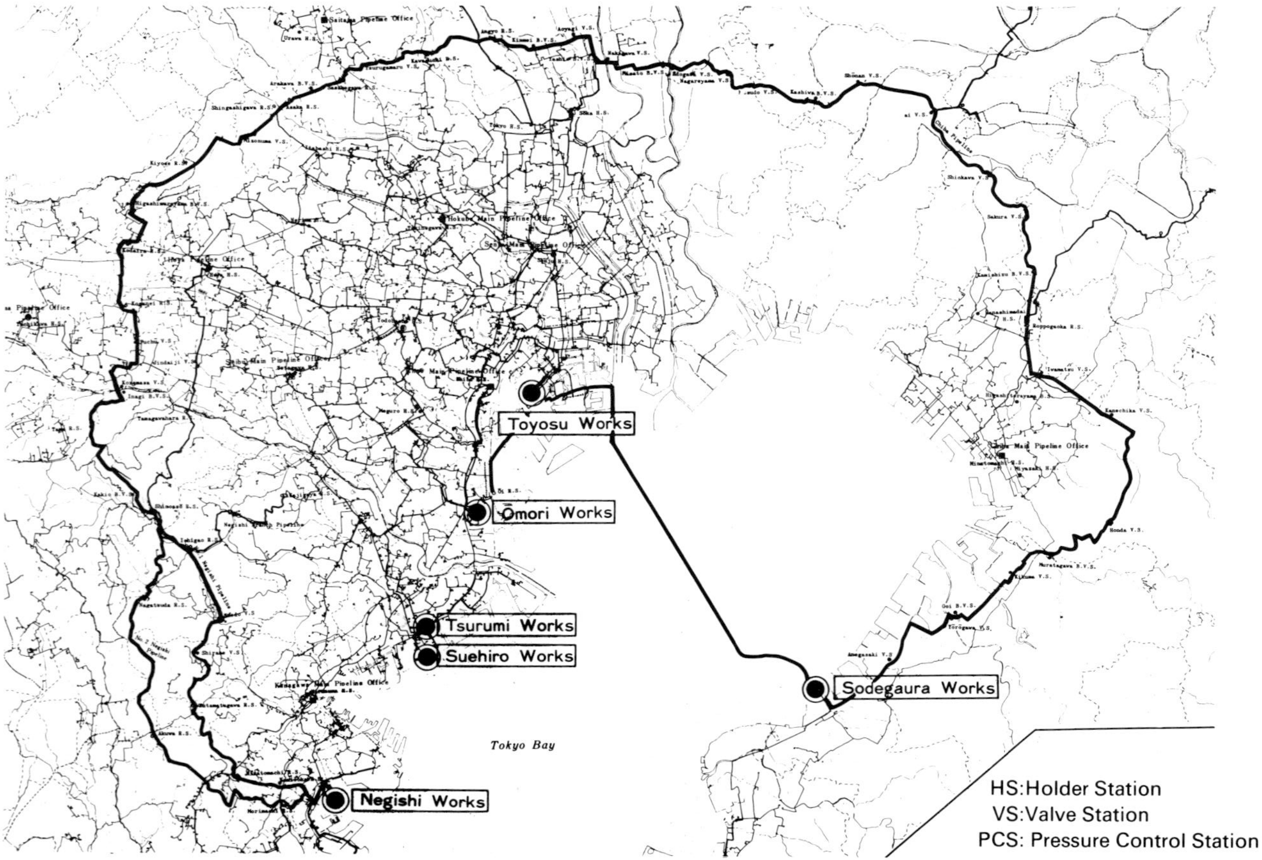

Fig. 7.3 High/medium pressure gas pipe network diagram.

formed into networks. Holder stations in the networks have the function of storage and pressure adjustment for city gas. There are also pressure control stations where the city gas is branched and distributed from the high and medium A pressure pipes to the lower pressure pipes.

The demand for city gas is, of course, seasonal and also fluctuates during the day.

7.2.3 TGCS software configuration

Table 7.1 shows the configuration of the TGCS software. It is divided into four sections: planning, coordination, control and service. In planning MOPS (Monthly Operation Planning System), WOPS (Weekly Operation Planning System) and PILOT (Planning Information System for LNG Operation and Transportation) are used for operation as well as for planning at the time of the yearly program appraisal.

Table 7.1, shows that there are functional and hierarchical structures existent in TGCS, which relate to the processing frequency.

7.2.4 TGCS hardware configuration

The hardware configuration of TGCS has a multilevel structure corresponding to that of the software to carry out the large-scale control of the distribution network stations (Fig. 7.4).

7.2.5 TGCS operation procedures

7.2.5.1 Planning procedures

The import, transportation and storage of LNG is important for both raw material planning and for cost reduction. We therefore calculate the yearly production policy for LNG with YOPS (Yearly Operation Planning System), PILOT and TOCS (Tank Operation Control System). The detailed optimal plant utilization program is computed using MOPS.

7.2.5.2 Performance program procedures

The performance program has as input the yearly planned capacity. However, in practice adjustment to the production-supply network status is neces-

Table 7.1 TGCS Software Configuration

Name of subsystem (Planning level)	Job	Program scale	Performance cycle	Processing method and computer
YOPS Yearly Operation Planning System	Yearly planning system Yearly supply forecast — Plant optimal utilization Optimal LNG — Various materials High/medium pipe network — Holder utilization	phase I . . .200 kB Pre-processing phase II . . . 400 kB Main body phase III . . . 150 kB Output LP editor reporter	Approx. 4 times/year	S/370 Batch process
MOPS Monthly Operation Planning System	Monthly planning system Monthly supply forecast — Plant optimal utilization Optimal LNG — Various materials High/medium pipes network — Holder utilization	Same as for YOPS	Approx. 2 times/month.	S/370 Batch process
WOPS Weekly Operation Planning System	Weekly planning system Weekly supply forecast — Plant optimal utilization Optimal LNG — Various materials High/medium pipe work — Holder utilization	140 kB × 6 8 kW × 2	One time/week	S/370 1800 On-line processing
TOCS Tank Operation Control System	LNG tank operation system Optimal LNG tank shift computation	13 kW	Approx. 4 times/year	1800 On-line background processing
PILOT Planning Information System for LNG Operation and Transportation	LNG control information system Estimating LNG tanker locations Estimating LNG tanker port arrivals LNG	6 kW × 20 14 kW	1 h cycle	1800 On-line processing
DOCS Daily Operation Control System	Daily production-supply control system Output forecast — Output pattern forecast Plant blackout simulation Pressure and governor stations utilization pattern forecast Optimal plant utilization — High/medium pressure pipe network simulation Optimal output operation guidance Output networks and weather information compilation	140 kB × 20	1 h–5 h cycle	S/370, 1800 On-line processing

(Coordinating level)

PPRS Port Performance Report System	LNG ship port entry control system Records and reports port entries	8 kW	once/day	1800 On-line background processing
PACS Production Analysis and Control System	Production control system Raw material information Difference between planning and performance Feedback to the planning operation	100 kB × 5 8 kW × 5	once/day	S/370, 1800 On-line processing
DEMS Distribution Equipment Maintenance System (Control level)	Supply equipment control system Pressure adjustment and governor stations Accident statistics and operation statistics Optimal maintenance control	150 kB × 2 8 kW × 10	once/day	S/370, 1800 On-line background processing
GPCS Gas Production Control System	Gas production automatic control system Process control Scanning and monitoring of process data Preparation of daily schedule and report on LNG transaction Operator information services	10 kW × 30	ms–1 h cycle	1800 YODIC-600 On-line processing
TACS Telemeter-telecontrol Automatic Control System	Daily production-supply automatic control system Production-supply network information compilation Central control Production-supply status automatic graphics Automatic control of pressure and governor stations Operator information services	8 kW × 30 12 kW, 14 kW	ms–1 h cycle	1800, S/7 On-line processing
Service program package	High/medium pressure pipe network analysis Gas combustion computation Miscellaneous statistics Miscellaneous report programs System maintenance program Data maintenance program	140 kB × 10 100 kB × 5	Optional time	S/370, 1800 On-line processing

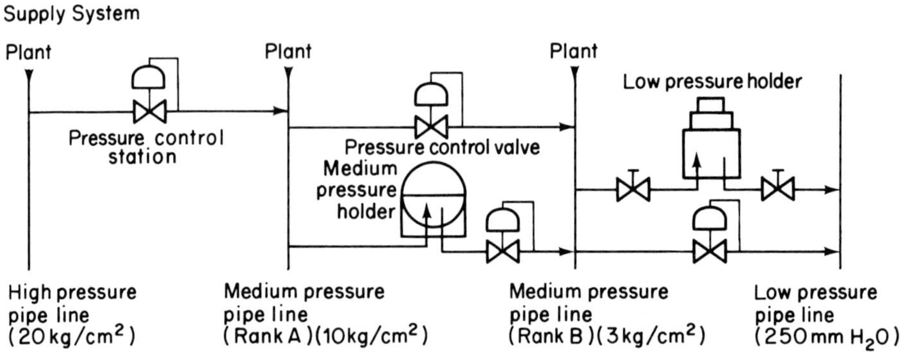

Fig. 7.4 TGCS Hardware configuration.

sary using on-the-spot estimates to resolve differences between the yearly program and the performance program.

The calculation algorithm is designed to meet on a realistic basis requirements under the TGCS hardware configuration.

7.2.5.3 Daily operation procedures

Operators in the central control centre issue hourly updated production-supply estimates. To support this operation, DOCS can be accessed interactively by the operators using graphic displays. This is done on scheduled basis five times daily, and at other times as necessary.

Using DOCS, the information retrieved from the data bank is selected and analysed (which holds information on production and supply over the last ten years) and, taking into account the weather data, predicts the daily gas supply. Historic data such as the hourly weather data for a particular day can also be retrieved to determine the hourly supply of city gas. After computing the necessary production for all plants utilizing the holders, individual plant requirements can be computed by the supply network administration.

In this way, after confirming the attainable production-supply, production commands are sent to the plants and the designated control values at the supply points to TACS simultaneously. Upon receiving these values, TACS automatically controls the supply networks.

The following information exchange programs can be utilized interactively by the operators of TGCS through graphic displays: multiple regression analysis, demand pattern forecasting, similar data information retrieval, emergency simulation, data sets for linear programming, production level simulation, linear programming analysis, plant production pattern decisions, holder storage pattern decisions, pressure patterns, as well as calculations on the medium pressure (A, B) distribution network and the Chiba-Negishi line networks. Approximately 100 graphic display images may be necessary for any one of the above operations.

7.2.6 Telemeter system

The first step in any centralization must be used for data. In the company, a telemeter system was introduced at the head office in Nihonbashi Tokyo in 1956, making the hourly observation of the gas supply trends possible at strategic points in the production-supply networks. The variables are gas pressure, volume flow, holder storage, operating valve opening and several digital signals such as valve positions, seismograph, etc.

All these variables are data processed by the computers, and some of them are displayed and recorded for use in a supervisory capacity by the operating personnel.

7.2.7 Telecontrol system

In 1959, installation of telecontrol apparatus was begun. With this equipment, direct operation of the remote pressure control stations and the holder stations became possible via the data communication network shown in Fig. 7.5, thus eliminating the need for manual operation at these stations. Complete control of remote operations at the key points in the supply networks then became possible.

Considerable care was taken (as in the telemeter system) in sequentially matching the logic in the command entry to avoid noise and maloperation caused by unreliable signal transmission.

Following the development of TACS in the TGCS, the remote stations are now directly set point controlled by computers.

7.2.8 Operation Control of Plants

7.2.8.1 Outline

The Sodegaura Plants may be used as an example. It is the main plant of the Tokyo Gas Company, receiving mainly LNG as raw material.

It also supplies all the fuel required by the neighbouring Sodegaura Power Plant of the Tokyo Electric Power Company. Changes in plant operational conditions are given audibly through a computer-edited automatic announc-

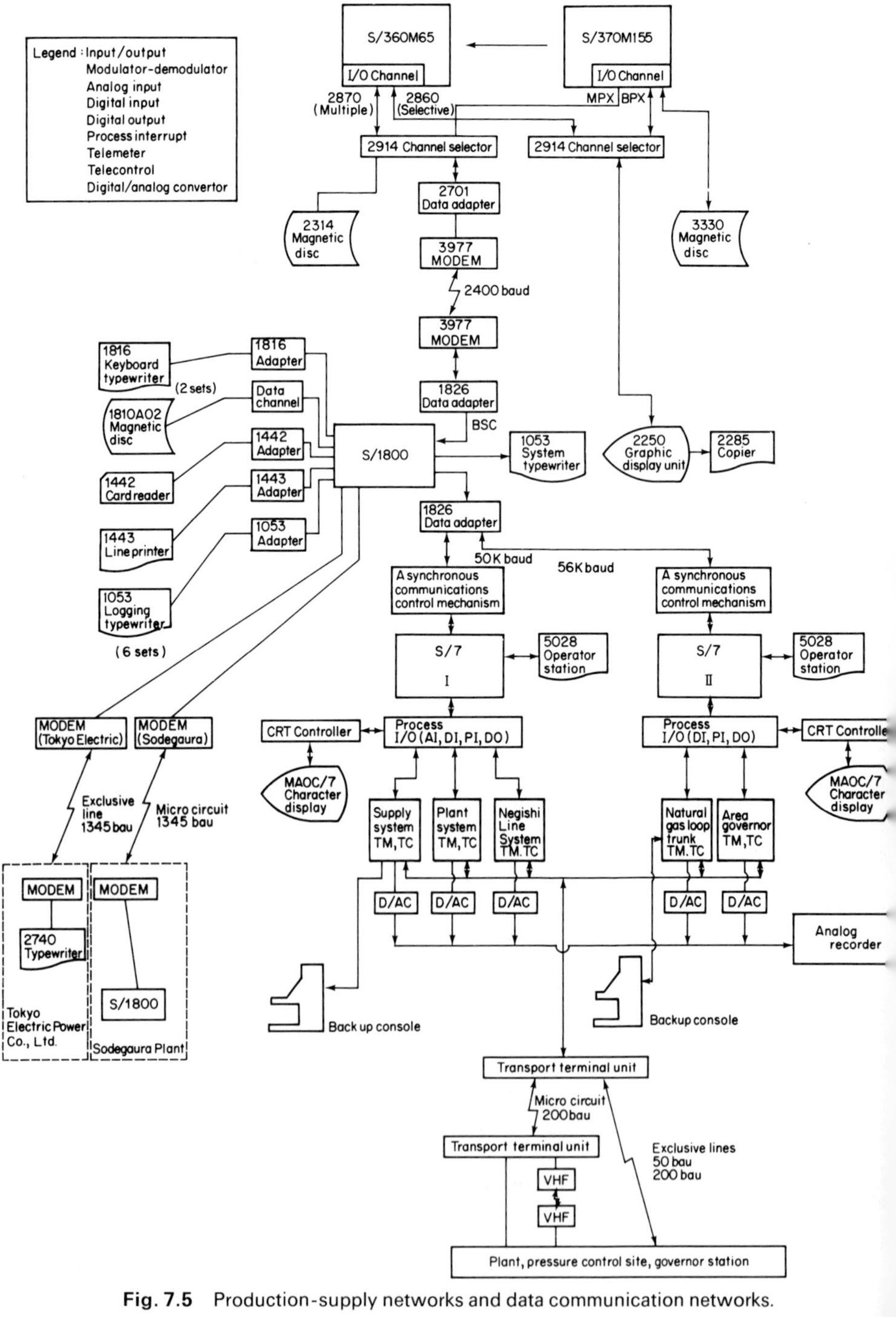

Fig. 7.5 Production-supply networks and data communication networks.

ing system. Visual displays are used only when detailed information is needed by the operator, or an emergency condition exists. (Under normal conditions, the plants are started up and shut down automatically.) In case of an emergency, most sections are dealt with automatically, and the operator only has to issue repair and spare equipment commands. To perform these functions, there is a hierarchical system at the plant level consisting of a SCC (Supervisory Control Computer) system which performs planning and control functions, a DDC which directly controls plants and a monitoring computer.

This system having been designed assuming the scope of future control may include:

LNG	30 units approx.
Primary pumps	60 sets approx.
Secondary pumps	10 sets approx.
LPG pumps	15 sets approx.
Boil off gas compressors	15 sets approx.
Vapourizers	25 sets approx.
Calorific value regulating systems	2 systems
Boiler, water and utility facilities	1 set

7.2.8.2 Software design

The SCC is directory-connected on a computer-to-computer basis with TACS and with the control computer in the Tokyo Electric Power Company Sodegaura Power Plant. Upon receiving information from these computers, it carries out an optimal automatic scheduling operation considering load variation forecast, safety margins in operating capacity, various LNG stocks, machine operating time, distribution route, equipment running costs, etc., and delivers the operation commands to the DDC.

In addition, the SCC carries out the following operations: supervision of DDC operations, detection and diagnosis of various instrument irregularities, information communication to the operators, to the Tokyo Electric Power Company and to the TACS, computation of LNG transactions based on the output gas, utility, electricity, storage quantity and received amounts, and data processing such as miscellaneous statistics, daily loggings and journal computations.

Upon receiving the above commands from the SCC, the DDC automatically performs all the necessary operations such as computing the necessary units of equipments, each unit optimal load distribution, start up and shut down of the machine and PID control. In addition, a system is set up to facilitate monitoring of the plant and the electric apparatus.

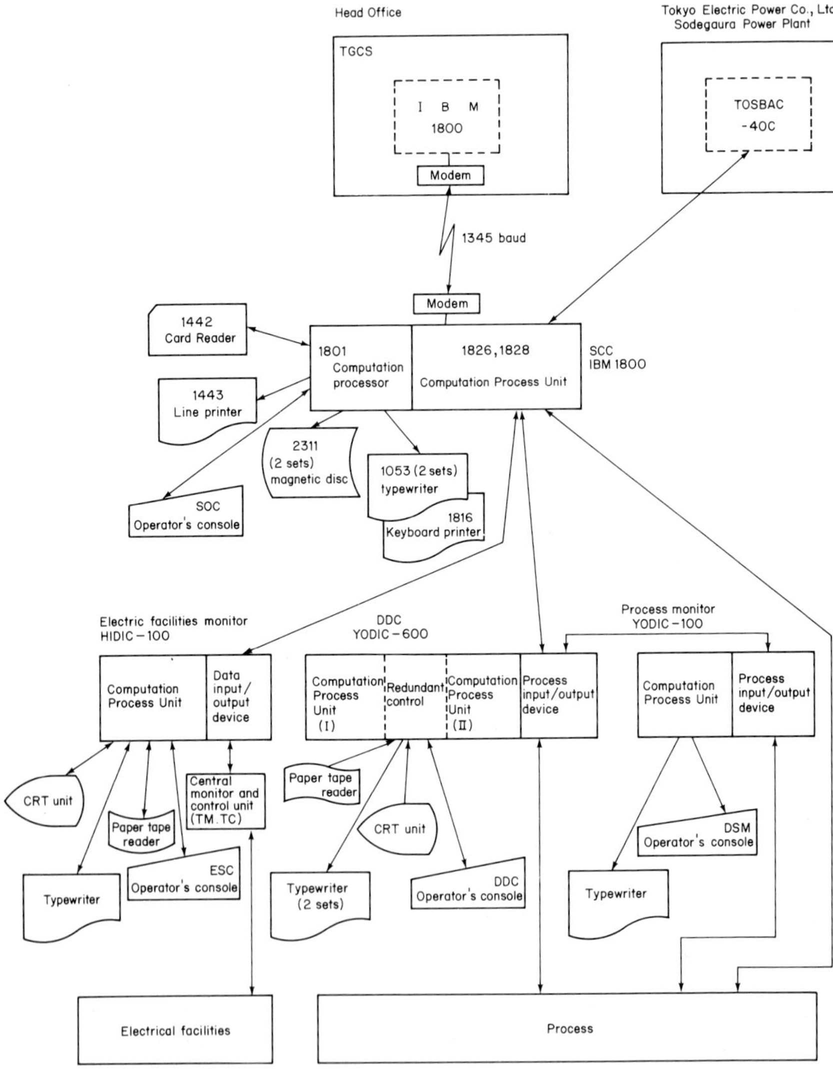

Fig. 7.6 Sodegaura Plant operation system hardware configuration.

7.2.8.3 Hardware configuration

Figure 7.6 shows the hardware configuration. In the DDC system, system reliability is the main priority, because auxiliary storage can then be avoided. All the programs required were written using assembler language so that efficient use could be made of the main memory units. A dual system was introduced for greater reliability.

7.2.8.4 *Effectiveness of the DDC*

The role of peripheral control equipment is particularly essential when considering the total system. The TGCS adopted the DDC at control level, with the following results:

Control

(a) Control function of a higher order
The current analogue control system is composed of integrated feedback routes of the tree type. However, to control complicated processes with good dynamic response, a control of the parallel feed-forward type is required. To achieve this, a control algorithm is introduced which is of a different level to the PID. This advanced control system allows compensation for security, minute control, and extremely accurate operation.

(b) Total control function
Current analogue instrumentation cannot perform effective control of such a complex system. However, the flexibility of digital computer techniques tenders makes co-operation between plants feasible.

Stabilization of special operations

Special operations such as shut down and start up present difficult problems for operators. By installation of DDC, the process operations involved in start up can be achieved at a speed much higher than previously possible by human control, and with much greater stability.

Maintenance factors

Since DDC was applied, problems of maintenance have decreased to about one tenth of the former rate. This result illustrates the remarkable improvement in overall reliability of the total control system.

Composite evaluation

As can be seen, the DDC system functions at a different level from the current analogue system. DDC cannot be compared directly with the analogue system; it should be evaluated by its performance in automated business transactions at the control level, receiving commands from both planning and co-ordinating levels in a complex hierarchical control system. The DDC introduced here as an example has all these functions and present daily operation is satisfactorily carried out by only three persons per shift.

7.3 Conclusions

Development of the TGCS was achieved by the combined efforts of more than 30 staff. It was inaugurated in September 1983, and has been performing satisfactorily since.

The multilevel hierarchical structure of TGCS suggests a way for similar systems to be established in the future.

8

Integrated Analysis System for Power System Planning and Operation

T. Minakawa, Senior Consultant Manager, Power System Planning and Operation Department, Tohoku Electric Power Company, Sendai, Japan

8.1 Introduction

In the course of making decisions and recommendations in system analysis and design, the power system engineer is required to analyse the dynamic performance of power systems in terms of transient stability, transmission facility over-load, short-circuit and voltage regulation. These studies are closely related to power system planning and operation involving network configuration, generator maintenance schedules, generation dispatch schemes and load forecasts.

These problems have usually been studied through individual analytical programs, the evaluation of the results and the exchange of data among individual analytical procedures being implemented manually. Because of the laborious work involved in data preparation and the time required for the detailed analysis of results, the time available for comprehensive assessment and decision making has been sacrificed. In order to improve this situation,

CAD/CAM and MIS in Japan
ISBN 0-12-682580-7

the establishment of overall methods for the assessment and improvement of power system planning and operation is urgently required by power system engineers.

Utilizing database techniques and power system engineering innovations, a comprehensive analysis system, designated as the Power System Analysis System (PSALM-73), has been developed in Tohoku Electric Power Company,[1] which satisfies the following requirements:

(1) Engineering data and information associated with power system planning and operation, such as facility, load forecast, operating constraint and expansion plan, are stored in the database and retrieved easily and immediately whenever needed. Confidential data are accessible only to authorized persons.
(2) Power system analyses are performed dynamically and evaluated automatically alongside the requirements of power system planning and operation involving network configuration, generator maintenance schedules, generation dispatch and load forecasts.
(3) The results of analysis and evaluation are compiled into the database, and intelligible reports are automatically presented.
(4) The system is provided with some optimization functions, including the ability to suggest practical improvements in power system planning strategies and operation. The system also allows the engineer more flexibility.

8.2 Description of Original PSALM-73 System

This chapter introduces the original Power System Analysis System (PSALM-73) designed as a tool for the assessment of dynamic performance of power systems. The PSALM-73's design has the following features:

(a) standardization of input and output data for various analytical programs;
(b) integration of data management and maintenance;
(c) data access and preparation by macro commands;
(d) intelligible reports summarizing the results.

The PSALM-73 consists of the following three subsystems:

(1) Master Database Management System
(2) Operating Condition Determination System
(3) Power System Analysis and Evaluation System

The overall information flow of the PSALM-73 is shown in Fig. 8.1.

The analysis and evaluation programs in the PSALM-73 are executed under the direction of macro input commands, retrieving necessary data from the master and intermediate databases, provided that these databases are produced once by previous procedures. The macro input commands include: study time (year, month, time), expansion patterns associated with transmission systems, thermal, nuclear and hydro generating facilities, states of stream inflows, generator maintenance schedules, generation dispatch schemes and other miscellaneous directives.

8.2.1 Master Database Management System (MDBMS)

The MDBMS initiates the master database, validates the data for the MDBMS master database and modifies the master database once the initial loading has been completed. Through MDBMS, the databases for power system analyses are established as follows.

(1) *Load forecast database.* Load forecast information for a period of ten years is loaded into the master database.
(2) *Interchange energy database.* Interchange energy information for a period of ten years is also loaded into the master database.
(3) *Generating facility database.* Generating facility information is stored in the individual databases corresponding to the respective expansion pattern data on thermal, nuclear and hydro facilities and generator types and characteristics.
(4) *Transmission and substation facility database.* Transmission and substation information is stored in the individual databases corresponding to expansion pattern data on transmission and substation facilities and transformer types and characteristics.
(5) *Hydro supply database.* Hydro supply information is stored in the database.
(6) *Bus information database.* This master database defines ratings and operating limits of bus voltages and reactive power injections including the equivalent impedances of external power systems connected to the bus.
(7) *System management database.* The MDBMS activates system management databases to supervise and monitor the data processing.

8.2.2 Operating Condition Determination System (OCDS)

The OCDS determines operating requirements and conditions prior to power system performance evaluation by simulation. After referral by the engineer

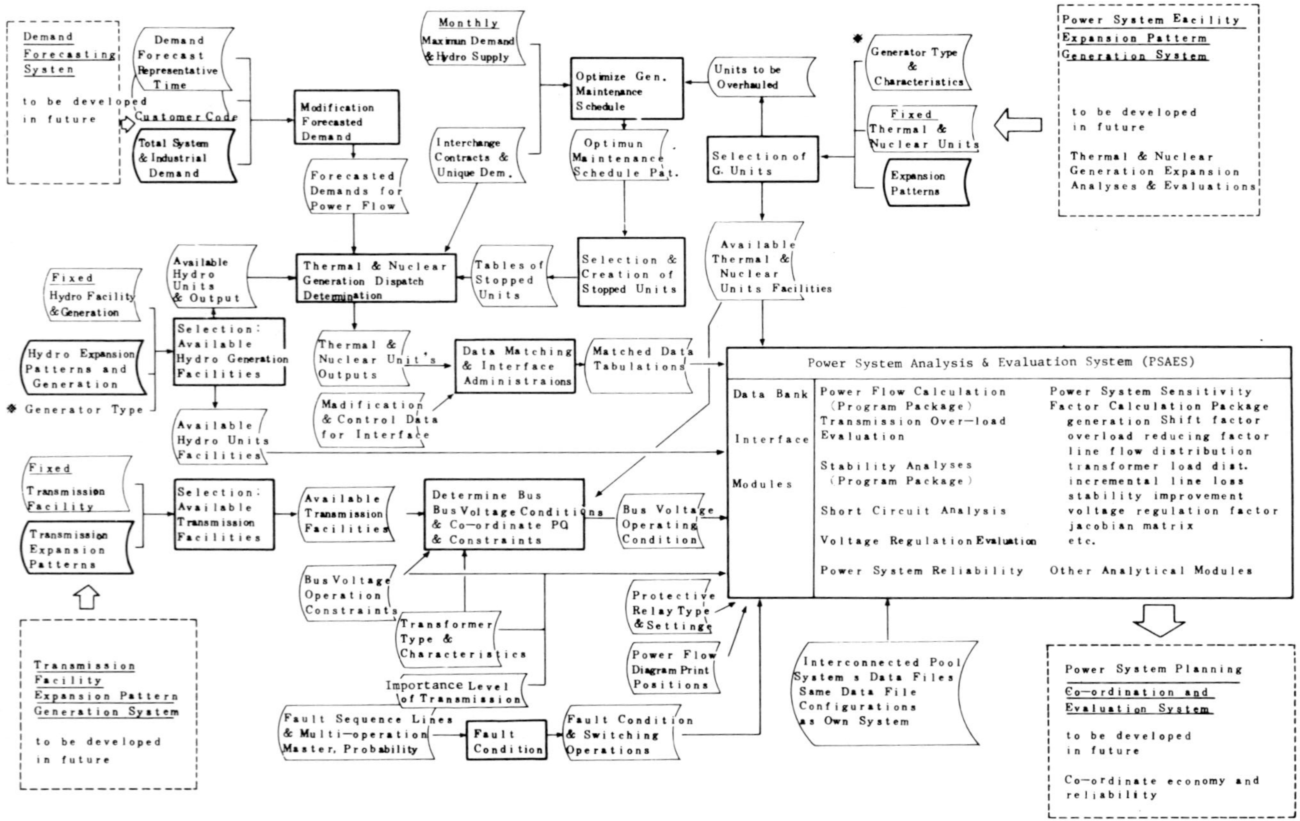

Fig. 8.1 Information flow in the power system analysis system (PSALM-73).

the results of these preliminary analyses and optimizations are transferred to the Power System Analysis and Evaluation System (PSAES). The OCDS has seven functions, whose features are briefly as follows.

(1) *Modification of load forecast*. The modification service on load forecast of substations concerning the total revised load of a power system is carried out by the interpolating method.
(2) *Selection of power system facilities*. Following macro input commands, the relevant power system facility data for the corresponding year are retrieved and tabulated on a monthly basis.
(3) *Hydro Supply Setting*. Hydro supply data are set in accordance with the macro input commands.
(4) *Thermal and nuclear generator maintenance schedule*.The OCDS can optimize the maintenance schedules of thermal and nuclear units, and retrieve the requested number of best schedule patterns. Employing the concept of equal risk, the generation reserve is distributed as uniformly as possible in terms of both geological allocation and period of year. This feature is useful as a supplementary method to study the problems of generation expansion. Without using the maintenance schedule optimization routine, engineers can specify any combination of overhaul-units.
(5) *Thermal and nuclear generation dispatch*. In order to undertake an ideal analysis of dynamic power system performance, various dispatch schemes, for security reasons, are required to simulate all possible operations from the optimistic to the pessimistic state. The following dispatch schemes can be selected by the user.

(a) units are successively loaded in the order of their efficiencies;
(b) units are loaded in proportion to their rated capacities;
(c) unit loading is subject to security control dispatch;
(d) unit loading is pre-determined by users;

(6) *Bus voltage determination and security improvement*. In compliance with operating constraints, bus voltage and reactive power conditions are determined in order to improve the security performance of power systems.
(7) *Data matching and interface administration*. In the use of PSALM-73, this function provides man-machine interaction, enabling the engineers to discover the optimum solution in power system studies logically.

8.2.3 Power System Analysis and Evaluation System (PSAES)

The PSAES is designed to provide engineers with a means of obtaining accurate and rapid solutions and evaluations associated with power system problems. All analysis programs are executed by macro input commands, retrieving pertinent data from the master and intermediate databases produced by the previous procedures. They can also be executed independently by manual card input without using the database. The results of performance studies are compiled in the database and reports summarizing the studies are printed. The components of the PSAES are as follows.

(1) *Power flow calculation program package.* The power flow program is indispensable in power system analyses. Various kinds of power flow calculation program modules are provided in the package.

(a) AC power flow models: Newton-Raphson method, Flow AC method, Fast load flow method, Ill-condition load flow method (Tamura-Iwamoto Method), etc.

(b) DC power flow models: Flow DC method, Reactance Matrix method, etc.

(2) *Transmission system overload evaluation programs.* The programs examine the line overloading in a transmission system in unforseen circumstances, simulating the available recovery operations within 30 minutes after the occurrence of the disturbance. The recovery operations considered in the program are OL relay operation, generation shifting, load shedding and switching operations. The time sequence of recovery operations within the 30 minutes and the overloading of transmission facilities are evaluated both before and after recovery operations. This is done by comparing the line power flow of each transmission facility with its time-varying ratings, such as emergency (5 minutes), 30 minutes and continuous ratings. The programs use the Flow DC Method for power flow computation.

(3) *Power system stability analysis and evaluation package.* This package makes it easy for engineers to study the stability problems of power systems, even though they may have little knowledge about stability analysis and evaluation. The package provides for several stability evaluation programs based upon approximate and accurate models. The following programs are selected by the user in his planning and operation studies.

Approximate models. Fast and useful for planning and on-line stability evaluation of security control: (a) Simplified fast transient stability; (b) Steady state power limit analysis. Steady state power limit is determined automatically.

Accurate models. Accurate and versatile for use in the planning, operation and system protection studies of large power systems; (c) Transient stability

(I), with AVR, Gov., load characteristics; protection relay excluded; (d) Dynamic transient stability (II), with AVR, Gov., load characteristics; relays include SOR, ZR, and OCR; (e) Dynamic transient stability (III), with AVR, Gov., load characteristics; relays include SOR, ZR, HOC, UFR, UVR, OLR, and pilot relays.

(4) *Fault current analysis package*. The package is designed to calculate fault currents of imbalanced and balanced network problems. The method used in the package is based upon the sparse matrix method,[2] which has the advantage of finding the solution faster, with less core memory.
(5) *Other analytical programs*. The PSAES provides for other analysis and evaluation programs, as follows:
(a) Flicker load voltage regulation evaluating program;
(b) Power system sensitivity factors determination package;
(c) Power system reliability analysis programs.

8.3 Models for Improvement of Dynamic Performance

This section introduces two typical models developed for the practical assessment of the dynamic performance of power systems. The models are components of PSALM-73 used in Tohoku Electric Power Company for power system studies.

8.3.1 Security-Oriented Power Flow Model

The problem of improving stability in a given power system is defined as an optimal power flow problem. To resolve this quickly decomposed equations and suboptimizations with respect to real and reactive powers are considered adequate.[3] The algorithm of the left-hand side pseudo inverse is used in the suboptimization, which minimizes the sum of squares of the differences between the left and right sides of linear equations. That is, under the equality constraint $H \cdot u = z$, where H is defined as an $(m \times n)$ coefficient matrix with $m<n$, u as an $(n \times 1)$ unknown vector, and z as an $(m \times 1)$ known vector, the proposition is to find $u = u^{\circ}$, which minimizes the objective function $J = \|H \cdot u - z\|^2$. Its optimal solution is given as

$$u^{\circ} = (H^{\mathrm{T}} \cdot H)^{-1} \cdot H^{\mathrm{T}} \cdot z$$

The decomposed power flow equations in terms of P and Q are:

$$P = X_p \cdot \theta + \Delta_p \quad Q = X_q \cdot V + \Delta_q$$

where X_p and X_q represent the matrices with regard to the P and Q vectors respectively, the vector θ and V represent the bus voltage angles and magnitudes, and Δ_p and Δ_q the correctors. The vector P is divided into two vectors P_G and P_L, where P_G stands for variable generation sources and P_L for constant loads. The vector Q is also divided into two vectors, the variable reactive power sources Q_G and the constant reactive power sources Q_L.

The secure condition necessary for stability may be incorporated into an objective function which is minimized for the angular differences among generator bus voltages, i.e. the node-branch incidence matrix C and the vector δ representing the angular differences between the terminals of branches, the equality constraint being stated as $\delta = C^T \cdot \theta$. For security as well as stability, δ is minimized. Whereas under normal power systems, the target bus voltage V_o is specified. If it is not specified, it may be assumed as $V_o = 1.0$ p.u. The procedure is to determine Q_G which satisfies the voltage condition $V = V_o$ as closely as possible. Consequently, as for the real power problem, the objective function which is minimized can be established as

$$J_p(\theta) = \|\delta\|^2 = (C^T \cdot \theta^T) \cdot (C^T \cdot \theta)$$

According to the left-hand side pseudo inverse algorithm, the solution of this equation is

$$P_G = -(A_G^T \cdot W_p \cdot A_G)^{-1} \cdot A_G^T \cdot W_p \cdot (A_L \cdot P_L - A \cdot \Delta_p)$$

where A_G and A_L stand for decomposed submatrices of $A = X_{pp}^{-1}$ corresponding to Q_G and Q_L respectively, and $W_p = C \cdot C^T$. Similarly, the objective function in the reactive power problem is formulated as:

$$J_q(V) = \|V - V_o\|^2 = (V - V_o)^T \cdot (V - V_o)$$

and the solution of the equation is:

$$Q_G = -(B_G^T \cdot B^G)^{-1} \cdot B_G^T \cdot (B_L \cdot Q_L - B \cdot \Delta_q - V_o)$$

where B_G and B_L stand for decomposed submatrices of $B = X_q^{-1}$ corresponding to Q_G and Q_L respectively.

These two suboptimization procedures are executed alternately until the solution is converged. That is, $J_p(\theta)$ is minimized assuming that it is related to real power P and phase angle θ and that bus voltage magnitudes are constant. As to the reactive power problem, $J_q(V)$ is minimized on the assumption that it is related to voltage and reactive power with fixed real power. The co-ordination between the two suboptimization problems can be taken into

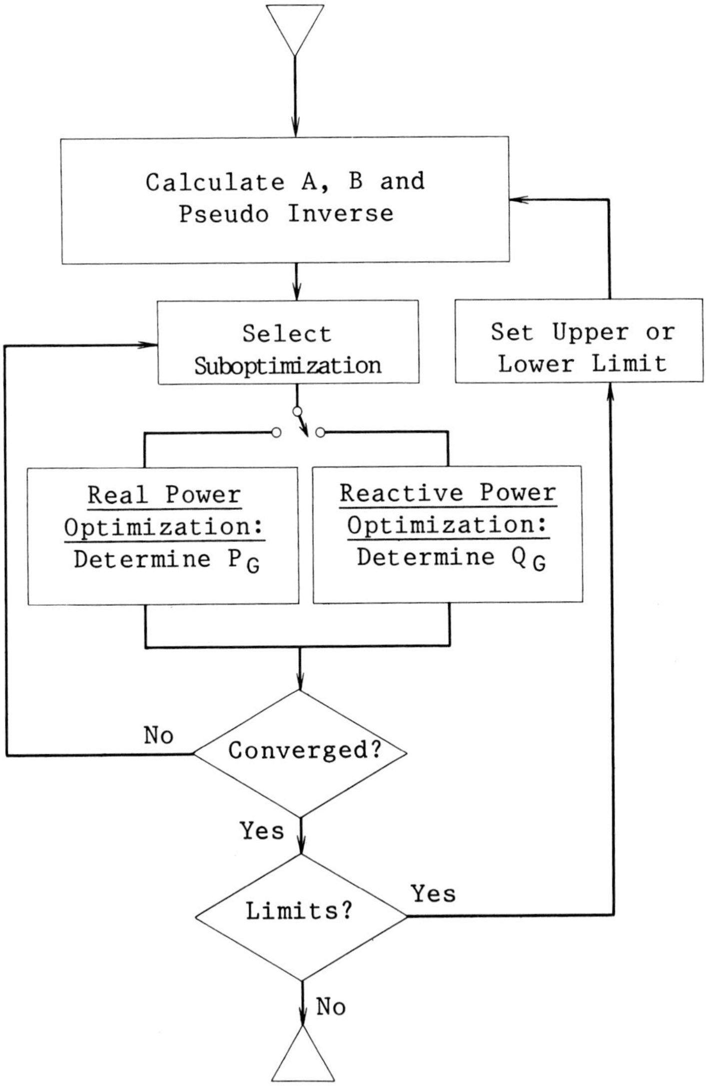

Fig. 8.2 Security improvement scheme by P and Q co-ordination.

account by considering correctors associated with the respective procedures, assuming that θ determined in the previous real power procedure is treated as constant in the subsequent reactive stage, and V determined in the reactive power procedure is treated as constant in the next real power suboptimization. Thus, the co-ordination algorithm demonstrates the accurate solution. Figure 8.2 represents the flow chart of the security improvement procedures.

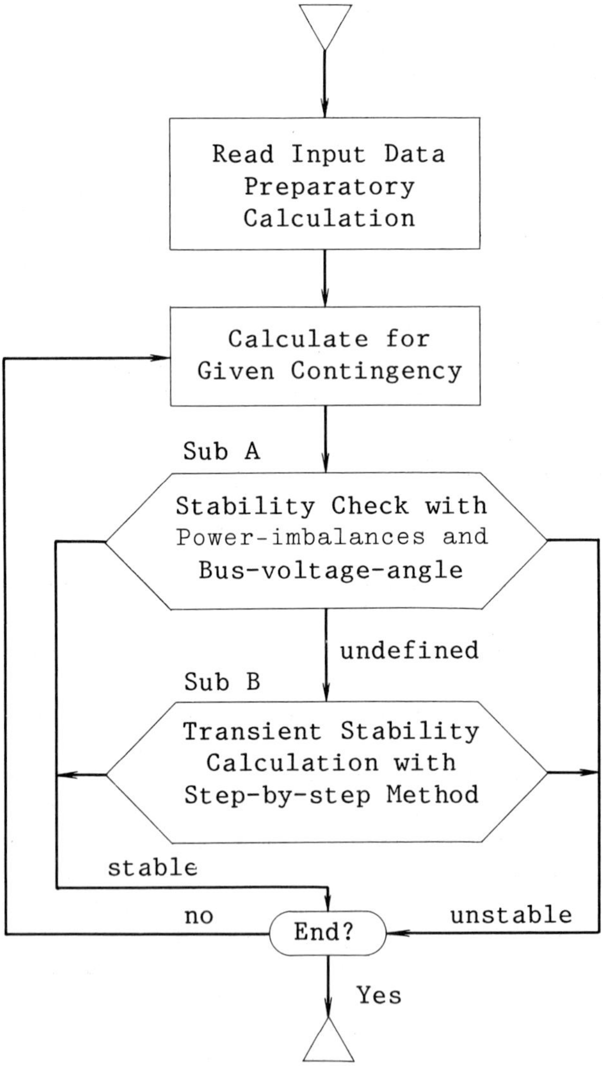

Fig. 8.3 Simplified fast transient stability evaluation.

8.3.2 Simplified Fast Transient Stability Evaluation Model

This model has the capability of rapidly determining whether a given power system is stable or not regarding any possible disturbances. As illustrated in Fig. 8.3, the proposed program consists of two parts labelled as Sub A and Sub B. Sub A gives a steady-state power flow solution for a given disturbance, and also a rough estimate of stability by calculating the supply demand power imbalance and generator voltage angular differences. Sub B performs the so-called step-by-step transient stability calculation with the linearized power flow model. This part is called for only when the system stability cannot be clearly determined in Sub A.

(1) *Feature of Sub A*. Sub A is to execute a rapid check for stability. It is first determined whether the system becomes separated under the given contingency. The following two measures are applied to each separated system for stability:
(a) ratio of total demand to total supply;
(b) maximum angular difference between generator bus voltages. The bus voltage angles are calculated with the linearized power flow model maintaining the assumed power balance within each system. These two stability measures are compared with empirical data, and the state of the system is determined to be stable, unstable or uncertain. The empirical data are predetermined from the engineering aspects for the given system.[4]
(2) *Feature of Sub B*. If the system stability is undefined in Sub A, the step-by-step transient calculation procedures are required. For fast execution, two linearized power flow networks and a simplified generator model are utilized:
(a) Real power distribution is computed at each time step with the first linearized power flow network designated as the P-θ network. This network expresses the relationship between real power injections P and bus voltage angles θ under constant bus voltage magnitudes.
(b) Bus voltage magnitude calculation is carried out every few time steps, because bus voltages magnitude present a relatively slow and small change for each increase in time during the swing of generators. This calculation is done with the second network known as the Q-V network, which gives the linearized relationship between reactive power injections Q and bus voltage magnitude V under constant line real power flow.
(c) The generator model assumes constant voltage behind constant transient reactance. The internal generator voltage behind transient reactance is calculated using the pre-fault condition.

In addition to the above simplifications, the following methods and models are used in Sub B:
(d) Runge-Kutta method of numerical integration is used for the solution for the generator swings.
(e) Matrix triangular factorization techniques are applied to the power flow calculations for both P-θ and Q-V networks.

(f) Load characteristics are represented either as constant impedance or as an expression of bus voltage magnitude and system frequency.
(g) Pre-fault power flow is determined on the assumption that all voltage magnitudes are specified.
(h) The contingencies given as system disturbances include three-phase and single-phase-to-ground faults at buses and generator outages.

As the step-by-step calculations proceed in Sub B, the following criteria are employed to determine whether the system is stable or not:

(i) If the maximum difference between the internal generator voltage angles $\Delta\delta_{max}$ exceeds $\Delta\delta_2$, then instability will occur.
(j) If $\Delta\delta_{max}$ is less than $\Delta\delta_1$ ($\Delta\delta_1 < \Delta\delta_2$) after t_1 seconds, the system will be stable.
(k) If $\Delta\delta_{max}$ lies between $\Delta\delta_1$ and $\Delta\delta_2$ but is decreasing in magnitude after t_1 seconds, the system will be stable.
(l) If $\Delta\delta_{max}$ lies between $\Delta\delta_1$ and $\Delta\delta_2$ but is increasing in magnitude after t_1 seconds ($t_1 < t_2$), the system will be unstable.
(m) As soon as any one of the above conditions is met, the step-by-step calculation procedure is terminated. The parameters $\Delta\delta_1$, $\Delta\delta_2$, t_1 and t_2 are empirically predetermined from the off-line study on a series of transient stability calculations.

This model calculates the generator swings of 2-second duration within 5 seconds for the main power system of Tohoku Electric Power Co.

8.4 Features of PSALM-73

The following information gives the scale of PSALM-73. Number of program modules: 126. Total number of program statement lines: nearly 150 000 (written in FORTRAN IV). Maximum number of JOB STEPS in catalogued procedures: 10 steps. Number of required data cards for master database initiation: nearly 8000. Load forecast data for a period of 10 years occupy about a quarter of the required data cards for the master database which is updated at least once a year. The data associated with network configuration patterns and system expansion facilities are also revised every year. Most of the other master databases usually required no regular modification once they have been established.

In comparison with the conventional procedures without the PSALM-73, about 100 times as much calculation service has been rendered with the same amount of man-day labour in carrying out power system studies. The significant contributions to the practices afforded by the PSALM-73 are summarized below.

With various system conditions such as network configuration, generator maintenance schedules and generation dispatch schemes, broader investigation of power system performance can be achieved as well as a significant improvement of accuracy in system studies. Moreover, PSALM-73 permits engineering interaction in the investigation procedures shown in Fig. 8.1, and enables the engineers to work out the optimum plan in the power system studies with the help of the suboptimization function of the PSALM-73.

Intelligible reports summarizing the results of analyses and evaluations are produced automatically in the form of graphical tabular formats, without utilizing any special I/O device (see Fig. 8.4). Graphic formats such as power flow diagrams and generator's phase angle swing curves can also be produced on a line printer. These functions improve not only the accuracy of studies but also communications between man and machine.

All analysis and evaluation procedures are carried out using a few macro input commands. These not only significantly reduce the laborious work involved in data preparation, but also eliminate the possibility of input errors.

The establishment of master databases results in unification of both engineering and management data associated with power system planning and operation.

In the design of the PSALM-73, the module structure of the component programs and the coding scheme of the relevant data are adopted so as to achieve a maintenance-free system. These techniques make it possible to incorporate easily additional programs in the PSALM-73.

8.5 Practical and Present Problems

The PSALM-73 has been in service in the routine practice of power system security studies in both the head office and seven branch offices. These studies have been done by batch-mode processing through the remote terminals in the company's computer network system. The batch-mode use of the PSALM-73 by RJE terminals (Remote job entry, IBM's computer system consisting of card readers and line printers) has now reached capacity level. This unfavourable situation arises from the constraints of batch processing, resulting in longer turn-around time due to the increase in computation frequency.

To reduce the waiting time for input/output in the batch-mode processing, the Company installed interactive CRT (Cathode-Ray Tube) terminals in its offices in 1982, thereby improving the efficiency of analytical computations, applying the PSALM-73 to the interactive terminals with no modification of its batch-mode-designed programs. Figure 8.5 shows the statistics of the PSALM-73 job processing during the past six years. However, interactive equipment of the computer system is expected to better operating performances by the the application of conversational-mode-designed softwares.

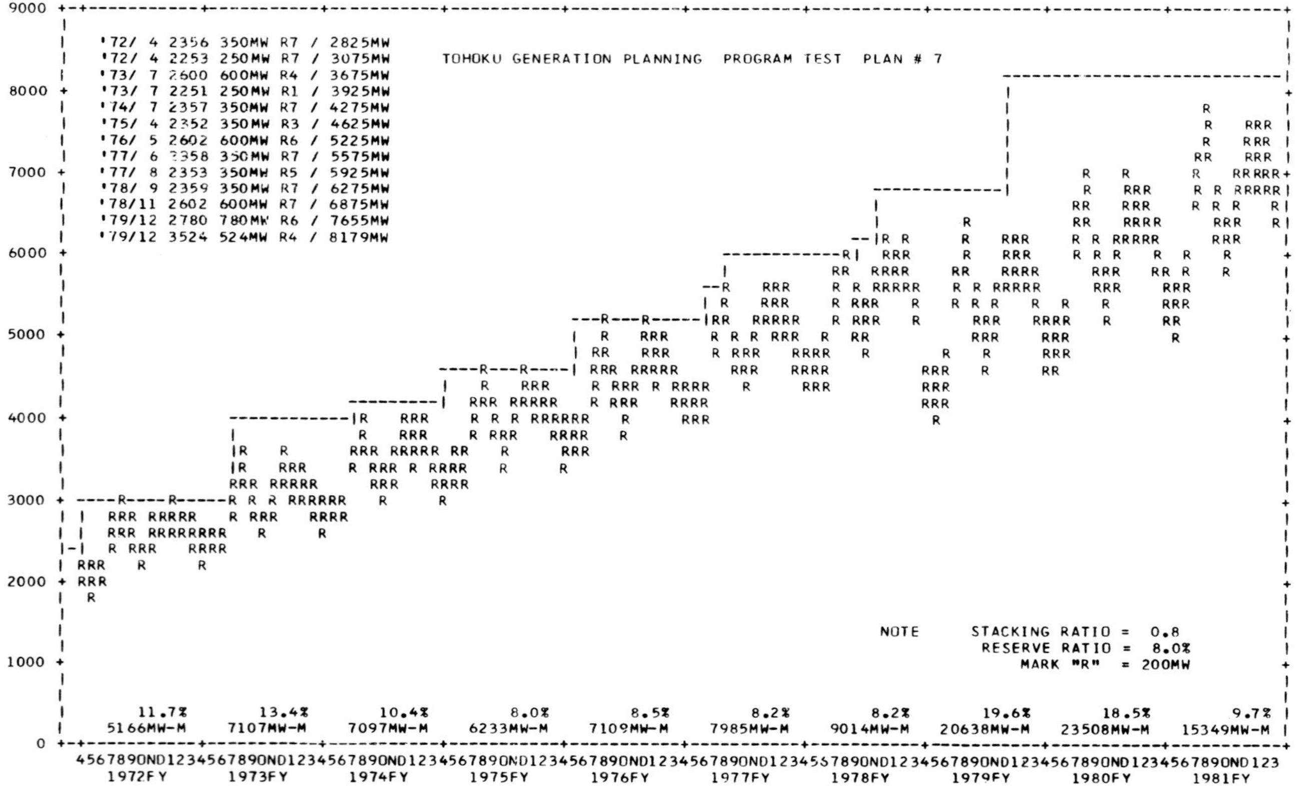
TOHOKU GENERATION PLANNING PROGRAM TEST PLAN # 7
'72/ 4 2356 350MW R7 / 2825MW
'72/ 4 2253 250MW R7 / 3075MW
'73/ 7 2600 600MW R4 / 3675MW
'73/ 7 2251 250MW R1 / 3925MW
'74/ 7 2357 350MW R7 / 4275MW
'75/ 4 2352 350MW R3 / 4625MW
'76/ 5 2602 600MW R6 / 5225MW
'77/ 6 2358 350MW R7 / 5575MW
'77/ 8 2353 350MW R5 / 5925MW
'78/ 9 2359 350MW R7 / 6275MW
'78/11 2602 600MW R7 / 6875MW
'79/12 2780 780MW R6 / 7655MW
'79/12 3524 524MW R4 / 8179MW
NOTE STACKING RATIO = 0.8
RESERVE RATIO = 8.0%
MARK "R" = 200MW
9000
8000
7000
6000
5000
4000
3000
2000
1000
0
11.7% 5166MW-M
13.4% 7107MW-M
10.4% 7097MW-M
8.0% 6233MW-M
8.5% 7109MW-M
8.2% 7985MW-M
8.2% 9014MW-M
19.6% 20638MW-M
18.5% 23508MW-M
9.7% 15349MW-M
4567890ND1234567890ND1234567890ND1234567890ND1234567890ND1234567890ND1234567890ND1234567890ND1234567890ND1234567890ND123
1972FY
1973FY
1974FY
1975FY
1976FY
1977FY
1978FY
1979FY
1980FY
1981FY

(a)

```
MAINT.SCHEDULE CODE     54005001                                                     54 FISCAL YEAR

 NO   NAME        APRIL  MAY    JUNE   JULY   AUG.   SEP.   OCT.   NOV.   DEC.   JAN.   FEB.  MARCH    NO CLASS AREA CAP(MW) MAINT(MW)

  1  S.SEND;2     ****    +      +      +      +      +      +      +      +      +      +   ********    1  ( 1)   4   600.0    582.0
  2  ONAGAW#1     +    ***********      +      +      +      +      +      +      +      +      +    +    2  ( 2)   4   524.0    500.0
  3  AKITA #1     +       +      +      +      +      +      +      +      +  ***********     +    +    3  ( 3)   3   350.0    336.0
  4  AKITA #2     +       +    ********   +      +      +      +      +      +      +      +      +    4  ( 4)   3   350.0    336.0
  5  AKITA #3     +       +      +      +      +      +      +      +      +      +   ********     +    5  ( 5)   3   350.0    336.0
  6  S.SEND#1     ***     +      +      +      +      +      +      +      +      +      +   +******    6  ( 5)   4   350.0    336.0
  7  N.KYOD#1     +  ********    +      +      +      +      +      +      +      +      +      +    +    7  ( 5)   7   350.0    336.0
  8  T1NIG.KY     +       +    ********   +      +      +      +      +      +      +      +      +    8  ( 5)   7   350.0    336.0
  9  T2NAUE#1     +       +      +      +      +      +   +*******      +      +      +      +      +    9  ( 6)   7   350.0    336.0
 10  T3NAUE#2     +       +      +      +      +      +      +      +      +      +      +   *******  +   10  ( 6)   7   350.0    336.0
 11  HACHIN;4     +       +      +      +      +   *******      +      +      +      +      +      +   11  ( 7)   1   250.0    237.0
 12  NIGATA#3     +       +      +      +      +      +      +      +      +  *******      +      +    +   12  ( 7)   7   250.0    237.0
 13  NIGATA#4     +       +      +      +      +      +      +      +      +      +  *******      +    +   13  ( 7)   7   250.0    237.0
 14  HACHIN;3     +       +    *******  +      +      +      +      +      +      +      +      +    +   14  ( 7)   1   250.0    237.0
 15  SENDAI#1     +       +      +      +      +      +      +   +*******      +      +      +      +   15  ( 8)   4   175.0    153.0
 16  SENDAI#2     ****    +      +      +      +      +      +      +      +      +      +      +  ****   16  ( 8)   4   175.0    153.0
 17  SENDAI;3     +   *******    +      +      +      +      +      +      +      +      +      +    +   17  ( 8)   4   175.0    153.0
 18  NIGATA#1     +       +      +      +   *******      +      +      +      +      +      +      +   18  ( 9)   7   125.0    118.0
 19  NIGATA#2     +       +      +   *******+      +      +      +      +      +      +      +      +   19  ( 9)   7   125.0    118.0
 20  HACHIN#1     +       +      +      +      +      +   ******  +      +      +      +      +      +   20  (10)   1    75.0     69.0
 21  HACHIN#2     +       +      +      +      +      +      +      +   ******     +      +      +    +   21  (10)   1    75.0     69.0
```

(b)

Fig. 8.4 Examples of intelligible graphic presentation of analysed results (a) Generation expansion analysis; (b) optimum maintenance schedule of generator units.

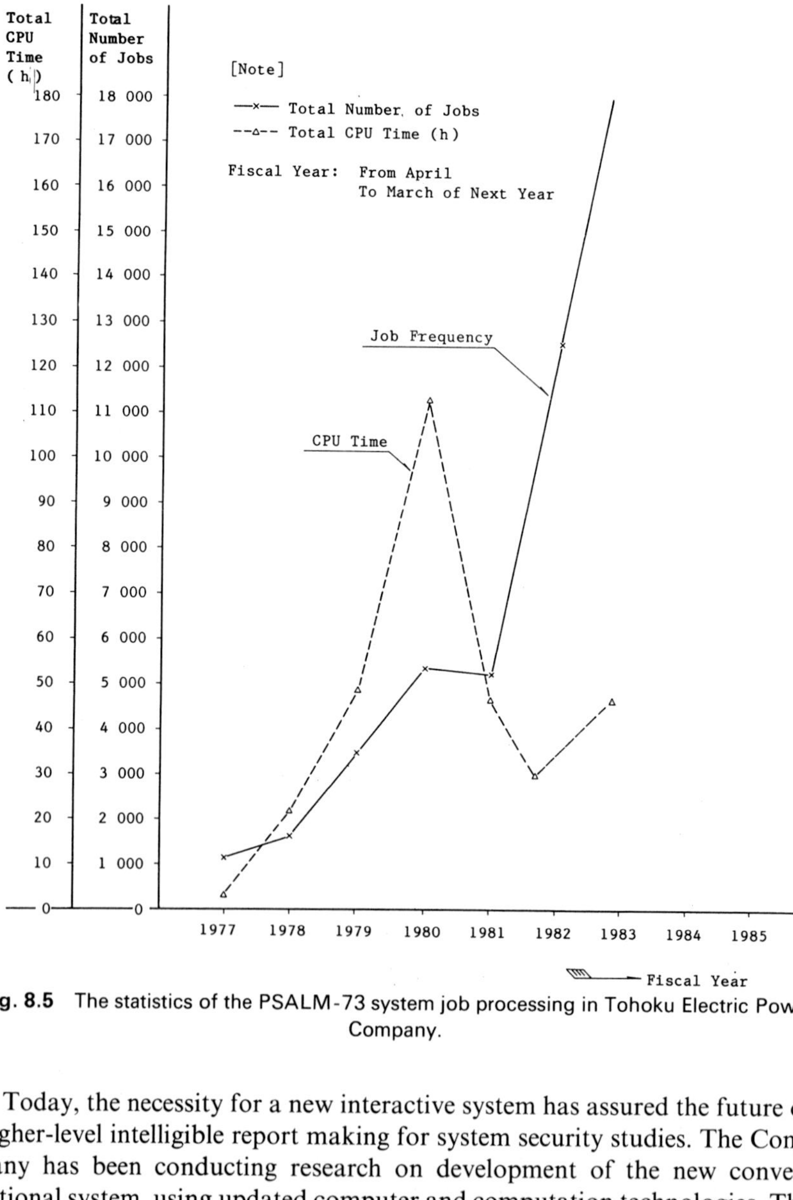

Fig. 8.5 The statistics of the PSALM-73 system job processing in Tohoku Electric Power Company.

Today, the necessity for a new interactive system has assured the future of higher-level intelligible report making for system security studies. The Company has been conducting research on development of the new conversational system, using updated computer and computation technologies. The new system is expected to deal efficiently with broader impacts arising from an energy-shifting strategy thereby lessening the problems related to oil-dependency and the environment.

In the following sections, the philosophy and strategy of the new interactive database for the improvement of security of the power systems are described, together with the design of the system and the corresponding measures. The specific procedures are also outlined with the requirements for the new interactive system.

Examples of the fast computation techniques applied to practical system studies are presented with an estimated evaluation on the cost savings offered by the fast computation programs.

8.6 New Interactive Database

The overall experience gained in operating PSALM-73 during the past decade were reviewed, and the design criteria included in the functions of the new system were studied. Prior to the system developing procedures, a feasibility study on the new interactive system was made. As a result, the following requirements were needed:

(a) curtailment of computing response time;
(b) simultaneous requests by many on-line CRT terminals;
(c) refined management for data and softwares;
(d) sophistication of graphic display functions.

The new interactive system has been designed to meet the requirements, as shown in Fig. 8.6.

The greatest difficulty in the process of developing the interactive system is the curtailment of response time. In order to ensure faster computing response, it would be necessary for the new system to adopt such measures as incorporating faster programs, separating programs into small modules, and the formation of high-level database.

As for requirement (b), small size "temporary sub-databases" have to be allocated to respective users to meet the procedures required. Protection of database from the interference caused by simultaneous access has also to be considered in OS programs.

For requirement (c), program products such as IMS, TSO, GDDM and GAS should be included in the system design, to help engineers reduce maintenance work on both data and programs.

For requirement (d), graphic display systems and plotter programs will be developed to improve cost and benefit.

8.7 Application of Faster Programs

Sample cost evaluations were made in the feasibility study which revealed

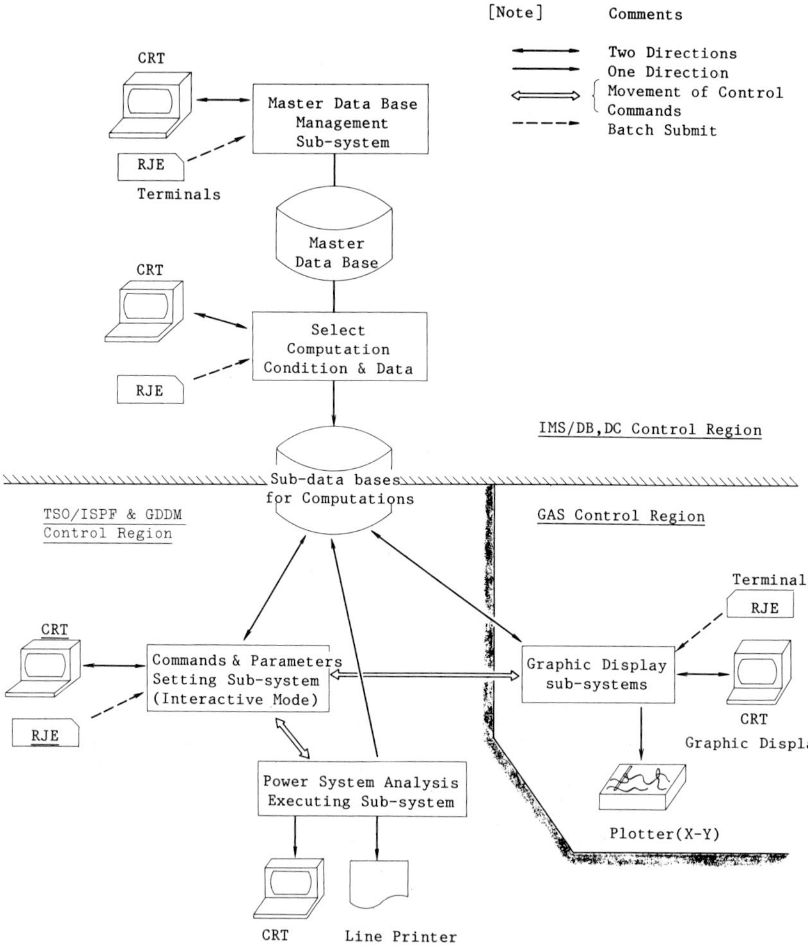

Fig. 8.6 The architecture of the interactive database for system security study in Tohoku Electric Power Company.

that about 20 million yen a year was already saved and 40–50 million yen a year is expected to be saved in the next couple of years by applying faster analytical computation programs. The following examples were used speeding up calculations:

Fast load flow program

System engineers often encounter the following difficulties in implementing load flow calculation: solutions for ill-conditioned data are not easily obtained; much time and labour are required to obtain adequate results. In

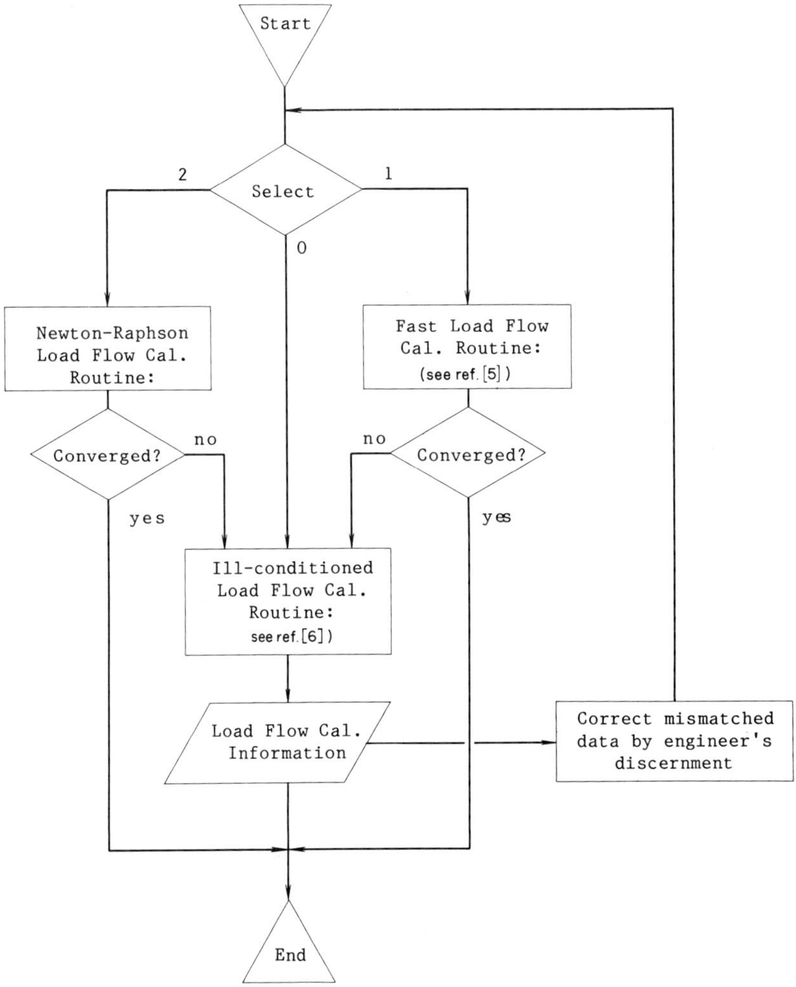

Fig. 8.7 The integrated load flow calculation program chart.

order to solve these an integrated load flow calculation program by the second-order Formation[5, 6] was incorporated into the new system, as shown in Fig. 8.7.

Speeding up fault current calculations

The Z-bus matrix calculation, used for fault current calculations and sensitivity factor analyses, was so far carried out using the building-up algorithm. However, this algorithm has the disadvantage that the computing time and memory storage increase drastically with system size. To solve the problem,

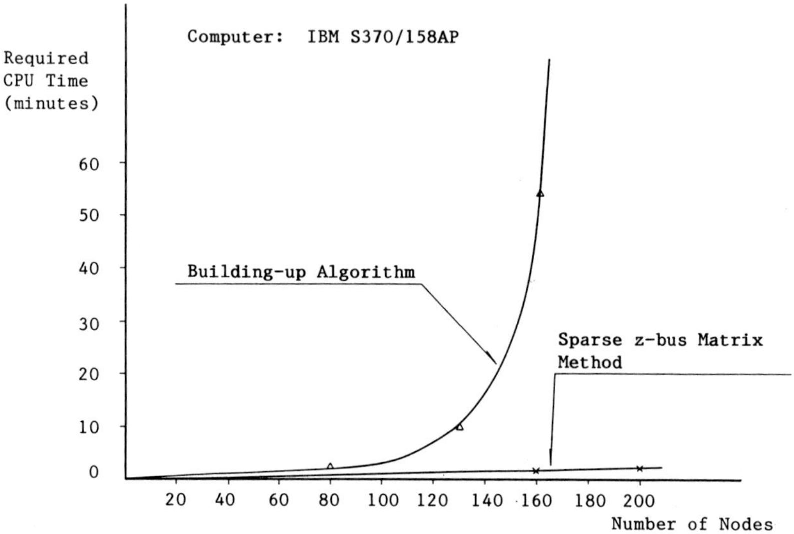

Fig. 8.8 The comparison of computing time characteristics on fault current calculations (building-up algorithm—sparse Z-bus matrix method).

the Sparse Z-bus Matrix Method[2] was applied to calculate only the elements necessary of the Z-bus matrix. By incorporating this routine into the conventional fault calculation programs, a significant speeding-up has been achieved. Figure 8.8 shows the comparison of the calculation speed between the two methods.

8.8 Schedule of Development

A task force was organized in May 1982 to promote this project under the sponsorship of Tohoku Electric Power Company in collaboration with Waseda University, Central Research Institute of Electric Power Industry, and Kaihatsu Computing Service Centre.

The first phase of the project was carried out before commissioning the new system in service in April 1984. The object of the first phase is to remodel the functions of the currently-used PSALM-73 into interactive mode systems.

In the second phase of the project are planned the sophistication of the functions on the graphic display programs, the grading-up of the database functions and the integration of other systems. The training simulator functions for system operators are to be developed in the second phase.

References

1. Minakawa, T., Mikuni, B., Takahashi, K., Masumo, I., and Tamura, Y., *Practical Assessment and Strategy for Improvement of Dynamic Performance of Power System*. CIGRE, 26th Session, 1976, Group 32, No. 32-02.
2. Takahashi, K., Fagan, J., and Chen, M. S., *Formation of a Sparse Bus Impedance Matrix and Its Application to Short Circuit Study*. IEEE PICA Conference, TPII-B, Minneapolis, pp. 63–69.
3. Tamura, Y., Narita, S., Yokoyama, R., Saito, A., and Kasahara, T., *Optimal Real Power Allocation Algorithm by Pseudo Inverse*. Information Study Committee of JIEE, Paper IP-74-2. 22 Jan. 1974, pp. 2-1–2-12.
4. Takahashi, K., Masumo, I., Minakawa, T., and Atsumi, I., Computing Method for Power System Expansion Planning (Report No. 1). CRIEPI Report No. 72008, July 1972, pp. 11–23.
5. Iwamoto, S. and Tamura, Y., "A Fast Load Flow Method Retaining Nonlinearity", *IEEE Trans. Power Apparatus and Systems*, Vol. PAS-97, pp. 1586–1599, Sept/Oct. 1978.
6. Iwamoto, S. and Tamura, Y., "A Load Flow Calculation Method for Ill-conditioned Power Systems", *IEEE Trans. Power Apparatus and Systems*, Vol. PAS-100, No. 4, April 1981.

Note: CRIEPI: Central Research Institute of Electric Power Industries, Japan.

9

Processing of Technical Information

Toshiaki Tabuchi, Research Laboratories, Ashigara, Fuji Photo Film Co Ltd, Minamiashigara, Kanagawaken, 250–01 Japan

9.1 Introduction

The driving force of an industrial firm lies in its technical capabilities. In today's climate of rapidly developing technology, one of the keys to developing new technologies and products efficiently is the ability to investigate quickly and accurately enormous amounts of advanced technical information.

The scientific literature and patent information that must be studied is increasing daily. For example, *Chemical Abstracts*, the representative abstract journal in the chemistry field, includes upwards of 500 000 papers and patents every year.

An old reference indicates that approximately half of all workers were conducting duplicate research, unaware of existing studies,[1] which was probably due to the fact that the volume of information has grown to such an extent that there would be no time left for experiments were one to read all that information! Further, it is impossible to evaluate deficiencies and problems precisely without fully studying the problem at first hand. Hence, duplicate

CAD/CAM and MIS in Japan
ISBN 0-12-682580-7

studies occur. When one considers that the volume of information has doubled since this study was published, the number of redundant research efforts must also have increased.

However, when starting any research, one must confirm existing research and patent rights in order to conduct research and development efficiently. For this, the collection, processing and classification of necessary information, as well as a system to retrieve information as needed, is necessary.

9.2 Total System for Information Processing

Fuji Photo Film Co recognized this need early on. It developed its own information retrieval system in the mid-1950s and has operated that system ever since. The total system for processing technical information, which gave birth to highly sensitive, high-quality colour film and a new drying technology for photographic film, is composed mainly of three search sub-systems.

(1) TI Card System (Technical Information Card System).[2]
(2) ITPAIS (Image Technology Patent Information System).[3,4]
(3) Commercial on-line search systems.

The TI Card System and ITPAIS are used for processing information in areas of technology that deeply concern the company's activities, but as each operates under essentially different classification schemes, investigations are often conducted using both systems in tandem. Processing all the required information by these two systems is naturally impossible, so for investigating basic information on science and engineering, commercial on-line search systems are also used (Fig. 9.1).

Both of these systems are computerized processing systems. In an informa-

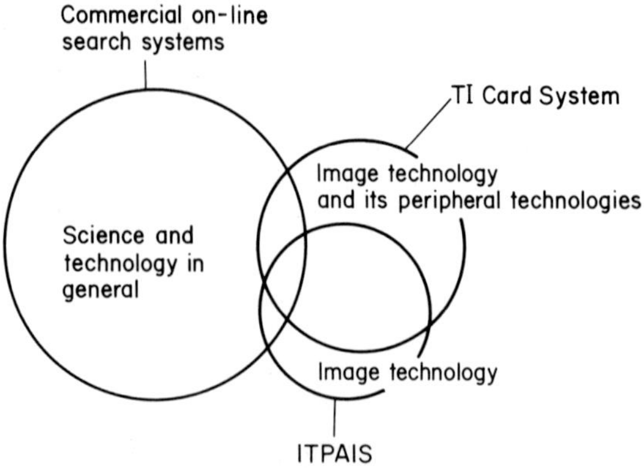

Fig. 9.1 Information retrieval systems and their areas of information.

tion system that processes textual information, the stages preceding computer processing, in other words, the collection, processing and classification of that information, are important; and without this processing efficient computer retrieval is impossible. From first starting to process technical information, emphasis has been consistently placed on the appropriate classification and indexing of information (Fig. 9.2).

9.3 Information Retrieval Subsystem

9.3.1 TI Card System

This information retrieval system was begun in 1958 for Fuji Photo Film's collection of technical literature and patent information in the field of image technology, and uses the UDC (Universal Decimal Classification) classification. In the twenty-six years since, newly generated information has been continually processed as well as retrospective information. Abstracts prepared outside the organization have also been added to the collection, so that by 1984 the total number of TI cards available for searching was an enormous 1 300 000.

9.3.1.1 The selection of information and the preparation of abstracts

With this system an organized group of researchers and technical people are responsible for collecting and abstracting the information. In order to ensure the comprehensiveness of the collected information, in the case of patent information, TPs (Technical Person for Information Science) are appointed for each area of technology, and these people (leading several group members) are responsible along with information scientists for the collection of information. For scientific literature, managers of these, by using the study group organization established for each technical area, are responsible for the collection of information. Also, in order to maintain the quality of abstracts, which is the deciding factor in the success or failure of a search system based on abstracts, the preparation is done not by the information scientists, but by the researchers themselves who understand the technical contents. The TPs and the study group managers are only responsible for their supervision. This system is now also used in subsidiary companies.

When these abstracts and patents are received, the technical contents are then discussed by each group, and, where appropriate, any opposition to patent claims examined. Abstracts which have been used as discussion material are all collected in the information section(s), and classified and stored for future reference.

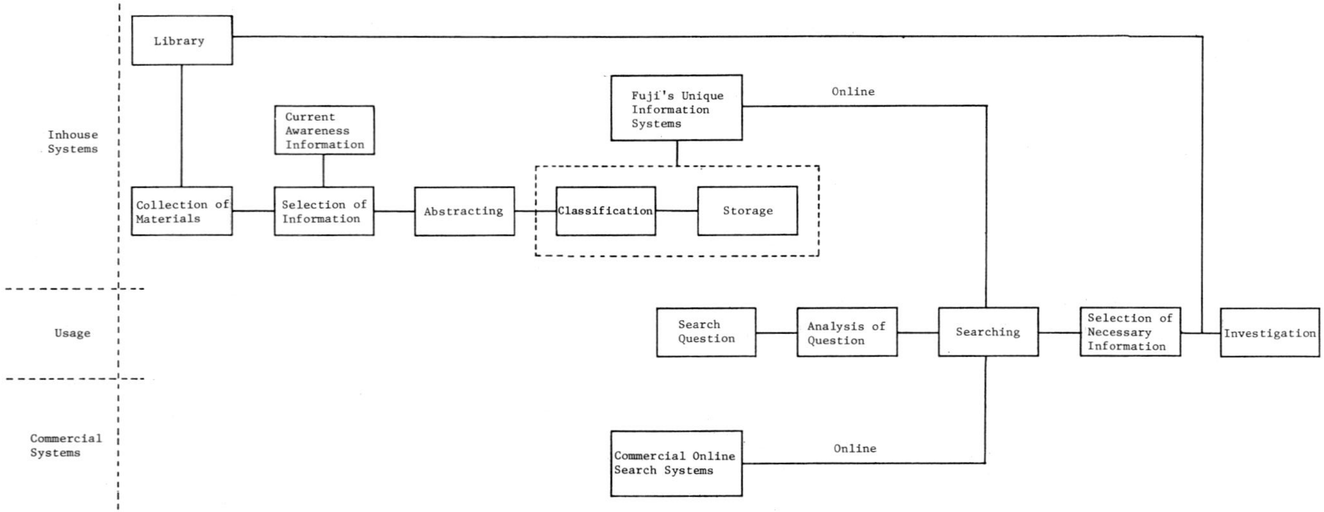

Fig. 9.2 Layout of the total information processing system.

9.3.1.2 *Classification of abstracts*

Abstracts of scientific papers and patents are classified using the UDC code. This code has been widely used for many years as a classification scheme, and the company has been using it consistently since it first started processing technical information.

UDC is a systematic classification scheme for concepts within a particular area. Moreover, since it is systematic, the appropriate classification is easy to decide, and differences between indexers are less common than in the case of indexing by keywords and descriptors. In the TI Card System, classification is normally done on the basis of the abstract, and not on the original document, so there is a significant advantage in using systematic classification.

Abstracts are always classified by a number of people: differences between indexers on the extraction of points to be classified, and the appropriateness of the classification, need to be resolved, and the final classifications decided. With this co-ordination, each person is able to comprehend the contents of the stored file and can determine the search statement for retrieval.

With the rapid development of technology, expansion of the classification has become necessary in order to make retrieval more efficient. As expansion and revision of UDC is often slow, it is sometimes pointed out that it is not appropriate; but at the company the subclassification scheme is expanded, based on UDC principles as needed, in order to enhance search efficiency (Fig. 9.3).

9.3.1.3 *Storage of abstracts*

Abstracts which have been assigned the UDC Classification are printed in card format, and distributed to those on circulation; at the same time, cards for searching are filed. Cards were originally used for searching, but as their number now exceeds one million, they are produced in microfilm.

Microfilming of these cards was implemented so that identical information could be used in several offices, to co-ordinate information, and to link the microfilm file to computer searching.

9.3.1.4 *Characteristics of this system*

Better quality information and search efficiency was planned by co-ordinating the activities of the researchers responsible for their own particular technical areas, and the information scientists. This resulted in researchers' better understanding of all the pertinent information and the technology and patentability, which in turn led to more accurate and efficient daily research activities.

772.F PHOTOGRAPHY USING PHYSICAL PHENOMENA
Auxiliary numbers for 77 are not applied here

772.F3 MANOCHROMATIC ELECTROPHOTOGRAPHY
Mostly the arrangement of the following auxiliary numbers occurs in the sequence .0 *and* -, *when both are necessary*

-1 SENSITIVE MATERIALS
Subdivided almost in parallel with 77.021.1; *Specify forms by* -4 *for* 62.4
-11 PHOTOCONDUCTIVE MATERIALS
-13 SPECIAL ADDITIONS TO PHOTOCONDUCTIVE MATERIALS
-132 Chemical Sensitizers
-133 Optical Sensitizers
Specify compounds by :54, *e.g.* 772.F35.2.062-133:547.674 *eosin as a spectral sensitizer for zinc oxids resin layers*
-139 Others, e.g. additions for reduction of fatigue, dispersant in photoconductor resin layers
-15 LAYERS OF SENSITIVE MATERIALS
Subdivided almost in parallel with 77.021.5
-151 Underlayers.
-152 Solvent hold-out and/or conductive layers
E.g. 772.F35.2.062-152:546.33'284 *conductive layer of sodium silicate for zinc* 772.F35.2

772.F3-53 PROPERTIES
Subdivided almost in parallel with 771.53
-533 PROPERTIES OF PHOTOSENSITIVE MATERIALS
-535 PROPERTIES OF LATENT IMAGE AND TRANSFERRED LATENT IMAGE
Subdivide as 772.F3-537, *e.g.* 772.F3-533.44 *electrostatic contrast of latent image*
-536 PROPERTIES OF THE DEVELOPED, BUT NOT YET FIXED IMAGE, AND ITS TRANSFERRED IMAGE
Subdivide as 772.F3-537
-537 PROPERTIES OF THE FIXED IMAGE
.43 Density. Density range
.44 Contrast
.61 Resolution
E.g. 772.F35.023.424-537.61 *resolution of magnetic brush developed image*
.62 Sharpness
.66 Development effects, edge effects
E.g. 772.F5-537.66 *image undercutting of reroradiograph*

-56 RECEIVING MATERIALS. TRNSFER PAPER
Subdivide as -17
e.g. -562 *receiving paper*

-6 APPARATUS. DEVICES
-7 CHEMICAL

Fig. 9.3 An example of the UDC subclassification developed independently at the Company.

As the classification scheme used is systematic, there are few misshits, and comprehensive searches that include related technology are possible. This is especially useful in preliminary searches preceding research, and in searches concerning technologies of any competitors that may infringe patent rights.

9.3.2. ITPAIS

This system is mainly used to index patent information in the image technology field, and can be searched by the computer on-line. Indexing is done by an item index using descriptors, a chemical compound index, and a corresponding patent index. It is a comprehensive patent information retrieval system for the image technology field. The Agfa-Gevaert Co, the Eastman Kodak Co, and the Fuji Photo Film Co have co-operated in preparing the indexes and revising the thesaurus since 1973.

9.3.2.1 The technical area for information collection

The scope of information that is processed includes patents, research disclosures and scientific literature in the image technology field, such as:

(a) photographic chemistry;
(b) photosensitive materials;
(c) photographic processing;
(d) non-silver halide photography;
(e) image-recording systems.

It is difficult to define these areas in detail, but apart from photographic equipment and the electrographic image recording system, 94% of past patents in these fields have been indexed. The number of patents indexed is 37 000, and if corresponding and related patents are included, 120 000 indexed items of data are filed in the computer.

9.3.2.2 The thesaurus and indexing[5, 6]

In contrast to the TI Card System, which uses the representative UDC systematic classification, ITPAIS uses strictly controlled descriptors as index terms in its classification. The thesaurus consists of 4800 basic terms (23rd edition), and for each descriptor synonyms, narrower concept terms, broader concept terms, related terms, and scope notes of descriptors are established (Fig. 9.4).

The ITPAIS thesaurus combines a systematic classification and classification by keywords, and incorporates the characteristics of both. This indexing method does not allow for variations between indexers, as in the case of uncontrolled keywords. Adequately controlled index terms must therefore be

```
ELECTROGRAPHIC CHARGING APPARATUS      011P25
                 NO NARROWER TERMS WILL BE ADMITTED
                 UNDER THIS TERM.  IF THE INDEXER WISHES
                 TO DIFFERENTIATE AMONG THE VARIOUS TYPES
                 OF CHARGING APPARATUS, HE SHOULD MAKE
                 USE OF THIS TERM AND ONE OF THE TERMS
                 FROM THE ELECTROSTATIC CHARGING METHODS
                 DISPLAY.  THIS DESCRIPTOR SHOULD BE USED
                 TO INDEX THE CONCEPTS OF A DEVICE FOR THE
                 DEPOSITION OR THE REMOVAL OF STATIC
                 CHARGES IN ELECTROGRAPHY.
  USE FOR
     ANTISTATIC DEVICES, ELECTROG
     CHARGING DEVICES, ELECTROGRAPHIC
     DISCHARGING DEVICES, ELECTROGRAPHIC
     ELECTROSTATIC CHARGERS, ELECTROG
  RELATED TERM
     ELECTROG PARTICLE CHARGING APPARATUS     011P64
     ELECTROSTATIC CHARGING METHODS     ******
  NARROWER TERM
     CORONA WIRE     026P74
  BROADER TERM
     APPARAT, ELECTROG     ******
```

Fig. 9.4 A portion of the thesaurus.

established and the thesaurus is the most important component of the system.

In the case of systematic classification, the concept to be classified can always be positioned, regardless of whether the subdivisions of the classification are lacking. However, when indexing by descriptors, this cannot be done if the corresponding descriptor is not included in the thesaurus, and a new descriptor must be established. By establishing a new descriptor, the positioning of it within the thesaurus, including its related descriptors, must be carefully selected.

Patents and scientific papers are indexed using this thesaurus. The entire document is carefully read, the contents considered in detail and descriptors selected. In order to limit the concept further after indexing by descriptors, roles may also be combined with the descriptors. These may be radiation sensitive element location within the recording materials (role 5), or parts of classification which include an agent in the processing solution (role 7). If there is mention of a chemical compound in the document, then the molecular formula or the substructure code (MCC-TSS: Mechanical Chemical Code-Topological Search Screen)[7, 8] is also decided.

With the TI Card System, co-ordination of opinions regarding indexing is necessary to decide the final indexing, because standardization using descriptors is more difficult than using a systematic classification. Indexers have both the knowledge of the technology being indexed and the search system.

9.3.2.3 The various indexes that comprise the system

In this manner various inverted files are produced for computerized searching, with the source file indexed by descriptors at the centre and supplemented by molecular formulae and substructure codes.

Here, in order to make the contents of ITPAIS more understandable, the output record contents will be used for explanation.

Indexing by descriptors

Descriptors and roles are assigned to each patent and paper. In the case of patents, the applicant and application year are input, as well as information on the patent family, including corresponding and related patents. The number of descriptors assigned differs according to the document but they average 20 to 30 per record.

Figure 9.5 shows the data output format, but in the computer are files which have patents and documents with descriptor codes, and files which correspond the descriptors to the codes. As ITPAIS does not have abstracts, the contents of the document are read from the set of descriptors assigned. To confirm further details of the contents, users are referred to the abstracts

```
P11283  ITPA-P      ITPA-P   G
                   JA..74/10,687 : RELATED TO  DT..2,051,930
                     FR..2,066,407 GB..1,302,688 *  DT2,051,930 =
                     FR2,066,407 = GB1,302,688 = JA74/10,687
                     JA74/10,687 = 065-002 TREATING HYDROPHOBIC
                     PHOTOGRAPHIC SUPPORT SURFACES WITH HIGH
                     FREQUENCY ELECTRODELESS DISCHARGE AT FREQUENCY
                     GREATER THAN 1 MHZ (ULTRASONIC TECHNIQUES) TO
                     IMPROVE ADHESION OF COATED HYDROPHILIC LAYERS
                   *  000B02        INDEXED BY FUJI
                   *  000C31        FUJI
                   *  000D69        PRIORITY FILING YEAR 1969
                   *  005P35 R01    ADHESIVES
                   *  000P52 R05    CELLULOSE ESTER SUPPORT
                   *  027P29 R05    CORONA TREATMENT OF SUBBING LAYER
                   *  000T01        ENGLISH TRANSLATION MICROFILMED
                   *  001P92 R05    HYDROPHOBIC LAYER
                   *  003P14 R05    POLYESTER SUPPORT
                   *  018P43 R01    RADIATION TREATMENT, SUPPORTS
                   *  003P21 R05    RESIN COATED PAPER SUPPORT
                   *  003P89        SILVER HALIDE COMPONENT
                   *  004P16 R05    SUBBING LAYER
                   *  004P94 R01    SURFACE FINISH
                   *  004P48 R01    ULTRASONIC TECHNIQUES
                   *  000010         CHEMICAL NUMBER
                   *  000095         CHEMICAL NUMBER
```

Fig. 9.5 A record of the basic reference file.

in the TI Card System or to the microfilm file, which include copies of the patent specification.

The chemical compound index

Many access points are prepared for chemical compounds. If there is a descriptor within the thesaurus that corresponds, then that descriptor is assigned. Also, the documents are indexed so that the compounds may be accessed from molecular formulae, substructures and trivial names.

The molecular formula file contains the molecular formula, the chemical structure and compound code number (Fig. 9.6). In searching from the molecular formula file the code number of the compound with the desired structure is used in the search statement.

In searches for chemical compounds the MCC-TSS is used, in which two-dimensional chemical structures are coded under a specific rule, and notated linearly. Figure 9.7 is an example of a chemical structure converted into MCC-TSS code. The MCC-TSS code for the substructure to be searched is determined, and from this the code number for the compound located. This code can be used to search for the compound, or it can be combined with other descriptors in a search statement.

The chemical trivial name file and the polymer substructure name file are kept together with the compound code numbers. The necessary files are displayed, and the code number for the compound confirmed. The code number can be used for searching the compound structure, or combined with other index terms for searching.

$C_{12}H_{21}N_3O_3S$ KAN: 016018-6 $C_{12}H_{21}N_3O_3S$ CAS REGNO: 73164-48-4

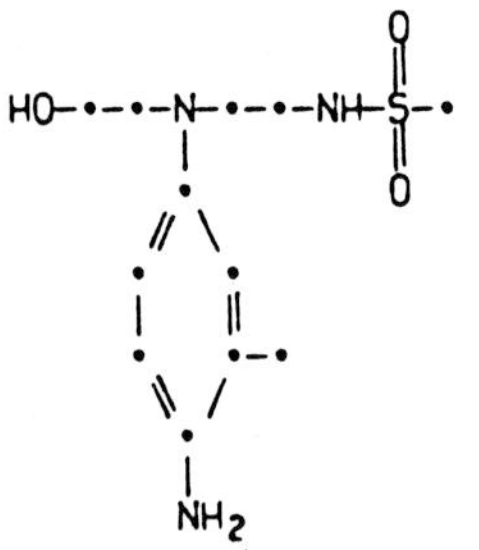

CAS NAME: Methanesulfonamide, N-(2-((4-amino-3-methylphenyl)(2-hydroxyethyl)amino)ethyl)-

Fig. 9.6 Record content of the molecular formula file.

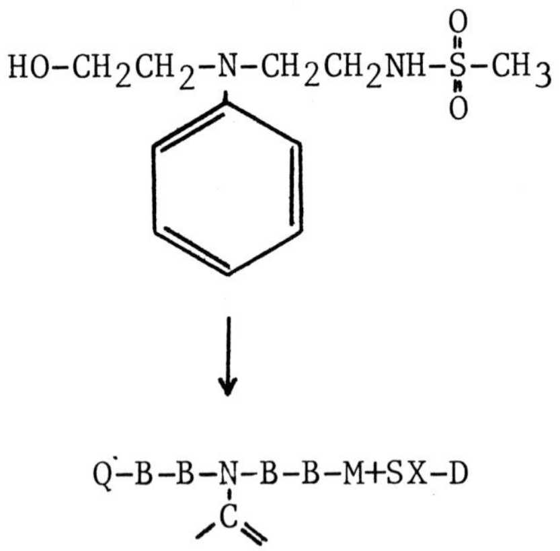

Fig. 9.7 An example of a chemical substructure converted into the MCC-TSS Code.

9.3.2.4 Characteristics of the system

The main technical feature of this system is that it uses descriptors in its classification. The individual descriptors represent concepts of narrower scope, compared with those in the systematic classification. These descriptors can be combined to search for those with an even narrower scope, resulting in an answer that is likely to be very close to the required item. ITPAIS is therefore appropriate for searching data suitable for opposition to a patent. However, on the other hand, if the concept is too finely narrowed, necessary information may be missed, making searching difficult. This is due to the fact that, as many descriptors are combined to represent the contents of a document, co-ordinating the descriptors that are selected during the indexing process, and those selected during the searching process is difficult.

Also, as previously mentioned, this system has many indexes, and searching can be done from many access points, and by combining various types of indexes. It is difficult to achieve perfection in indexing using descriptors, but as ITPAIS is designed as a collective index system, this increases its recall.

Since 1973, Agfa-Gervaert, Eastman Kodak and Fuji Photo Film (which are companies competing in technology and product development) have been co-operating to produce the indexes as well as expanding and revising the thesaurus. For competing companies with common interests, the technical information that must be processed is similar, so that co-operation in processing this information is profitable to all. Joint processing of information with competing companies has reduced development costs for the system as well as effort required for preparing the indexes. The system has been used not only by the three participating companies, but has also been available publicly for a fee since 1977 for the progress and development of image technology.[9]

9.3.3 Search System Using Free Keywords

This system does not use a particular classification scheme, as with the TI Card System or ITPAIS, but the person who collects the information decides on keywords to express those concepts which he considers will become access points, and uses them for indexing. As with the title of the record, all the words except stop-words are designed to be available for search.

With this system, there is less effort needed for classification, as indexing is by free keywords and the text can be searched, and unique data fields such as author names are used for searching. However, selecting the appropriate keywords is difficult, and recall in the search is not high compared to searching by descriptors. Due to this fact, this system is used more for processing transitory information, such as newspaper articles.

9.3.4 SPHINCS[10]

SPHINCS (Structure-Property Handling Information Network for Chemical Substances) is a chemical structure searching system for handling the structures, properties and reactions of chemical substances.

If searching by names is required, then it is necessary to index compounds using a systematic nomenclature. Even then, it is difficult to identify chemical compounds, especially if one wants to retrieve a group of compounds that have a common substructure.

For this reason, large-scale chemical structure search systems such as CAS ONLINE and QUESTEL/DARC have been developed, and are available commercially. The Fuji Photo Film Co has also developed a search system for chemical compounds used in the company by inputting chemical structures represented by two-dimensional images, and searching for compounds which have the desired structure.

9.3.4.1 The input of chemical structures

Component structures that frequently appear (sets of atoms) are first registered. On a graphic display, chemically significant alphanumeric codes (such as 6 for Benzene rings, 6N for a Pyridine ring) are retrieved and the desired chemical structure is drawn by specifying bonds, erasing or exchanging atoms or bonds, rotating the chemical structure or producing the mirror-image of the chemical structure. At the same time as drawing the structure, the chemical bond between atoms is registered in the computer in a connection table.

9.3.4.2 Searching

In order to ensure efficient and precise searching, primary screening and follow-up checks of the bonding relationship between atoms in a set reduction method are combined. In the primary screening, ring extraction, ring analysis, and chain analysis are done from the connection table, and a search file prepared. No matter how efficient the primary screening, as the access points are finite, the set of resulting compounds includes noise.

Therefore, by using all the information on bond relationships in the connection table, the bonding relations between atoms are checked one by one, the sets are reduced and noise excluded.

In processing the search query, the desired chemical structure is drawn on a graphics screen following a menu-driven set of commands, exactly as in the case of registration. The chemical structure judged to be appropriate after searching is displayed on the graphic display as a two-dimensional representation.

9.4 Total Computer Search System

The first computer application for the various search systems was to quickly and accurately process enormous amounts of data to form a response, no matter how complex the combinations of classifications or indexes in the search.

The systems are all on-line because of the ease of performing searches, and the locations in which they are used. The input of information for all of them is done directly from the terminal, at the point that information is generated, so that the systems may contain all the latest information.

The files for each system exist independently, but cross-system searching can be done in order to increase the coverage of a search. The results of searching the TI Card System using the UDC and the results of a search on ITPAIS using descriptors can be cross-referenced when outputting information.

9.5 The Use of Commercial On-line Systems

The Fuji Photo Film Co has independently collected and processed information in the major technical fields, including scientific literature and patent documents. However, it is impossible for it to process all the information in the scientific fields that form the foundation of these technologies, and all information in peripheral areas related to the company's activities.

Commercial on-line systems include not only the major abstract journals that have been used, but also abstract journals of a particular specialized technical nature, and numerous files including business information, so that these systems hold vast collections of data. Commercial systems give instantaneous access to any necessary information on-line, without purchasing these abstract journals. Apart from the systematic classifications and index fields used in the printed abstract journals, keywords within titles and abstracts and various supplementary index fields can also be used. The company started using the JOIS, DIALOG, and SDC on-line systems in 1979, and is presently using 12 on-line services. These systems include the world's major scientific literature and patent information files. Depending on the type of search needed, either in-house or commercial systems are used in tandem, or the commercial systems only are used. Use of the commercial systems comprises 65% of all searches conducted in both modes combined.

In looking at the most frequently used files in terms of collective usage hours, the order is CA SEARCH, JICST, WPI, CLAIMS, INSPEC, and the usage of Chemical Abstracts and the patent information files is extensive.

The effect of using these systems is that the scope of searches has expanded. The use of commercial search systems can also be combined with searches using the in-house systems for better recall. Problems in using these commercial systems arise in the selection of keywords, and the construction of the search statement combining those keywords, which is difficult.

These systems and files were produced by third parties, so that the information covered and indexing policies is not always clear. In order to resolve these problems, we have run study groups for over 21 years in the use of indexes of abstract journals, and have introduced their use among information scientists. The search topics processed daily are also gathered every week, and the net value assessed. The combination of these activities contributes to a better knowledge of systems and files, to learning the techniques for the selection of appropriate search terms, and the establishment of appropriate search statements.

9.6 Investigation System

Finally, the actual method of investigation using these systems will be described. In the case of searching, several information scientists analyse the contents of the investigation; the reason for more than one, is so that the topic of investigation can be judged from various viewpoints, and the most appropriate search system and classification used.

Where the in-house system will be used in the investigation, the person producing and the person selecting the classification for searching will be the same. With commercial systems, if more than one person selects the index terms and sets up the search statement, then the search will be more precise than if it were done alone.

When the system(s) to be used, and the classification and indexing terms have been decided, the information scientists search each system. The person who extracts the information desired from the output is naturally the one who requests it, because he has the clearest idea of what information is needed. Moreover, by reading all the information that has been output, hints for further research may arise from information that was not the direct object of the investigation.

9.7 Conclusion

The in-house technical information processing systems, operated as on-line computer systems, has been described. However emphasis is focused on classification and indexing and their use, instead of on the hardware and software, because this is what is required by the people who produce the indexes and classifications, and because they are considered to be the most important aspects in the processing of information.

Once the data are properly classified and indexed, design of the search system is not difficult. However, in fully utilizing the completed system to conduct precise searches, human factors again play a large role; e.g. the ratio of the human factor in information proccessing is large. Without sufficient consideration of this aspect, highly precise searches cannot be expected, no matter how rapidly the computer can search huge amounts of data. On the other hand, human contribution is limited by effort and time. In an age when the information to be processed is generated in vast amounts, the human effort required for abstracting and preparing indexes presents a problem. The company constantly evaluates the balance between computer processing and human effort, and is proceeding with the total systematization of technical information processing by computers.

In 1978 the Union of Japanese Scientists and Engineers awarded the Ishikawa Prize to Mr. Kotaro Sato, Dr. Yoshitada Tomoda, Mr. Kameji Nagao and Mr. Toshiaki Tabuchi in recognition of their work in modernizing the management of the Fuji Photo Film Company as a result developing these technical information systems.

Acknowledgement

Acknowledgement is made to the great contribution of Dr. Kenzo Hirayama for establishing and developing technical information systems, and the guidance and advice from Mr. Yoshikata Sakaguchi regarding further computerization since the Ishikawa award. Additionally, thanks is due to Mr. Hirozo Ueda, the senior Managing Director of the Fuji Photo Film Co Ltd, for approving publication of this paper.

References

1. Anon., *Japanese Scientific Monthly* **21**, 49 (1968).
2. Aoki, T., Discussions on the TI Card System.
3. Graham, R. W. and Roberts, H. E., *Phot. Sci. Eng.* **17**, 278–281 (1973).
4. Keller, K., *World Patent Information* **1**, 24–29 (1979).
5. Tabuchi, T., 179th ACS National Meeting, Houston, Texas 23–28 March, 1979.
6. Tabuchi, T., 3rd ITPAIS Symposium, Rochester, N.Y. 30 April, 1981.
7. Lefkovitz, D., *J. Chem. Doc.* **7**, 186–192 (1967).
8. Lefkovitz, D. and Gennaro, A. R., *J. Chem. Doc.* **10**, 86–94 (1970).
9. 2nd ITPAIS Symposium organized by the Graphic Arts Research Center, Rochester Institute of Technology, Rochester, N.Y. 21 April, 1977.
10. Hanai, S. and Miyakawa, M., 186th ACS National Meeting, Washington D.C. 28 August–1 September, 1983.

10

Management Decision Support System

Toshio Sunada, Management Systems Development Department, Systems Engineering Group, Fujitsu Ltd, Tokyo, Japan

10.1 Introduction

Generally speaking, computer systems have become indispensable for data processing. Based on experience in computer usage and the progress of computer technology, more advanced applications in the area of management decision-making are being developed in various organizations.

The Japan Development Bank (JDB), which is a governmental institution established to supply long-term loans to the private sector for economic and social development, has been applying software tools for decision-making activities since the introduction of a computer in 1966. The JDB has developed various computer models, mathematical and statistical programs, and databases, which are among the most sophisticated systems in Japan. However, the expansion of computer-aided operations has generated problems, such as the limitations of batch processing, shortage of computer specialists, the high cost of modifying the system and the time-lag between data gathering and presentation.

In order to solve these problems, the JDB organized a joint research project with Fujitsu in 1972 for the development of an advanced system named Management Decision Support System (MDS). The ultimate goal of MDS

CAD/CAM and MIS in Japan
ISBN 0-12-682580-7

was a man–machine interactive system providing effective support to all staff.

To realize this objective, the following features were included in the MDS software.

1. Graphical output for visual problem-identification, and the improvement of decision-making capability through interactive graphic terminals.
2. An interactive system to provide flexible and comprehensive combinations of problem-dependent analytical tools.
3. A general purpose system to satisfy various user needs.
4. A powerful modelling system to enable non-programmers to build and modify simulation models instantly through a terminal.

In addition, the following objectives were added when MDS was redesigned and enhanced to MDS-IV in 1979 by Fujitsu:

(a) Japanese character handling capability to provide managers with the more readily comprehensible output;
(b) two-dimensional table-oriented data structure to handle more complex information, together with tabulating and reporting functions;
(c) menu handling capability to provide a more flexible man-machine interface allowing tutorial operation, direct commands and customization;
(d) interface to management database systems which are based on both network and relational database systems.

Today, MDS-IV is being used not only by the JDB but also by many FACOM users in Japan, Australia and Korea. FACOM is the brand name of Fujitsu-made computers.

The sphere, design methodology, and system specifications of MDS-IV and MDS are discussed in the following sections.

10.2 Requirements of Decision Support System

As mentioned above, recent years have witnessed increased use of advanced computer programs for a variety of management functions. However, each traditional system has a restricted scope of application.

In this age of complicated and rapidly changing environments, improvements need to be made in the efficient use of computers, especially in management planning and control—mainly because they are so diverse that no single support system can fully meet the needs of its users.

Management
The recession, a high yen-to-dollar ratio and resource and energy problems have brought about rapid changes in corporate activities. Under such circumstances, structural changes due to streamlining of business operations are inevitable, and information rapidly becomes out of date. Consequently, existing systems are insufficiently flexible to keep pace with the rate of change, thereby precluding the computer from accomplishing its proper role.

End-users
Timing is essential for successful business negotiations. Deadlines are often missed because of slow progress in programming resulting from large backlogs and insufficient business knowledge. Recently, "office automation" has become an important element of business policy, and the computer facilities provided to end-users include packages, user oriented languages and personal computers having user-friendly man–machine interfaces.

EDP staff
Having no practical business experience, the EDP staff find themselves groping in the dark in an attempt to design a new system for management planning and control. For example, they are faced with the problem that they don't know what the skill of an end-user is, and for what purpose the system is used. Moreover, when the system is finally completed, the requirements of the user are found to have changed.

These comments are common to most corporate EDP systems. MDS-IV, the current version of MDS, is designed taking these points into consideration. Its aim is to develop a system to cope with the unexpected, and with changes in business operations.

10.3 Scope of Application

MDS-IV can be applied in a variety of operations to support management planning and control:

Planning and research. Analysis and forecasting of business achievements, economic environment, and industry trends.
Financial and accounting. Preparation of financial statement; financial analysis.
Sales. Demand forecasting; sales planning; comparative analysis of sales results versus sales planning.
Production. Production analysis; preparation of cost management table; comparative analysis of divisional output versus production planning.

Personnel. Analysis of wages and allowances; personnel planning and simulation; personnel expenditure versus expense budget.

The business procedures shown above are diverse and each requires professional knowledge and analytical skill. But, from the computer system point of view, they may be classified into facilities for data retrieval, manipulation, statistical analysis, tabulation and graphic presentation.

It is interesting in this context that a survey of senior management staff at the First National Bank of Chicago in the United States showed that 86% of their office hours is spent in data collection, 10% in data analysis, and 4% in report preparation.[1]

These staff members are well paid mainly because of their high analytic capability. However, they are spending 86% of their time turning over the pages of thick binders and ransacking file drawers in search of data to be used in their analytical work. As a result, only 10% is left for their professional activities.

Report writing is an important task for effective presentation of the analytical results, but only 4% is available for this job, because too much time is spent in collecting the data.

Similar situations are probably common in most corporate organizations. By supporting such tasks as retrieval, manipulation, analysis, tabulation and graphical representation, MDS-IV works indirectly to increase the efficiency of various business operations involved in management planning.

10.4 System Specification

MDS-IV has six important functions consisting of more than 90 commands. It is operated interactively through an alpha-numeric character display, a colour graphic display, a Japanese character display and a personal computer. Output is normally available for display, but comprehensive output (such as a set of several graphs and sentences) can be printed onto a laser printer at the DP site. Figure 10.1 shows the functions of MDS-IV.

(1) *MDS database.* Two-dimensional data structure is the key element of MDS-IV. According to our survey into the types of data structure used among management staff, tabular data (two-dimensional data) is the most widely and frequently used.

(2) *Tabular data manipulation (TDM).* This provides the basic functions for the manipulation of tabular data including file input/output functions such as simple retrieval of tabular data, editing of time series data, copying/deleting tables and locking/unlocking security functions.

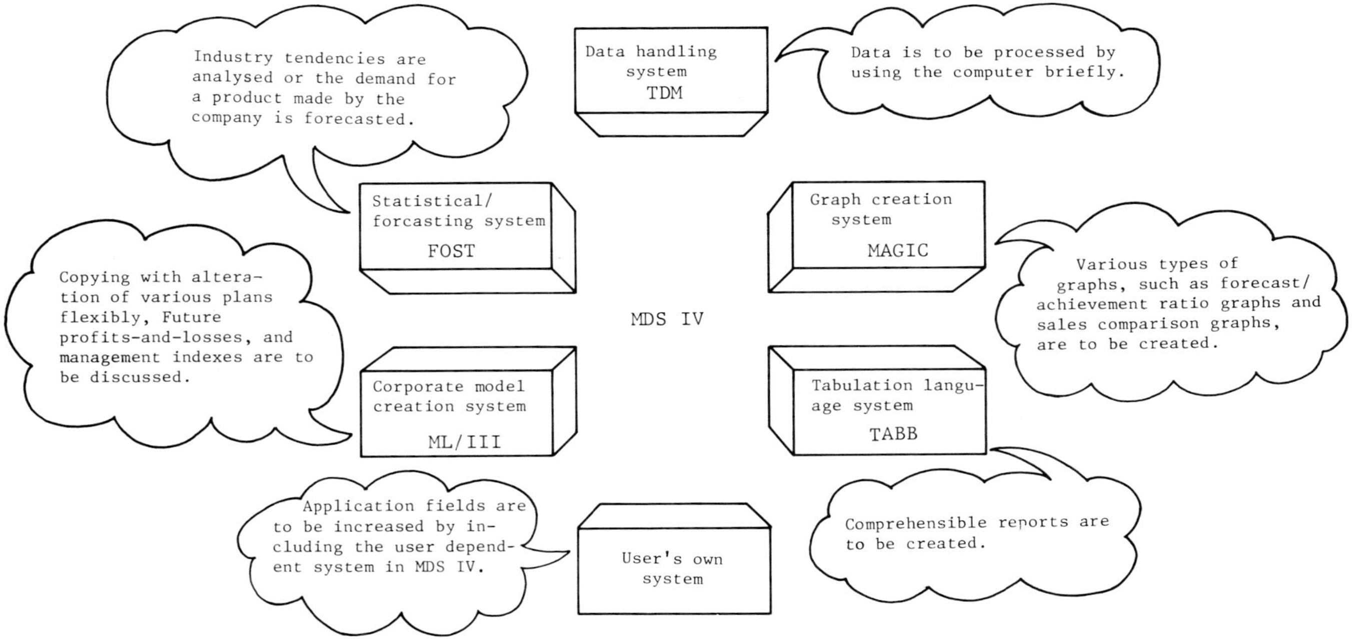

Fig. 10.1 MDS-IV functions.

(3) *Table generation language subsystem* (*TABB*). This is a simple and powerful language for end-users, which operates on tabular data with the row-wise or column-wise arithmetical and functional formulae to generate tabular type reports furnished with headers, ruled lines, footnotes, etc.
(4) *Graphical generation subsystem* (*MAGIC*). This is a system for preparing a variety of visual ouput including line graph, bar charts, scattergrams, composition ratio graphs, polygon graphs and portfolio graphs. Figure 10.2 shows a sample graph of break-even point.
(5) *Statistical and forecasting subsystem* (*FOST*). Forecasting by regression analysis, or by fitting actual data to theoretical curves, statistical analysis and confidence testing can be carried out. More sophisticated economic forecasting is done using another econometric system.
(6) *Corporate modelling language subsystem* (*ML/III*). This subsystem allows the building of simulation models (such as corporate financial models) and the use of various techniques including sensitivity analysis, risk analysis, and goal seeking.

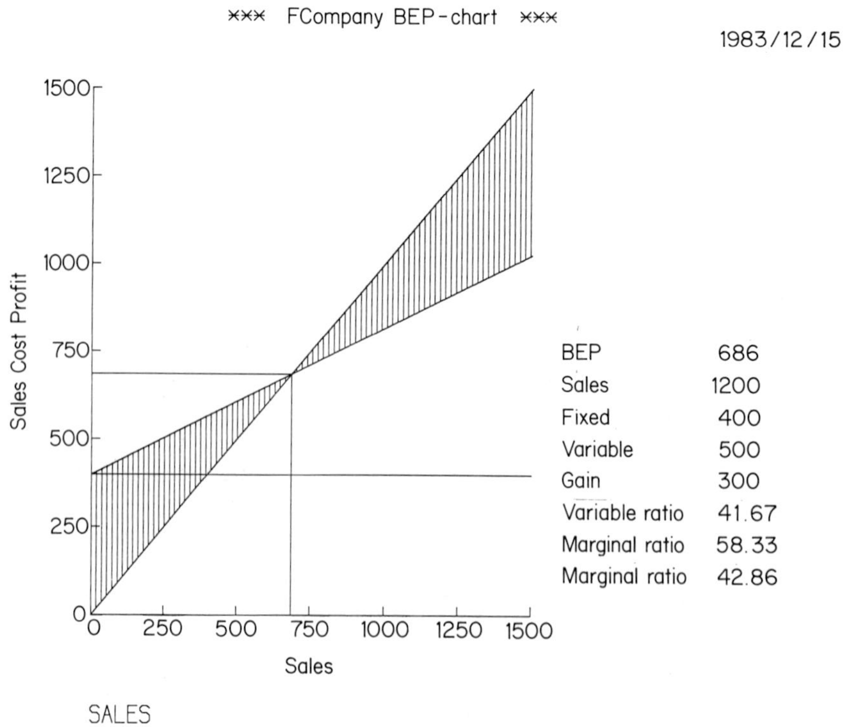

Fig. 10.2 A sample "Break-even Point" Graph.

(7) *Integrated interactive software*. The above-mentioned functions provided by MDS-IV are performed through commands or screen-oriented menus using an integrated interactive software architecture called INTERACT to interface to other programs such as the query languages of traditional network and relational database management systems (Fig. 10.3).

All the commands in each subsystem of MDS-IV are based on the same man–machine interface. And if one wishes to use more detailed data, it can be accessed in a network database system (or VSAM files) using the query language, and can be transferred to the MDS database by a simple command. One can then manipulate them using MDS-IV functions.

10.5 Management Analysis System and its usage in the JDB

10.5.1 Effectiveness of the System

The target of the management analysis system in the JDB is to support loan officers and improve their judgement.

The following are some examples where the system has been applied effectively.

(1) Evaluation of capital investment and of long-range plans using simulation models, and assessment of client's applications for loans and their pay back schedules.
 (a) The month-long effort required for manual calculation is reduced to only a few minutes by computer-based simulation.
 (b) Improved risk analysis, sensitivity analysis, and analysis of each project as a whole.
 (c) Evaluation of the JDB's financing policy to each project under the governmental budget.
 (d) Management services and consulting work for clients on big projects.

(2) Quantitative corporate analysis based on a large amount of data to establish more objective decision taking.
 (a) To avoid misjudgement arising from individually biased experience.
 (b) To avoid increase in staff by improving office productivity.

(2) Management analysis system in the JDB.
 The key features of the management analysis system in the JDB. are:
 (a) data banks;
 (b) software library;
 (c) corporate models.

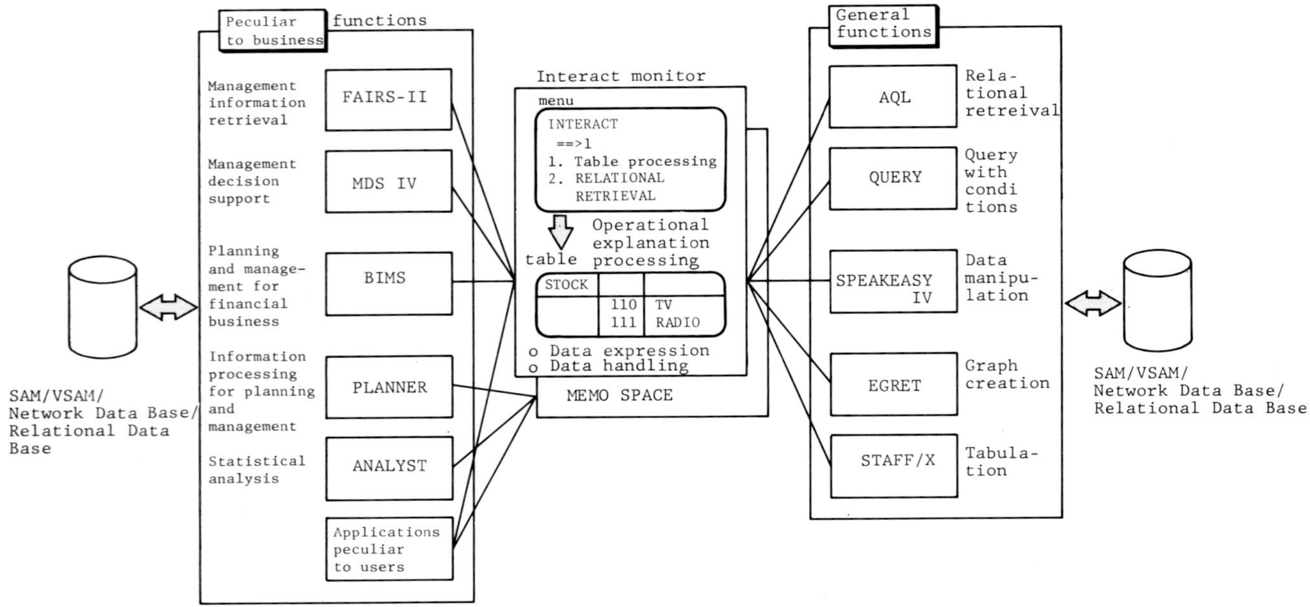

Fig. 10.3 Structure of Interative Systems.

MDS-IV, and other management science-oriented software jointly developed by the JDB and Fujitsu Ltd are playing an important role. The JDB has been maintaining its data banks, which are the nucleus of the system and which contain financial data on around 4000 companies.

In the early stage of computer usage in the JDB, these services were in batch mode, but since 1974 on-line services were available.

In 1980, the following items in the system were redesigned and improved on the basis of six years' experience of this on-line interactive system.

(1) Improved man–machine interface. More user friendly systems and better response time.
(2) Implementation of Japanese character processing and colour graphics displays.
(3) More integrated data banks using new database management systems rather than individual data files.

In 1981, a new management analysis system was developed on FACOM M-180 II AD, using the OS IV/X8 operating system, the AIM database management system, MDS-IV and other software with Japanese character processing features and colour graphics displays (Fig. 10.4).

The heart of the JDB's management analysis system are data banks, software libraries and corporate models.

Data banks
There are two data banks, one of which is a financial data bank, and the other is a macro economics data bank. The former contains more than 4000 domestic and overseas companies. It is used not only to produce formatted financial reports but also *ad hoc* financial analyses with an interactive query system. The latter contains economic time series data, and is mainly used in economic forecasting and analysis using the econometric system.

Software library
Not only MDS-IV software, but also the JDB original financial reporting and analysis system called 'FAST' are held in the library. Additional packages included are an integrated econometric system (KEMPF/X), to analyse economic fluctuations (SIECOM), and KACET/F, which is a quantitative corporate analysis tool based on multi-variate analysis.

Corporate models
In the JDB, various corporate models have been developed and used not only to evaluate clients' applications for loans but also to establish its own management policy. The following are some examples of the models and simulators developed.

```
OS IV/X8
   |
   |
   |--- INTERACT
   |        |
   |        |
   |        |- QUERY              (query language)
   |        |
   |        |
   |        |- MDS IV             (management decision support system)
   |        |
   |        |
   |        |- FAST               (financial analysis through terminal)
   |        |    |
   |        |    |
   |        |    |- FIS           (financial statement)
   |        |    |
   |        |    |
   |        |    L- COA           (corporate appraisal)
   |        |
   |        |
   |        L- STAFF/X            (report writer )
   |
   |
   |--- DYNAMIC MODELS
   |
   |
   |--- MATHEMATICAL PROGRAMMING
   |
   |
   L--- STATISTICS AND FORECASTING
            |
            |
            |- KACET/F            (corporate appraisal)
            |
            |
            |- KEMPF/X            (econometrics system)
            |
            |
            |- SIECOM             (simulator for econometric model)
            |
            |
            |- MULVA/X            (multi variate analysis)
            |
            |
            L- CLUSTER            (cluster analysis)
```

Fig. 10.4 Software Library in JDB.

(a) long-range planning models for manufacturers;
(b) models for shipbuilders and shipping companies;
(c) models for hotel planning;
(d) models for urban development;
(e) models for electric companies.

10.6 Conclusion

Ten years ago, management information systems were said to have failed in Japan. However, innovation in hardware and in system design methodology now makes it possible to apply computers more effectively in management. As explained, the key lies in providing user-oriented software to improve the user friendliness of the system.

The Japan Development Bank and Fujitsu are continuing their efforts in applying computers to more advanced management systems.

Reference

1. Nash, D. R., "Building EIS, a Utility for Decisions". *Database, SIGBDP of FCM*, Vol. 8, No. 3 (1977).

11

Kimitsu Integrated Information System

Nippon Steel Corporation, Tokyo, Japan

11.1 Introduction

Kimitsu Works, built in 1967 during the period of Japan's rapid economic growth, was conceived and designed as a competitive and modern plant with emphasis on applying computers throughout its operation. The computer-aided operation of an integrated steel plant was not a novel notion for the company. The basic ideas and thoughts of the Kimitsu Integrated Information System (KIIS) evolved out of the company's vast pool of practical knowledge and experience in production control, various standard values and computer applications already in use for some years. KIIS realizes an integration of systems for processing information and data on steel plant operations, from the receiving of orders to the shipping of products by assigning various functions of production control and operating control to a hierarchical structure of business computers and process computers.

The prototype of KIIS was completed in 1970, Since then, this system has been playing an indispensable role in the production control of Kimitsu Works. At that time Kimitsu on-line real time system was said to be ahead of time and to be a prototype of the new age of steel works.

During the high growth period which continued up to the oil crisis of 1973,

CAD/CAM and MIS in Japan
ISBN 0-12-682580-7

the computer systems were expanded with the construction of new plants. After the oil crisis, however, restructuring and modification of the existing systems became necessary to adapt to changes in the economy and plant-operating conditions, and to meet changes in manufacturing processes. Against such a background, computer applications have been further expanded to fit on-line systems to administrative areas in addition to the production control area, and new computer systems have been developed to facilitate the control and analysis by staff departments.

In addition, efforts have been made to utilize computers to improve the efficiency of various operations and activities in the computer system division.

11.2 Outline of Kimitsu Works

The Nippon Steel Works started the construction of Kimitsu Works in 1967 and four blast furnaces started operation one by one in 1968, 1969, 1971 and 1975. By 1971, after four years of operation, the production capacity has been increased to 10 million tonnes of crude steel per year to cope with the rapid economic growth.

From a systems standpoint, beginning with the Heavy Plate System, each one has been developed one after another, synchronizing the cut-over of operations in each facility.

Today, Kimitsu Works has the largest production capacity in Nippon Steel. It is located in the Kanto area where over 30% of all steel products are consumed in Japan (Fig. 11.1).

The transition of crude steel production at Kimitsu is shown in Fig. 11.2.

By establishing Kimitsu Works close to the metropolitan area, it became possible to provide good customer service with effective product distribution.

The total site of Kimitsu Works is about 10 million m^2, and it is equipped with the facilities to produce various kinds of plates, wide flange beams, wire rods, and pipes. The plant layout consists of three main blocks. One is the group of blast furnaces, converters, and iron resource plants located adjacent to the raw material quarry. The rolling facilities are separated into two other blocks, East and West. The plate and sheet producing facilities are located on the West side, and the rod and pipe making facilities on the East side. This layout makes it easy to move products from the arrival of raw materials to the shipment of finished products (Fig. 11.3).

The facilities of Kimitsu Works have a high production speed and capacity, and can produce almost every kind of steel product except specialized steel (Table 11.1).

The company considers it very important to utilize these facilities to make

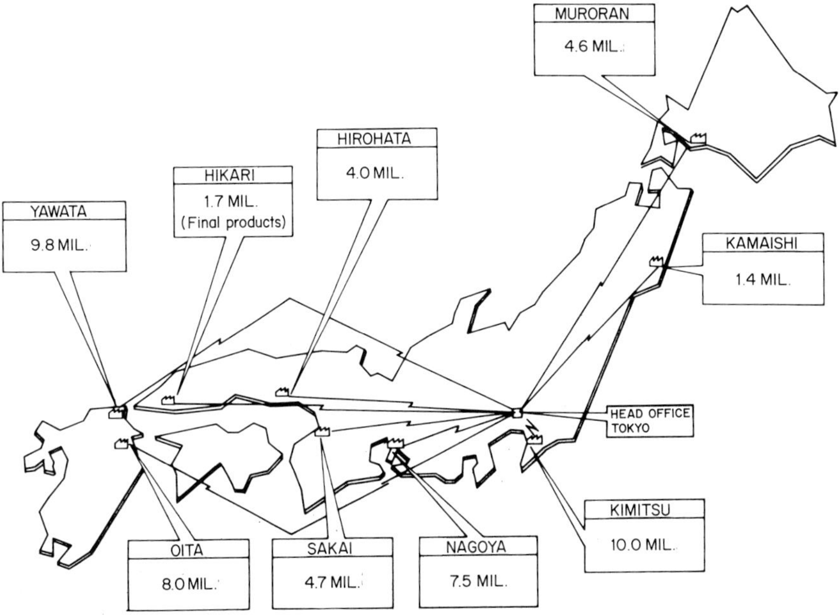

Fig. 11.1 Locations of NSC Steelworks.

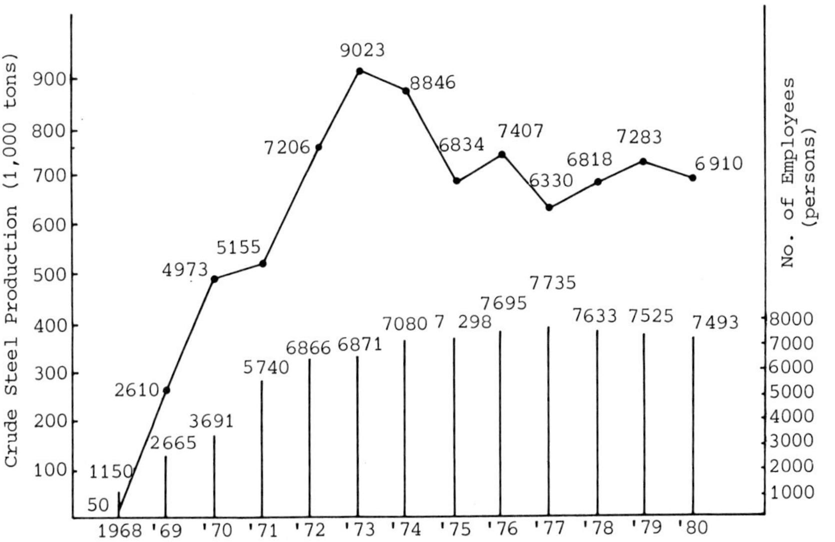

Fig. 11.2 Transition in crude steel production and number of employees. *Note*: In addition to the above figures for employees, approximately 9000 persons are employed at the works as contract labour.

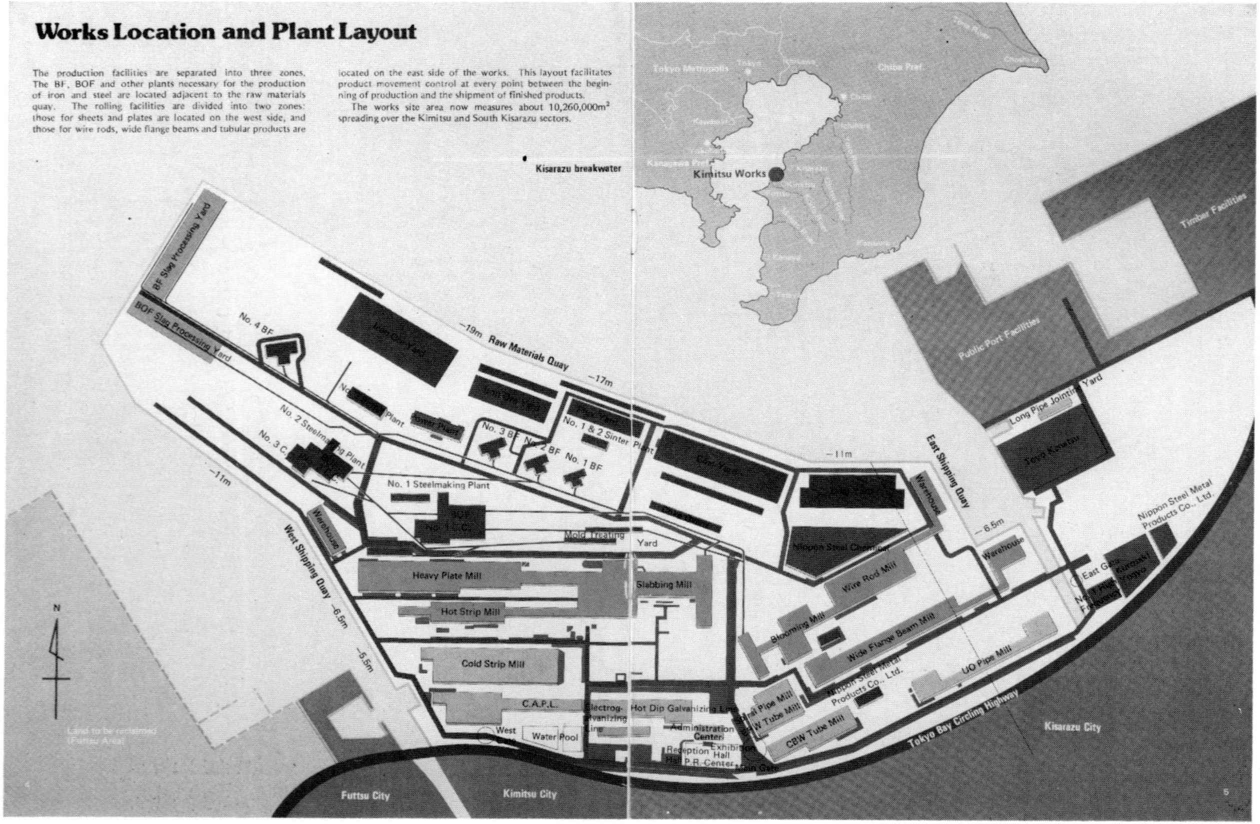

Works Location and Plant Layout

The production facilities are separated into three zones. The BF, BOF and other plants necessary for the production of iron and steel are located adjacent to the raw materials quay. The rolling facilities are divided into two zones: those for sheets and plates are located on the west side, and those for wire rods, wide flange beams and tubular products are located on the east side of the works. This layout facilitates product movement control at every point between the beginning of production and the shipment of finished products.

The works site area now measures about 10,260,000m² spreading over the Kimitsu and South Kisarazu sectors.

Fig. 11.3 Plant layout.

Table 1 Production capacity

Equipment	Plant	Start-up	Capacity
Blast furnace	No. 1 BF	Nov. 1968	6000 t/d
	No. 2 BF	Oct. 1969	7000 t/d
	No. 3 BF	Sep. 1971	10 000 t/d
	No. 4 BF	Sep. 1975	11 000 t/d
BOF	No. 1 BOF	Dec. 1968	400 000 t/m
	No. 2 BOF	Sep. 1971	360 000 t/m
Continuous caster	1CC	Sep. 1970	120 000 t/m
	2CC	Mar. 1980	210 000 t/m
	3CC	Feb. 1982 (scheduled)	210 000 t/m
Primary rolling mill	Slabbing mill	Dec. 1968	500 000 t/m
	Blooming mill	Apr. 1971	240 000 t/m
Plate and strip mills	Plate mill	Mar. 1968	160 000 t/m
	Hot strip mill	Jan. 1969	480 000 t/m
	Cold strip mill	Apr. 1965	360 000 t/m
	Coating mill	Jul. 1968	115 000 t/m
Shape and wire rod mills	Large shape mill	Apr. 1972	93 000 t/m
	Wire rod mill	Apr. 1972	85 000 t/m
Pipe and tube mills	UO	Jun. 1970	43 000 t/m
	SP	Jan. 1968	18 000 t/m
	ERW	Jan. 1968	7000 t/m
	CW	Jan. 1969	23 000 t/m

profitable products which meet the customers' needs, while maintaining a balance between cost, quality and productivity (Fig. 11.4).

11.3 Characteristics of a Modern Steelworks

Before explaining about Kimitsu computer systems, the characteristics which today's modern steelworks must have are first described. For example, the works must have high speed, and a large capacity which adopt to any new technologies in order to satisfy the many users' requirements for quality and delivery.

The manufacturing processes of steel products have become more complex because of the wide variety of the products being produced. There is thus a strong need not only to control each process, but to control the entire works in order to satisfy customer requirements.

The necessity for total control is increasing in today's economic environ-

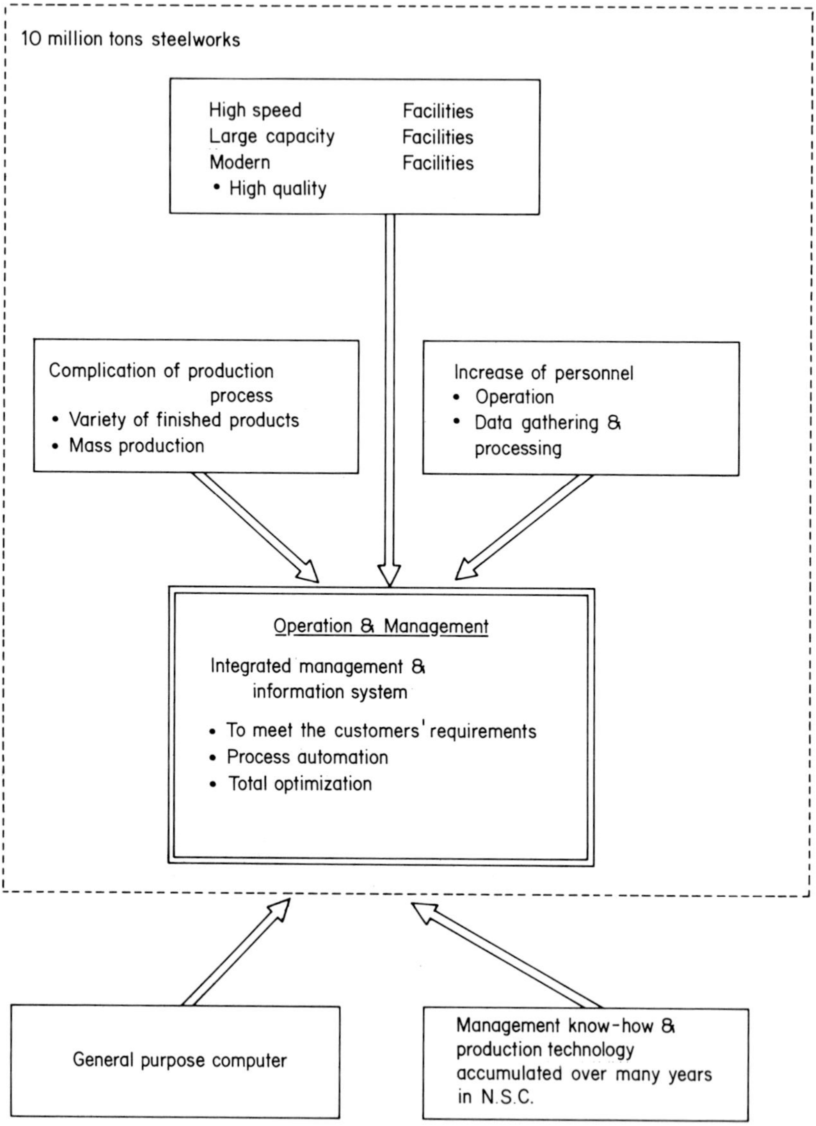

Fig. 11.4 The characteristics of a modern steelworks.

ment. Therefore a modern steelworks needs large volumes of information, people to act on it and the people to operate large-scale facilities in the most efficient manner.

Great importance is attached to what is believed to be a unique management system to achieve total optimized production in a large-scale steelworks like Kimitsu. In addition many mechanical and process automation techniques as are practical have been adopted.

Many computer systems are effectively used to implement the management system, close attention being paid to achieving the optimization of large-scale steelworks the management system was established.

11.4 Production Management and Control (Fig. 11.5)

One hundred per cent of steel products in Japan are made to order. In addition, trading companies are used as a unique business form in the steel products market. Almost all of the orders from customers are sent to headquarters through these trading companies by magnetic tape media. Headquarters creates a long- and mid-range production plan from the sales forecast. The plan is then authorized by top management. After receiving this plan, the steelworks makes hierarchical production plans, in which higher level plans control the lower ones. A common and definite agreement between lines and staff have thus been established. This is contained in the master schedule approved by management and used to make the hierarchical production plan, which is called "planned value", and which establishes the various standards and conditions for production.

Management's desires for the production authorized in the plan are executed by agreement in every section between headquarters steelworks. Production in a steelworks can begin once it has management's go-ahead in the production plan and the customer's requirements through the order entry system.

As described, products are made through the optimized plan and schedule in each mill. The total optimization of steelworks can be achieved by adhering to such a plan in each section. This approach is called "Plan Directed".

Another way to control production is by authorization delegated to the mill where production is controlled based on a partial optimization. Here the production is controlled by each event; this approach is called "Event Driven".

When work instructions are given to the mill, it is also important to include information on where to move the materials and products, in addition to the necessary information on quality, techniques and operations. It is a principle of the management system that the mill operators work to the instructions and plan which have been established.

This information is usually complete, and makes it possible to check the operations accurately in each step of the production process. Almost 100% accuracy in data gathered from the works has been achieved. This has enabled the production control system to function with high reliability.

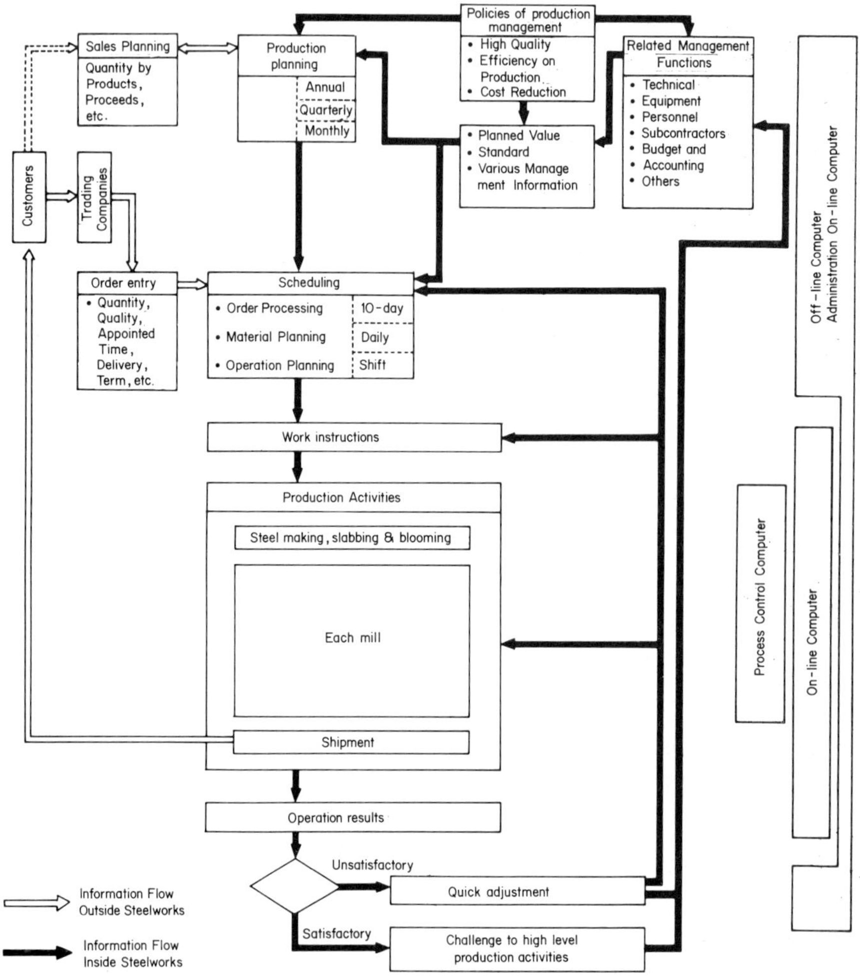

Fig. 11.5 Concept of production management.

Sometimes exceptional situations in production occur, but these are usually dealt with without jeopardizing the plan.

When exceptional cases do happen, decisions can be made by the staff and fed back to the operator and reflected in the schedule for the following processes. Accurate working results are fed to each function for reporting to management. These results are also used as base data to revise planned values and standards.

The quality of the management system is improving daily through these feed-back activities.

11.5 Kimitsu Integrated Information System

The Kimitsu integrated information system (KIIS) is a computer system hierarchically organized to cope with the structural production management concept (Fig. 11.6).

Business computers perform both off-line and on-line processing functions. Installed in the central computer room of the administration centre, they form a centralized processing and radial telecommunications system to control process computers and terminals installed in production plants.

This computer system is operated twenty-four hours non stop to support steel production. The computer in the headquarters and Kimitsu Works are connected directly by telecommunication line, order information is transmitted from the headquarters computer to Kimitsu Steel Works and all information necessary for production is included with the order information. The order is allocated to each mill after the material requirement calculation. This calculation is performed every five or ten days, depending upon the product class.

In this process, the scheduler pays attention to creating a well balanced plan in terms of production efficiency, cost and quality from the beginning of the steel-making process to the final shipping yard. Daily schedules for each mill are transmitted to the on-line computer by communication line or shared disk files.

This on-line computer creates production schedules and conveys the work instructions to the terminals and to the process computers which are attached directly to the production facilities in real time mode. Approximately fifty process computers and 1750 terminals are used, such as printers, CRTs and custom-made operator panels depending on the use and the response time required.

Operators in each mill act according to the work instructions, and they feed back the results with additional data only when there is an exception to the instructions already received from the computer.

In normal cases, with "Plan Directed" approach, the operator only sends the completion signal to the computer centre which moves the planned data to the completion file. The process computers are used mostly for automatic operation of the facilities and control of the process.

Completion is processed by one product and reported by on-line and off-line computers.

Application systems are becoming large scale. In April 1983 11.2 million statements were used in a business computer system. The ratio of use on a on-line application system is about 60% (Table 11.2).

Forty per cent of the total application code is for normal work, and 60% of the code is for the handling of exceptions and emergencies. Because these

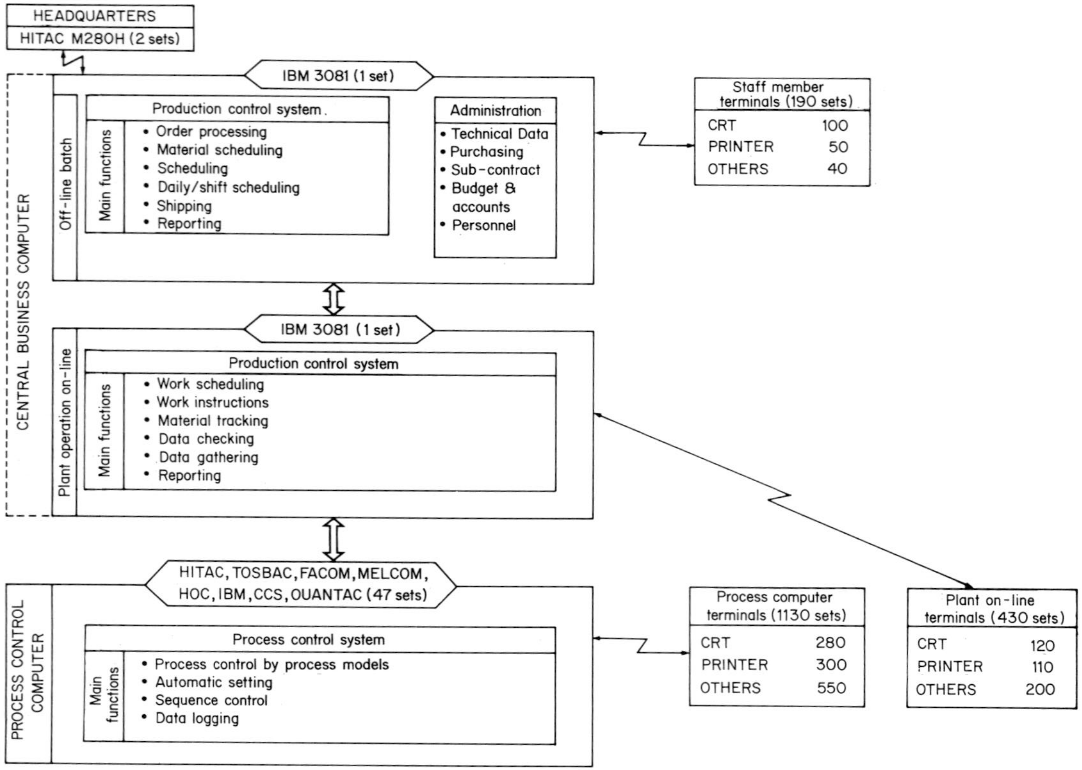

Fig. 11.6 Computer system configuration.

Table 11.2 The Size of Kimitsu Computer System (as of April 1983)

	Business computer system		Process computer system	Total
	Off-line	On-line		
Statements (K steps)	4815	6428	3500	14 743

have been provided for, the availability and reliability of the applications can increase, and thus extend the systems' life.

Although much benefit has been obtained through KIIS it is very difficult to calculate the profit contributed by computer systems, because of the many factors which affect profit.

A cost justification of KIIS is as follows. First, it was calculated how many people would have been necessary to carry out the management system in Kimitsu without computerization; then a comparison of the cost for these people with the present computer cost. Computer systems saved the cost of more than 2000 people once needed at Yawata Steelworks for the information processing to execute a similar management system. Much benefit was derived from the processing speed, in the complexity of calculation, and in the accuracy of data which cannot be accomplished without computers.

There was a tangible benefit of 3.5 times total DP costs. If the management system in the steelworks becomes more complex and sophisticated, the value of information, will become more important and, will, we think, bring more profit through computerization.

The benefits of the business computers are summarized as:

(1) Accomplishment of integrated management with minimum personnel.
(2) Accurate production control with improvement of customer services.
(3) People are freed from the simple and routine calculations to do creative work.

The main object in utilizing process control computers is to increase the total production of the facilities. Most of the savings are made through production increases. Total return of investment is over 3.5 times even for process computers. The advantages of process control computers being:

(a) accomplishment of process automation;
(b) improvement of work environment;
(c) high quality products through closer adherence to specifications.

In summary, the characteristics of KIIS are:

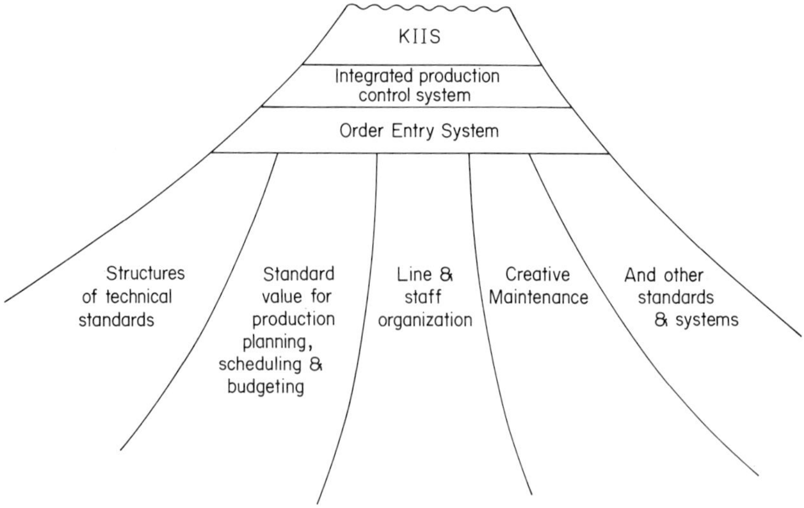

Fig. 11.7 KIIS Based on various control techniques.

(1) Integrated management system.
(2) Plan oriented system.
(3) User oriented system.
(4) Computer hierarchy and network system.
(5) Centralized processing and radial tele-communication system.
(6) Process automation.
(7) High data accuracy.

One final point which is emphasized is that KIIS is like the top of a high mountain: for support, it is necessary to have various functions at the base (Fig. 11.7) These functions are:

(a) basic agreement to the value of planning and standards;
(b) the management facility necessary to operate the total plant with stability;
(c) the organization which supports and executes the system.

These functions are the basis upon which an integrated production control system can be constructed. Only then can a KIIS system be possible.

11.6 Historical Background (Fig. 11.8)

This management philosophy has not appeared suddenly. In 1962, this revolutionary approach began in Tobata Works which had been integrated with Yawata Works sometime before. At that time, the approach for controlling the steelworks was changing from a partial optimization approach to the total optimization approach.

Various new methods were introduced such as a line-staff system, a planned value system, and a standards system in order to establish the unique management system at Tobata. The management system in Kimitsu was designed from the experience of the system in Tobata. Complete separation of line and staff made it possible for staff to control whole production activities. This new organization frees the production line managers and operators from pressure when the staff develop the computer system.

When the construction of Kimitsu Steelworks was started, it was very important to reduce the manpower required for production control and scheduling, for then there was a large labour force—about 600 employees in the production control staff and about 2000 for on site scheduling and data checking.

This objective was achieved because at that time general purpose computers became available which made it possible to construct Kimitsu Works with a complete computerization plan, including process automation.

11.7 Development and Operation of KIIS

One of the considerations when Kimitsu Works was built, was the computer systems, as was the management system which was to be used. The latter was therefore designed to meet the requirements of the facilities, and many ideas and opinions were applied in the design, so that the works were easily managed.

In order to construct Kimitsu Works, the Kimitsu Project Bureau was established, which was in charge of administrative work for the total construction and development plan of the management system.

Then, DP functions were a responsibility of the Construction Bureau the management system designed by Kimitsu Project Bureau using computer systems was developed.

During construction and expansion of the system the system development personnel worked in the nearby production control department. This allowed the system to be stabilized in a short time.

In the early 1970s, the separation of DP functions from production control functions and the establishment of standards for jobs in the DP department were promoted, in order to achieve better personnel and program development.

Another characteristic of the Kimitsu system is that both business com-

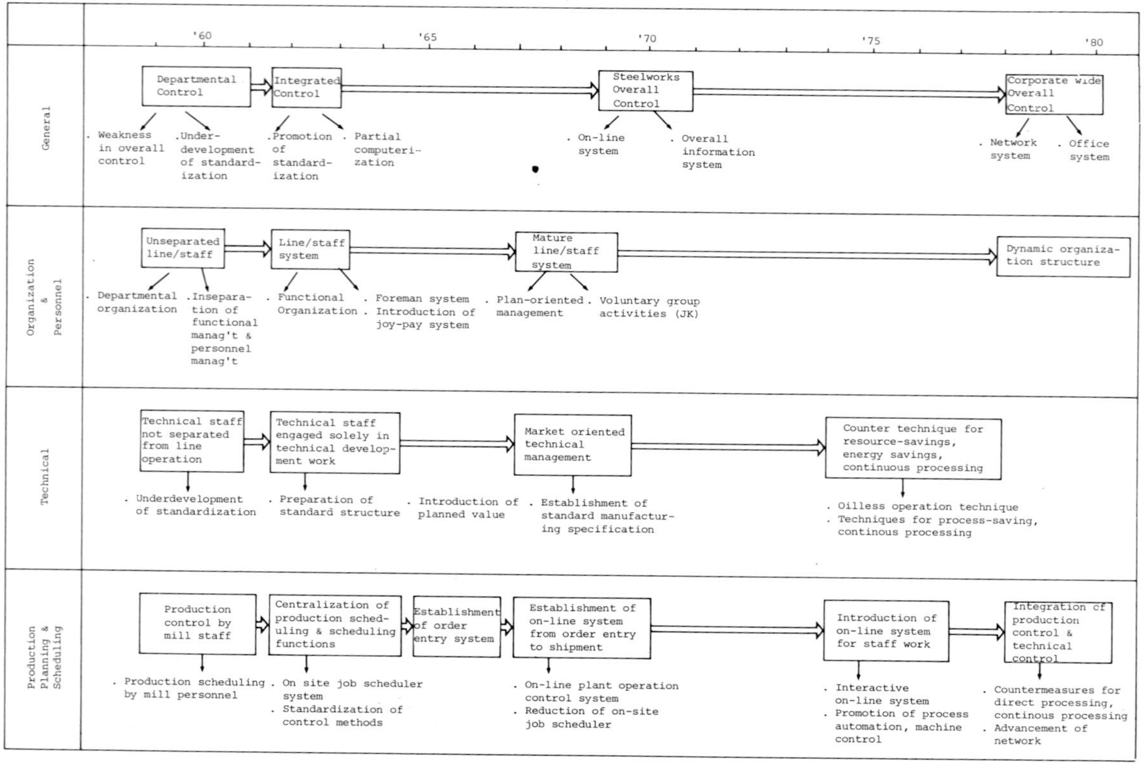

Fig. 11.8 Historical development of management structure.

puter and process computer systems are designed and operated by the computer system division, while the management of electrical and instrumentation equipment is managed by the relevant sections. The organization of the computer system division is also shown in Fig. 11.9.

General Superintendent
Deputy General Superintendent
Deputy General Superintendent

- Personnel Dept.
- General Administration Div.
- Computer System Div.
  - System Coordination
    - System Coordination
    - Computer Operation
    - Program Development
    - Standardization
  - (General Administration System)
  - (System Planning & Design)
  - (Ironmaking, Steelmaking, Slabbing & Blooming, Hot Strip Production Control)
  - (Plate, Shape & Wire Rod, Pipe & Tube Production Control)
  - (Cold Strip Production Control)
  - (System Engineering/Process Computer System)
    - System Engineering
    - Program Development
    - Equipment Technology
    - Application Development
- Environmental Control Dept.
- Labour Relations Div.
- Finance & Accounting Div.
- Contracting Office
- Technical Div.
- Production Scheduling Div.
- Blast Furnace Div.
- Steelmaking Div.
- Shape & Wire Rod Div.
- Hot Coil & Plate Div.
- Cold Rolled Sheet Div.
- Pipe & Tube Div.
- Equipment Div.
- Overseas Technical Cooperation Office

Fig. 11.9 Present organization for development, maintenance and operation of KIIS.

At present approximately 240 people are working in the division including about forty operators and fifty process computer personnel.

11.8 Future Direction

Internal and external conditions surrounding the industry have been rapidly changing after the oil crisis: great difficulties lie ahead—economic fluctuations, lack of energies and raw materials, soaring labour costs and pollution.

Any one of these problems is very difficult to solve, and as a group they represent a monumental challenge. The management of a steelworks is required to have flexible strategies to make decisions in order to cope with rapid changing conditions. Therefore the system always needs to be improved to meet these changing conditions.

The management system of the steelworks is fairly stable. And the computer systems which support the management system have been improved greatly over the past 15 years, and these kind of "creative maintenance" activities make the systems both active and flexible, for the larger the systems become, the more important the maintenance will be.

In addition it is necessary to make the systems less expensive and more efficient in order to increase the profitability of the entire organization. Great attention also has to be paid to the newest technologies of both hardware and software, so adopting them when prudent to do so. It is intended to make major efforts to improve techniques for development and maintenance of new systems.

In summary, future directions are:

(a) adaptation to changing social and economic environments;
(b) improvement of application systems;
(c) standardization of system planning and programming techniques;
(d) utilization of the latest hardware and software;
(e) security of data and systems.

It is clear to the management of the steelworks that the system is highly important. At the same time it is almost impossible to carry the management system to effect in today's complex, large-scale steelworks without a well-functioning computer system.

12

Management Information and Control System in a Job Shop

Tozo Hikichi, General Systems Manager, Hitachi Ltd, Tokyo, Japan
(Now Board Director, Hitachi Information Network Ltd)

12.1 Introduction

The Hitachi Management Information and Control (HIMICS) was developed in 1968 to produce a radical improvement in management. Because of the diversification of products manufactured, it was first designed individually for each works, and then followed by an overall management system. The system had three main goals:

(1) Realization of more dynamic management.
(2) Improvement of technology.
(3) Development of smaller groups of highly skilled staff personnel.

Dynamic management means flexibility to environmental change; it is necessary to adjust production capacity in response to changes in the market so that higher productivity is possible. Production forecast and control are the bases for this, especially in a large factory.

Improvement of technology means improvement in the productivity of engineering design work, which has high emphasis in a job shop. An effective

CAD/CAM and MIS in Japan
ISBN 0-12-682580-7

automatic design and information retrieval system for technical information has been developed and put into use.

12.2 Search for Standardization

As outlined in the previous section, each works began to develop their own HIMICS independently, concentrating on an operational level system which, it was expected, would bring more practical results than a management level system.

The shop's products are varied; for example, there are various types and capacities of generators. Nevertheless, it was realized that adherence to present production methods would mean that no improvement in productivity would be possible, and that it could not compete with advanced American and European companies. It is also true that mass production lines have a higher productivity than job shops, and after much discussion "group technology" (GT) was decided upon (Fig. 12.1) and "information retrieval" (IR) (Fig. 12.2) as the means to increase productivity.

GT is a technique using classification, which was originally invented for the rationalization of production; it seeks similar methods or similar tools used in processing parts for different products.

This concept has been extended to drawings, so that engineers now write down GT codes on every drawing. Further, all drawings that had been completed in the past and were still available, had GT codes added to them. This information is stored in the computer and used for the centralized continuous production of similar parts, and for parts standardization. Engineers can retrieve drawings designed in the past and make full use of data by accessing the stored information using the correct GT code.

"Information retrieval" is another useful concept, but if it takes too long to retrieve the required data it becomes ineffective. Furthermore, since the situation changes very quickly in a job shop it is necessary to maintain updated data. Therefore a character display unit was set in every department of the works. The device is connected to the main computer and makes it possible to get the required data without difficulty. For example, in the manufacturing division those in charge of process control can see the status and location of all parts at the terminal without walking around the shop floor to locate the problem.

12.3 Improvement in Engineering Efficiency

As explained above, engineering design work has an important place in the job shop. Whilst however the productivity of the manufacturing division has

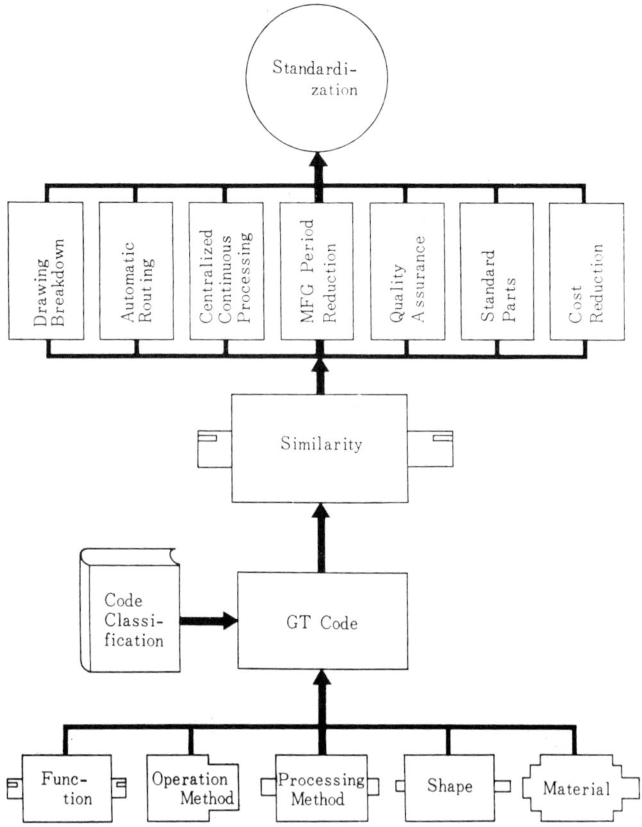

Fig. 12.1 Standardization through group technology code.

been much improved by the development of new machines and tools, there has been little improvement in the speed of making a drawing or in the design process.

This lack of improvement has naturally pushed up the cost of engineering and hence the price of the products, which naturally caused a problem that had to be solved. The cost of some parts, half in fact, was made up merely by engineering. Modern techniques were adopted such as a large computer and automatic draughting to change the engineering process.

Automatic design is done as follows. The computer is used to execute all the calculations during the design process to obtain the best drawing and to prepare the data for automatic draughting; a series of engineering tasks can thus be accomplished by the computer. The problem of rising engineering cost and shortening the period necessary for engineering design was therefore lowered.

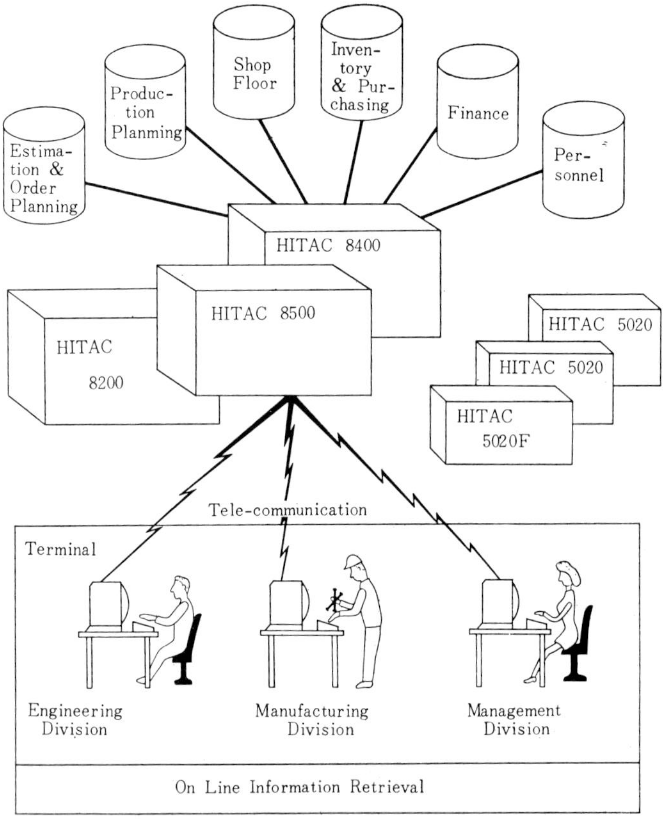

Fig. 12.2 On-line display system flow.

HIMOTOR (the Hitachi Induction Motors Ordering with Telex and On-line Response system) is a good example. The general flow of HIMOTOR is shown in Fig. 12.3. This system is applied to one of the main products—induction motors. From Fig. 12.3 it is seen that engineering sheets are produced by the on-line system according to the customer's specifications, and drawings are produced on an X-Y plotter. The effect of this system is remarkable in that the period necessary for engineering has been reduced to one day, though it took a week or more before the development of HIMOTOR.

12.4 Establishment of the Total System

As shown in Fig. 12.4, HIMICS consists of three major parts: an engineering system, a production control system and a management system. Through standardization it proved possible to computerize all jobs from order receipt

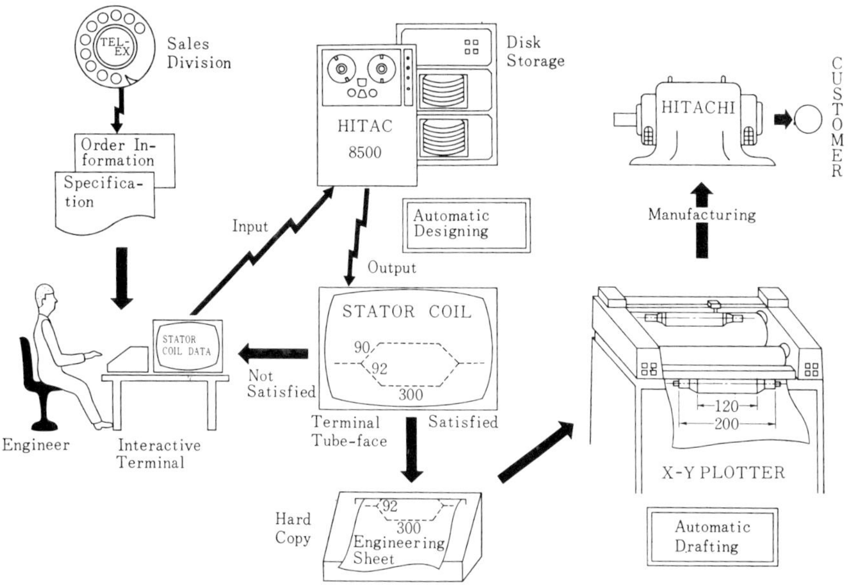

Fig. 12.3 General flowchart of HIMOTOR system.

to shipment. By reviewing all jobs, standardizing them and computerizing all routine work, an organization was able to be built where only creative jobs are left to humans.

GT and IR are the techniques used from order receipt to shipment; GT is the key for computer control of production activities, and IR makes it possible to see the situation on the shop floor and to take action.

Products have recently become larger in size and more complicated in specification, so production control has become correspondingly more difficult as is the need to control from a wider point of view than before. The production information combined effectively by GT and IR is condensed into the Master Sheduling System which reflects the actual status of the works. The Master Scheduling System represents in the network that which is prepared for each plant at every checkpoint, from order receipt to shipment. It uses network techniques, holds information on due time and shop load. The future shop load can therefore be forecast and the production schedule can be planned at the time of order receipt.

12.5 Conclusion

The characteristics of the management information system of our works are as follows.

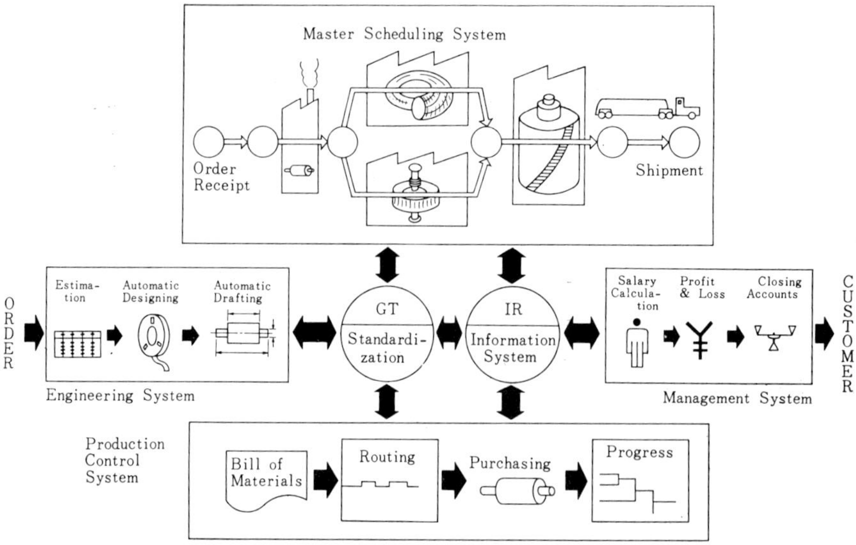

Fig. 12.4 General flowchart of Hitachi Works total system.

(1) The construction of a computer system in which standardization is blended with the diversification peculiar to a job shop. HIMICS is constructed on the basis of GT (which is adopted for standardization of parts and products), and IR (which allows quick response to changes in information).

(2) The automation of engineering work. Although engineering work has weight in job shops, little improvement in productivity has been made hitherto. All routine aspects have been computerized, such as automatic design and draughting, technical information retrieval, and progress control.

(3) The setting up of a total system in which all information from order receipt to shipment is involved. The Master Scheduling System embodies 10 years of computer experience and acts as the manufacturing control centre.

The system has just been put into practice, so there is much room for improvement. From the point of computer utilization, it is still at the elementary stage, and there is a need to improve man–machine communication.

It is now agreed that the most important merit of the management information system is that it can cope with rapidly increasing demand for fewer staff. Comparing now with 10 years ago, total production has tripled, whilst the number of employees has been reduced to half. This improvement of productivity would have been impossible without a computer.

Another major advantage is enlargement of the scope and area of management. This is conspicuous especially in a large-scale job shop where there is a tendency to have difficulty in exercising detailed control. It would not have been possible to keep techniques advanced without computer aids because the products have been becoming larger in scale and the pace of technological changes has become ever more rapid.

12.6 Supplement

The main part of this chapter, which was awarded the Ishikawa Prize, was written in 1971. Since that time computers have developed considerably in scope and technology, and the users' market has greatly expanded. The following is a summary of the latest developments.

12.6.1 Achievement of the HIMICS Project

During the 16 years since 1968, the HIMICS project has been thoroughly implemented and has not only produced marked energy savings, but also reduced the budget by 2700 million yen.

Over the same period Hitachi's production expanded 250%, whilst the number of clerical staff remained at the same level, although some departments such as business administration, purchasing and finance succeeded in reducing approximately 30% of their personnel. Other departments, such as design and sales, increased almost 30%—this represented structural change of personnel within the company. This increase in staff was dictated by the necessity of developing new projects indispensable for its future prosperity.

This change was successfully realized during the period of high economic growth. Its success has also strengthened the foundations of the company, and facilitated a quicker recovery from the first and second oil crises. After these, the Japanese economy moved into a slow growth period, and markets began to show a drastic change. In order to survive in such fluctuating markets the company had to adopt a more user-oriented strategy. Throughout the company it was necessary to set up a system to support the sales division.

It was in the context of this harsher economic climate that Hitachi Ltd started the new HIMICS Project in 1978.

12.6.2 Promotion of the New HIMICS Project

The New HIMICS Project can be divided into three main components, summarized as follows.

(1) Flow and connection—their more rapid combination: the vertical (= sales information) line must be rapidly connected with the horizontal (= business information) line.
(2) Establishment of a database. The unification of common information systems for everyone to access.
(3) Clerical work simplification. "Simple and speedy" was the motto.

This project was aimed at smoother and faster execution of business. The diverse manufacturing interests of Hitachi Ltd include computers, semi-conductors, electrical appliances, automobile parts, precision instruments and medical engineering equipment, and in each market the company has been faced with stiff competition from rival manufacturers.

In order to overcome the challenge from competitors, Hitachi, as a conglomerate, needs to acquire more efficient and faster methods of handling business. This is the gist of the so-called "Information System Design".

So far 13 teams have been organized and a three-year project promoted to establish a system to support sales activities. The results have been very encouraging.

12.6.3 The Promotion of Office Automation (OA)

Since 1981, office automation projects have been developed throughout the company. This is an extension of the first HIMICS Project which was awarded the Ishikawa Prize in 1971.

The company has been making great efforts to combine clerical business improvement with the formation of an information system making full use of computers. In other words, the establishment of "simplified business" comes first, and the next stage is then a thorough business review with the joint aims of halving all paperwork and at the same time doubling the speed of business.

A new target has been set annually, and this movement has taken root as the very foundation of office automation (OA). The essence of the information system in the OA era is a large-scale database and a network system. The network database, which covers orders, receipts, customers, cost estimation and also personnel, has been implemented throughout Hitachi.

The necessary data can now be retrieved from each branch plant. As far as the semi-conductor division is concerned, not only domestic but even overseas information is now available by means of artificial satellites. Orders, receipts, shipment or sales data can now be acquired on the spot.

On-line machines have spread throughout the company. Statistics for September 1983 show that on-line machines have been installed at the rate of one for every seven office-workers. OA has had an unexpected impact in that the introduction of these machines has proved indispensable to those rapidly developing business fields requiring very quick response. Throughout the recent company campaign, a trend to make full use of OA machines to improve business efficiency from the top management down to clerical staff is clearly evident. This movement is backed up and supported by numerous small group activities which have been fostered by Hitachi Ltd's traditional management campaign.

OA is expected to contribute significantly to the development of both hardware and software, and the company firmly believes that OA will continue to play an important role in business innovation.

13

LOGOS (Logistic On-line System)

Toray Industries, Inc., Tokyo, Japan

13.1 Introduction

Toray Industries, Inc. established in 1976 a large-scale hierarchical computer network, using an information processing centre as its core. The company developed a comprehensive on-line business system (LOGOS) for all Toray products as well as for 200 related Toray companies, enabling the firm to improve efficiency, quicken response and savings in intra-company sales activities.

In technological terms, Toray leads the industry in developing High Level Data Link Control (HDLC) procedures for centralized and decentralized control. The company has also developed, and is using, high level systems technology in preparing sales decision models, including automatic shipment instructions.

13.2 The Background of Systems Development

In 1973, Toray began to re-examine whether its information systems fitted the company management and control, particularly in the light of the general acceptance and use of the firm's on-line system. Realizing from this study that the establishment of a new information processing system for the Toray

CAD/CAM and MIS in Japan
ISBN 0-12-682580-7

group was urgently needed for coping with post-1975 sales activities, the company decided to set up an Information Processing Centre (hereafter referred to as the Centre) at its Shiga Plant, and a Toray Group Information Processing Network based upon this Centre. The Centre was completed in April 1976 so that the information system would serve to meet such basic needs as (a) increased sales efficiency in the face of general adoption of a five-day working week; (b) improved job and control efficiency in the face of a low-growth economy; (c) overall business group management.

The company implemented the overall business on-line system (LOGistic On-line System, or LOGOS) in October 1976. Since then effort has been spent to improve functions, to raise automatic shipment instruction levels, expand data transmission inside and outside the company and upgrade the computer.

13.3 The Information Processing Network

13.3.1 A Single Centre

As Fig. 13.1 shows, the LOGOS network is a typical large-scale hierarchical network composed of a large computer at the Centre (first level); front end processors (second level) for relay and terminal control at the two major business offices in Tokyo and Osaka; and about 260 terminals for the company's sales divisions and factories plus another 200 terminals outside the company, all controlled by large computers and front end processors.

The large computer systems previously installed in Tokyo and Osaka have been reorganized into the single independent computer centre for the following reasons.

(1) Centralized control of databases supporting the firm's information system (which is oriented toward decentralized processing under a centralized control system);
(2) Improved job efficiency through centralization of facilities and personnel.
(3) Shiga area is close to Osaka where the bulk of the data occurs.
(4) Location of the Centre in a special building away from the city centre, and the establishment of protection for a major part of the information system from earthquakes, fire and other potential disasters.

13.3.2 Functions of the Front End Processor

Advanced computer and communication technologies have facilitated development of centralized processing. Ideally, input/output data functions

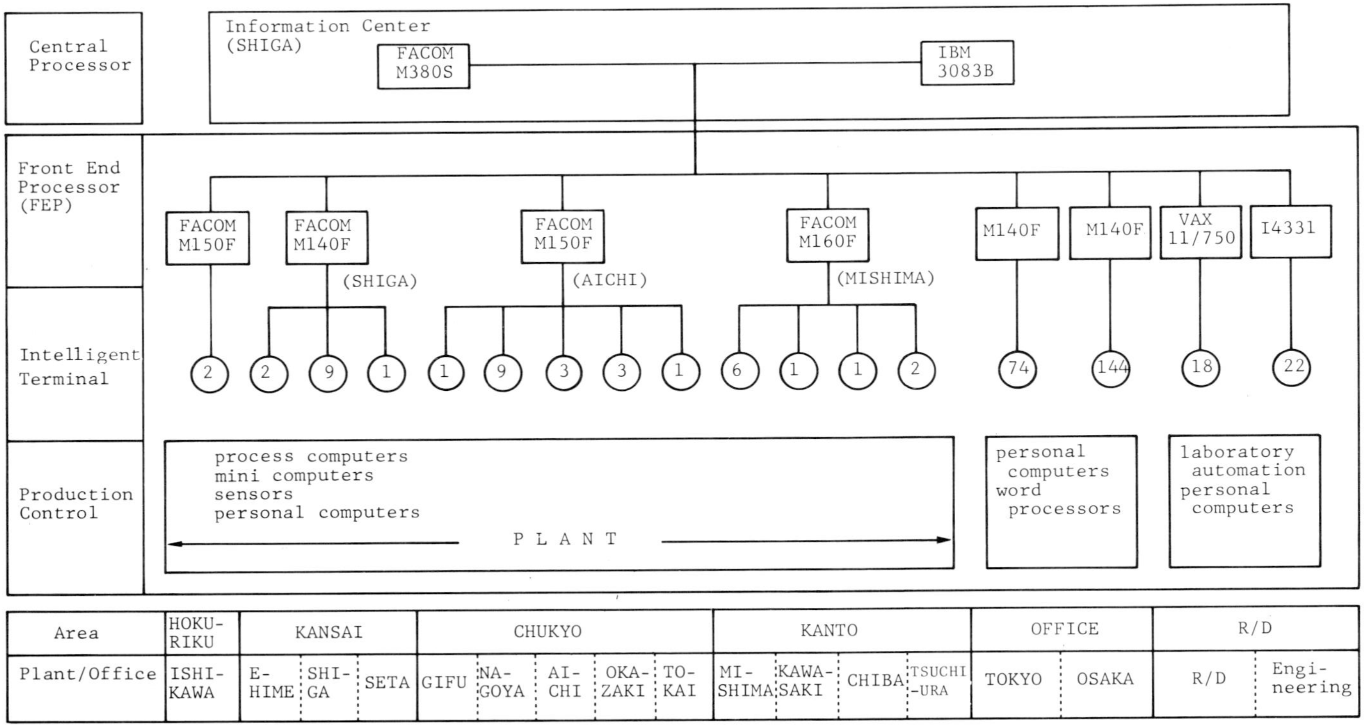

Fig. 13.1 Toray Data Processing Networks.

should be located where the end-users are, and data should normally be input at its place of occurrence whenever possible. Basic processing requiring immediate handling should be done on a real-time basis to improve business speed and customer service. For this purpose, numbers of Centre-controlled, sales-related terminals have been installed, especially in Tokyo and Osaka. Front end processors have also been installed in these areas to handle the other functions required on Tokyo and Osaka such as:

(a) central control of sales-related Toray terminals (Local type);
(b) control of sales-related Toray Textile terminals (Remote type);
(c) receipt and printing of data centrally processed at the Centre;
(d) dispatch and receipt of data to and from terminals at factories and outside the company;
(e) remote job entry (RJE).

Because multiple applications such as order entries, inquiries, collection and distribution of data and RJE requiring efficient transmission control are exchanged between the Centre computer and the front end processors, they are linked by a high speed (48 KBPS) communication line using HDLC control.

13.3.3 Types of Terminal Device

Table 13.1 shows the types of terminal device and their application and Fig. 13.1 shows the actual locations. The optimum function, performance and cost of these terminal devices are selected in accordance with the job site requirements, but devices of the same type and manufacturer are selected whenever possible.

Each terminal has the following main functions.

(1) Sales division terminals process order entries, inquiries (display terminal), and the printing of bills and display information requiring immediate handling (printer terminal).
(2) Factory terminals process input and output data to be exchanged between the company's factories and the Centre via leased lines.
(3) Terminals outside the company process input and output data to be exchanged between the Centre and related offices outside the company via public lines.

13.4 Application System

13.4.1 Purpose and Scope

LOGOS, which was developed to control all products produced and sold by

Table 13.1 Terminal types and applications (as of 1983)

Location	Terminal device		No. of terminals	Application				
	Type	Maker		Order entry	Inquiry	Collection distribution of data	RJE, local batch	ICS (see section 13.5.1)
Office	Alpha-numeric display	Fujitsu	156	+	+			+
		IBM	10					+
	Kanji, Graphic display	Fujitsu	35					+
		IBM	3					+
	Intelligent terminal	Toshiba	2			+	+	
	Personal computer	Fujitsu	12	+	+	+	+	+
Factory	Intelligent terminal	Toshiba	42			+	+	
Research and development, engineering	Graphic display	IBM DEC	40	Scientific calculation, computer-aided design				
Outside company	Personal Computer	Fujitsu	About 120			+		
	Telex	—	About 80			+		

Note: + = ???

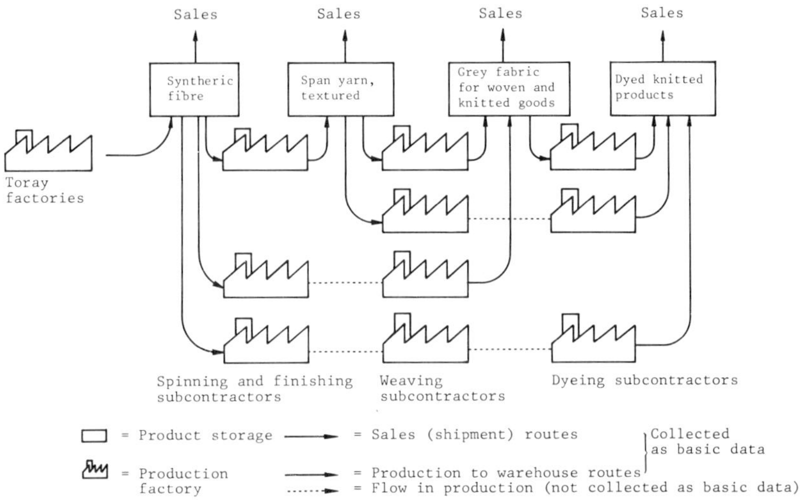

Fig. 13.2 Textiles production and Sales route.

Toray and Toray Textiles, is first of all intended to collect basic product flow data quickly, accurately and efficiently for daily sales activities (Fig. 13.2). Secondly, it is intended to use the basic data to draw up sales strategies and plans such as sales route input plans, and multiple stage production and inventory plans.

Products covered by LOGOS may broadly be classified as textiles, plastics, chemicals and newly developed items (artificial kidneys, sunglasses, carbon-fibres). These items have in common product diversity and complicated distribution routes, as Fig. 13.2 indicates. Synthetic fibre produced in the factory may be sold directly as unprocessed product; or it may be spun, and textured yarns and woven and knitted fabrics produced by other companies and sold as value-added products.

Thus, the products for final sales, production and processing are diverse with the same products produced or processed in many ways by different companies.

LOGOS handles all the firms concerned with these products and all the basic production data.

13.4.2 Basic Data

The basic data are classified broadly into:

(1) An order receipt group, which represents the flow of products from

storage, and is composed of data on orders received, shipment instruction data and data on delivery from the warehouse.

(2) An order placement group composed of data on orders placed and entry into the warehouse.

A description of basic data is given in Table 13.2. The data are totalled and used to show the product flow, including sales for each product and shipments for each destination. Some data are combined to show the position of stocked products, including order backlogs, unfulfilled orders and inventories.

Because, as mentioned before, the products are diversified in form, and have many production and sales routes, the basic data are divided into categories according to production stage, and these data are used to control each stage's product flow. Data are also used to carry out "multi-stage inventory control" throughout the production and processing stages.

13.4.3 Basic Function

Order entry

Data on orders received and placed, and shipment instruction data, are input when the salesman calls the menu screen from the display terminal of the sales department. Depending on the form of the product and the method of transaction, the salesman selects data with a light pen (menu system, two minutes or less, several screens for each case). Some data are keyed in, but the percentage of the key operation is very small, vastly simplifying the operation. During this process, errors are checked and only corrected data are sent to the Centre.

The Centre updates both the individual contract history for each order receipt (data based on orders received), and the position between order receipt and placement groups. The Centre sends rush data to the company sales division printer terminal, or to terminals outside the company, in the form of shipment instruction documents.

Collection and distribution of data

The Centre's computer processes shipment instructions and orders placement data to be sent to related terminals outside the company. It also processes shipment and warehouse entry data (including data occurring in company factories and warehouses) to be sent to the company. These data are exchanged between terminals inside and outside the company (collection and distribution of data).

Using collected data, the Centre updates the above databases overnight so

Table 13.2 Description of basic data

	Factors for classification					Feature		
Factor	What	Who	Where	When	How	Quantity	Unit Price	Price
Contract number	Product name, product type	Salesman, Section, Department	Contractor, Warehouse location. Stored location. Application/ destination	Contract date. Date of entry and delivery from warehouse, date. Planned shipment date	Settlement terms. Transaction form. Production method	Quantity of sales. Quantity of production	Sales price. Purchase price. Wages	

they may be processed on-line the following day. To assure efficient and uniform processing, the company uses its unique data collection and data distribution system. This has a variety of functions, such as a simultaneous data distribution, fixed-time data distribution and terminal initiated data distribution.

Inquiry

Inquiry is classified into (a) individual contract history inquiry (to check contract job progress using the contract number as a key) based on the data for orders received and placed, and (b) position inquiry (to check possible sales, shipment, in-proces inventory and product inventory using the product and stored position as keys). In both cases, inquiry is initiated from the display terminal.

The relationship between the basic data (flow) and balance (stock) for these inquiries is generalized according to the balance matrix shown in Table 13.3. This system has the virtue of both versatility and expandability when the data are to be subdivided into further categories, or when new balances are to be added.

Batch processing

The system updates the database overnight on the basis of the order receipt and placement groups. The system also prepares various daily statistics and management reports, and processes payment requests, receipts of money, and payments. Results are sent to the relevant sites by the data collection and distribution processing system by the time work starts the following day.

13.4.4 Extended Functions

Automatic shipment instructions

The system not only computerizes clerical work, but incorporates many routine decision-making jobs. A typical example is the shipment instruction program for the distribution division. This is the result of thorough analysis and standardization of the actual jobs performed, and the experience and knowledge of the personnel being gathered together and is embodied in an algorithmic formula (the algorithm assures optimum shipment by calculating the necessary evaluation function, based on such factors as freight required, the separation or merging of shipment lots and the similarity of shipped lots). This method has already been put into use for all synthetic fibres, spun yarn, plastic products, etc., to ensure shipment date and product quantity.

Report preparation on request

The system functions to prepare reports on request so that only the relevant

Table 13.3 Example of a balance matrix

Balance		In-group balance				Balance between groups		
Group	Basic data	Backlog	Unspecified	Unshipped	Unfilled orders	Sales position	Instruction position	Inventory position
Order receipt group	Order receipt	+	+			−		
	Shipment instruction		−	+			−	
	Shipment	−		−				−
Order placement group	Orders placed				+	+		
	Warehouse entry				−		+	+

Note: Calculation is made, for example, as follows: Inventory position = (current stock) + entry − delivery.

information is obtained at the desired time, thus eliminating routine management reports as far as possible. Specifically, the system produces routine reports for sales management (such as sales status for each sales route, sales progress status and material input status for each processing stage, as well as the daily schedule), and submits the required reports on the date specified by each sales section.

13.5 Project Results

13.5.1 Improved Data Accuracy

LOGOS has substantially increased the speed of clerical work both inside and outside the company, and has assured quick rectification of errors and greatly improved accuracy. The number of errors has also been reduced to half.

Previously, five to seven days were required from data occurrence outside the company to processing completion. This has been reduced to about two days. As a result, requests for payment, for example, can be calculated quickly, achieving a monthly profit of about 10 million yen in terms of interest alone.

13.5.2 Effective Manpower Use and Savings

The number of indirect sales personnel has been reduced by preparing reports using the on-line order entry, inquiry and request system. Automatic shipment instruction has eliminated the need for delivery personnel in the physical distribution division, and the number of data punchers and error checkers in the Systems Division has been reduced. Furthermore, the considerable cost saving achieved in many sections is evaluated in conservative terms as follows:

Fifty indirect sales posts eliminated;
Sixty delivery posts eliminated;
Eighty posts eliminated in the administration, accounting and systems divisions.

The Centre is run by the Systems Division as well as LOGOS, thus improving systems maintenance and operational efficiency, cutting the number of systems personnel by half.

13.5.3 Other Benefits

Use of the system has improved the efficiency and management quality particularly in the Sales Division. Furthermore, the efficiency of monthly settlement procedures, and the quality of various management and control functions has also improved.

13.6 Current Topics

Seven years have already passed since the LOGOS system was developed. During this period, remarkable advance has been made in hardware and software, giving rise to a considerable change in computer utilization. Direct computer utilization by end-users is also rapidly expanding. This is because conversational computing systems and databases have been developed, and the appearance of business-oriented personal computers has familiarized end-users with computers. In these changing circumstances, this report presents supplementary information concerning current topics on computer utilization.

13.6.1 Development of Toray's Conversational Computing System

Computer utilization where end-users communicate with systems through terminals is in gaining popularity. To meet a variety of end-users' requirements, Toray has developed a conversational computing system called ICS (Improved Conversational System), designed to provide a powerful aid to the on-line development of conventional processing systems, and interactive computing systems. It is based on software produced by computer manufacturers and software houses, and much consideration has been given to provide adequate expandability and to standardize utilization-related processes, including those for data transfer between different programs, security and terminal operation.

Systems developed using ICS currently cover systems for the entire company, individual local systems and personal computer-based systems as well.

The database Management System service as the heart of ICS is based on ADABAS relational database (developed by Software AG in West Germany) and uses simple languages, including FOCUS and EASYTRIEVE. These software components are combined to ensure standardized security and terminal operations, and to provide compatibility between FACOM and IBM equipment.

With the popularization of ICS, users' demand for higher terminal functions has increased. As a result, the implementation of high-performance terminals with kanji and graphics features, and business-oriented high-grade personal computers with adequate intelligence, is rapidly proceeding.

The ICS design philosophy is summarized below.

Higher efficiency and maintainability in application development
From procedure-oriented to function-oriented design, and use of an ICS paradigm for more efficient development (aimed to reduce the development period by 10% to 20% compared with conventional development methods).

Establishment of more flexible software systems
Establishment of a data link between software components such as ADABAS, FOCUS, and EASYTRIEVE, and personal computing based on simple languages, including PSP (IBM) and PLANNER (Fujitsu).

Standardized terminal operations.
These include standardized basic screen flow, standardized function keys and higher terminal manoeuvrability.

Application to office automation (OA)
Function allocation between the main system and personal computers.

Standardization of data security

13.6.2 Standardization of Terminals at Associated Companies in the Industry

In the synthetic fibre industry, processing such as weaving and dyeing are performed by specialized outside companies. Individual synthetic fibre manufacturers have developed their own on-line systems, and began data communication with outside companies through their own systems. As a result, synthetic fibre manufacturers were anxious to choose one standard terminal so that outside companies could communicate with different manufacturers through one type of terminal. As a result, terminal specifications, including transmission protocol, have been standardized. Toray selected the business-oriented personal computer FACOM 9450 as a standard, and is installing this terminal in outside companies. Most of whom have now completed the implementation of over 120 terminals. These are used not only for data transmission, but also for issuing basic documents, such as shipping instructions and delivery notices.

13.6.3 Data Communication with Trading Companies—Inter-company System

As on-line systems within companies became more extensive, the need for

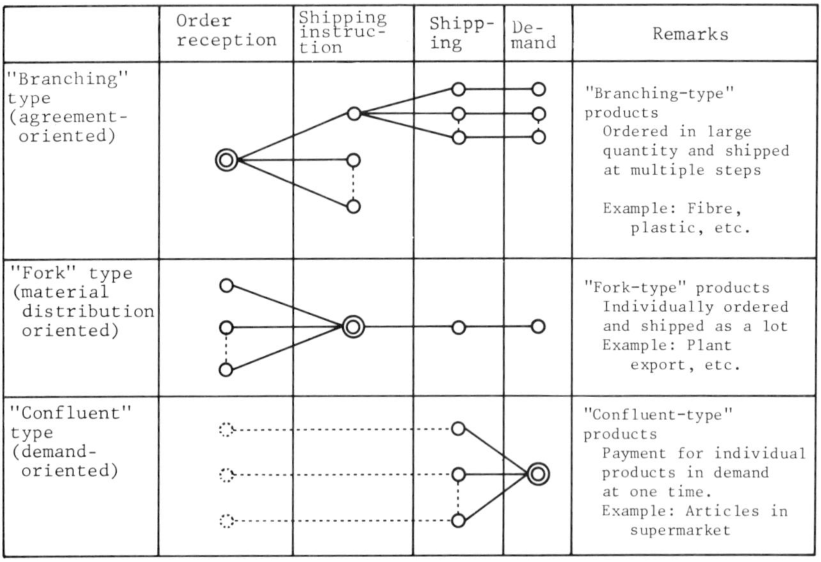

Fig. 13.3 Various transaction forms.

more laborate data communication between them arose. In general, the relation between an associated manufacturer and a trading company is such that sales by the manufacturer correspond to purchases by the trading company, and payment by the trading company corresponds to receipt by the manufacturer. If the manufacturer sends sales data to the trading company, and the trading company sends payments data to the manufacturer, this brings about a rationalization because the number of processes for sales/purchase and payment/receipt is halved. In the past, this data exchange was performed via magnetic tapes; recently, computers in the two companies are connected through a communication line to exchange data daily on-line. This practice is contributing to more rapid business processes and making office work more efficient.

13.6.4 Response to Diversified Products—Construction of Mini LOGOS

The conventional LOGOS system sometimes encounters difficulties in responding to diversified business and the appearance of products from new fields. The reasons for these difficulties include the type of new business and the nature of new products, but the main point is that a mixture of different fields or products involves different transaction forms. This is illustrated in Fig. 13.3.

The conventional LOGOS system is designed to fit to the transaction form for "tree-type" products, including fibre, plastic and other material products. On the other hand, many products from new fields, which are destined for final consumers, are "fork-type". With these, one agreement specifies that multiple products corresponding to different orders be shipped as a single lot. Since the conventional LOGOS system finds it difficult to respond to this situation, a LOGOS system oriented to "fork-type" products has been developed. Currently, the company is endeavouring to improve its features, and has begun to study the possibility of an inexpensive mini LOGOS system based on personal computers.

14

Development of a Comprehensive Management Information System

Okamura Corporation, Yokohama, Japan

14.1 Introduction

In the past emphasis has been placed on raising plant productivity, but increasing attention is now being paid to productivity improvement in office procedures (office automation), as an important tool of future management. Okamura Corporation has long made efforts to rationalize and modernize its management through measures to increase plant productivity by means of mechanization, and by the introduction of new management strategies based mainly on the use of computers.

These endeavours have led to the creation of a comprehensive management system, which enables the company to make timely responses to a changing corporate environment. It also helps to link outside information such as economic and market data with the efficient use of manpower, materials and money to aid the basic functions of corporate management, including marketing, production, purchasing, distribution, and technological decisions, accounting and personnel affairs. At present, individual subsystems achieve efficient management in their respective fields of function, but these subsystems work together, producing multiple and complex effects.

CAD/CAM and MIS in Japan
ISBN 0-12-682580-7

This is possible only when the system is a comprehensive, total one: the system is a step forward in realizing the concept of productivity improvement in offices; that is, more creative work with less manpower.

The major characteristics of the management information system are as follows.

(1) The nucleus of the system is on-line, which links all the company's sales, manufacturing, distribution and staff divisions, including those of some affiliated companies.
(2) The on-line system operates using a common database controlled from a single centre.
(3) The company's activities, from sales and order acceptance to production, distribution, manufacture and collection, are linked and controlled in an integrated manner.
(4) In addition to the information system which links each functional sector of the company, the company also has an electronic mail and electronic public relations system. Thus, a company-wide internal communication system is established.
(5) Computer power is used not only for control-centred affairs but is also utilized for planning sales strategies and tactics.
(6) New management techniques such as computer graphics have been introduced.
(7) A means for developing software programs for the system by the least manpower has been developed as a support system.
(8) Development of a decentralized processing system (which enables participation by all company personnel) is possible using the central main frame (on-line) system.

14.2 History of the Management Information System

14.2.1 Background and Circumstances

In anticipation that wage levels in Japan will soon approach Western levels, the company has long worked on modernization of its management, principally in rationalizing office work.

As a first step, the company inaugurated in 1966 a nation-wide on-line system for marketing, inventory control and delivery. Since then, computers have been introduced into each of the Company's manufacturing sectors, and efforts to achieve efficiency have been made in each location.

In 1973, the company produced a plan for a long-term comprehensive management information system. Behind this was the following philosophy: if we are to enjoy stable management in an age of slower economic growth, we should unify those systems which have been separately developed in each field, and combine them into a single concept to increase the overall efficiency of the company. Only then will we be able to make advanced use of computer power.

The company began to develop individual systems on the basis of the plan referred to above, and in 1974 began operation of a management information control system (MICS). This system links all sectors of the company by an on-line communications network.

The currently used management information system is an improved, upgraded version of MICS. The basic concept of development is, however, an extension of the original system. Currently, in addition to the traditional centralized use of the system, efforts are being made to introduce a decentralized processing method utilizing the less expensive equipment that has recently become available as a result of technological innovation. Improvement in corporate management will be promoted by encouraging more company employees to use computer techniques. This will facilitate both the achievement of the system target, and more effective use of equipment.

14.2.2 Basic objectives of the management information system

The company established some objectives for its future management information system when it was designed in 1973. These objectives are still valid, eleven years later. They are as follows:

(1) Corporate management at minimum cost through minimum personnel costs and improvement in the asset turnover ratio.
(2) Contribution to improvement in productivity in the field of production and delivery through appropriate reduction in the backlog of unfilled orders and efficient work plans (lots, orders).
(3) Contribution to improved customer service through response to diversified and sophisticated customer tastes, information on precise delivery dates and shorter delivery schedules.
(4) Correct and timely provision of data and analysis of the results required for correct decision making by management.
(5) Its usefulness not only for internal management, but also for sales activities meeting market trends.

14.3 Basic Concepts for Building the Management Information System

In an attempt to attain the objectives mentioned above, the company constructed its management information system on the following basic concepts.

To establish, first of all, a feasible system. A system having no firm foundation is just a castle in the air, and however carefully planned no useful output will be obtained. The more important the application, the more necessary it is to build a system that is feasible and which will provide the correct basic data.

To achieve this, it is necessary to reproduce the situation in the actual workshops as correctly as possible on the computer. It is also necessary to define and limit clearly both the locations that should provide information, and the content of that information.

To provide all personnel with information which is as correct as possible, and in sufficient detail for them to act upon it. To survive future corporate competition, all company personnel must be given the correct information and must act in concert towards the same goals. As the size of the company grows, it will become more difficult to provide everyone with correct data unless computer power is used to solve this problem.

To create a system that will encourage all people to participate in management (the bottom–up system). When only some personnel use computer power, things will not go smoothly. Consideration must therefore be given to enable everyone to use it. In addition, efforts to rationalize each computer-using section should be reflected in overall corporate management.

14.4 Basic Outline of the Management Information System (Fig. 14.1)

Since its introduction of computers in 1966, Okamura has endeavoured to develop and introduce systems through guidance from top management. A most important role in implementation and follow-up was played by the Business Improvement Committee. This body constituted a steering committee composed of representatives from each sector and top management. Under it, subcommittees were established for the promotion of computer introduction in each field.

The subcommittees were valuable—for not only did they participate in the design stage of systems to be introduced, but they also provided education and guidance on the introduction of these systems to each section concerned, and acted as a forum for opinions about the system.

14.5 Product Line, Distribution Routes and Plants

14.5.1 Product Line

This compiles furniture for office and public use, equipment for building and

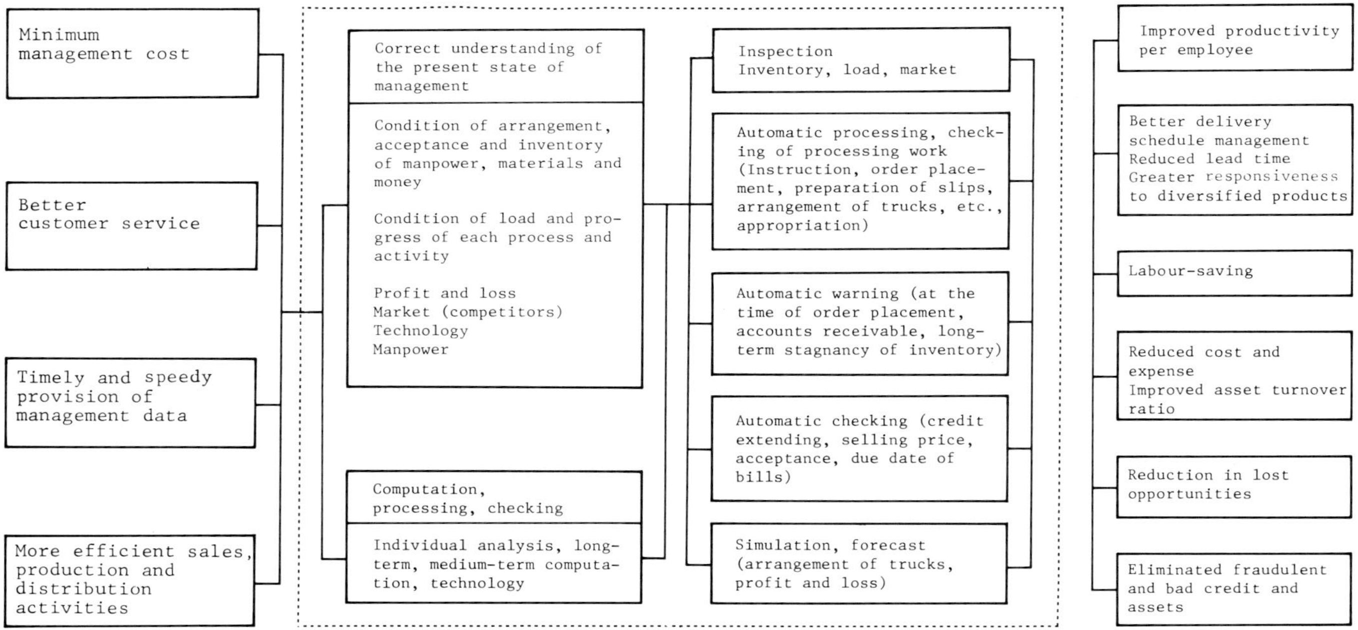

Fig. 14.1 Objectives, basic conditions and benefits of the management information system.

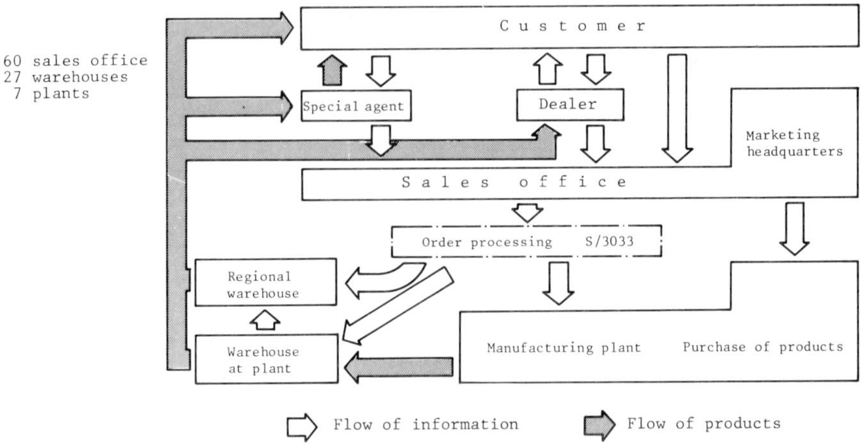

Fig. 14.2 Distribution routes.

air-conditioning, stores use and distribution, household furniture, industrial machines.

14.5.2 Distribution routes (see Fig. 14.2)

14.5.3 Manufacturing plants and their products

The Company's manufacturing plants and their products are as follows.

Oppama Plant, Yokosuka. Office chairs, furniture for reception and for conference rooms, tubular furniture, air-conditioning equipment.

Fuji Plant, Gotemba. Article control shelves, display shelves for stores, steel partitions.

Totsuka Plant, Yokohama. High-class desks, fixtures for banks, steel partitions.

Takahata Plant, Yamagata. Wooden high class desks, system kitchens, wooden partitions, students' desks, wooden display shelves and storage furniture.

Tyler Plant, Gotemba. Tyler refrigerated and cold storage showcases, counters, article control shelves, steel partitions.

Oppama Machinery Plant, Yokosuka. Shredders, automatic warehouse systems, Vulcan motor cleaners, industrial robots, torque converters.

Kansai Okamura Corporation, Higashi-Osaka. Office desks, drawing boards, counters, removable libraries.

14.6 Outline of the Management Information System

14.6.1 Main Frame System

The IBM 3081-K system is the main frame used for the company's entire system.

14.6.2 Sales Information

The prime purpose of the sales informatiom control system is to help salesmen spend as much time as possible on their main job, that is, sales activities. In this system, support groups consisting principally of women employees are available for assistance in such sales-related work as order processing, and control of and inquiry into delivery dates.

In addition, customer management and sales activity management are also systematized. These activities will become more and more important as markets mature in the future. Figure 14.3 shows the relation of each function.

14.6.3 Production Information Control System (Fig. 14.4)

The company is engaged both in manufacturing and selling. Because of this, the production information control system plays the most important role in sales support activities.

The principal characteristic of this system is that all instructions and responses related to production are through the computer. This provides salesmen with responses through terminals concerning the progress of individual orders. In addition, it has a "program system" which is made up of a combination of data on plant operation and information on inventory, inquiries and orders received or supplied to the on-line system from the sales division.

The company has made efforts to standardize and mechanize all management work. Consequently, even employees who have no specialized knowledge can now carry out business management work at an operational stage.

14.6.4 Company-wide Accounting Information System

This system handles monthly and yearly settlement of accounts, analysis of payment, accounts receivable and cash position, analysis of accounting statistics, item-by-item and sector-by-sector profit and loss simulation, and short-term and long-term profit analysis. Monthly and yearly settlement of accounts is carried out by the on-line and off-line system by unifying such subsystems as purchase of materials and parts, completion of processes,

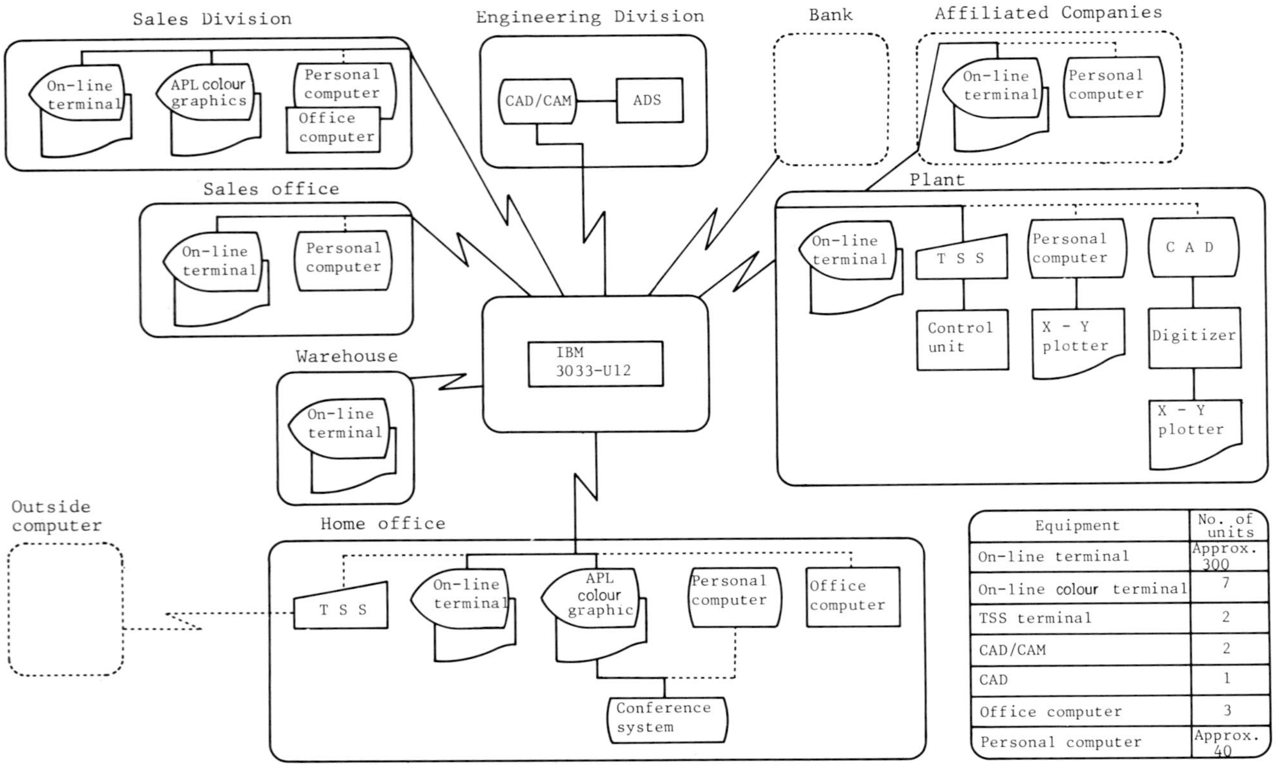

Equipment	No. of units
On-line terminal	Approx. 300
On-line colour terminal	7
TSS terminal	2
CAD/CAM	2
CAD	1
Office computer	3
Personal computer	Approx. 40

Fig. 14.3 Outline of the computer network.

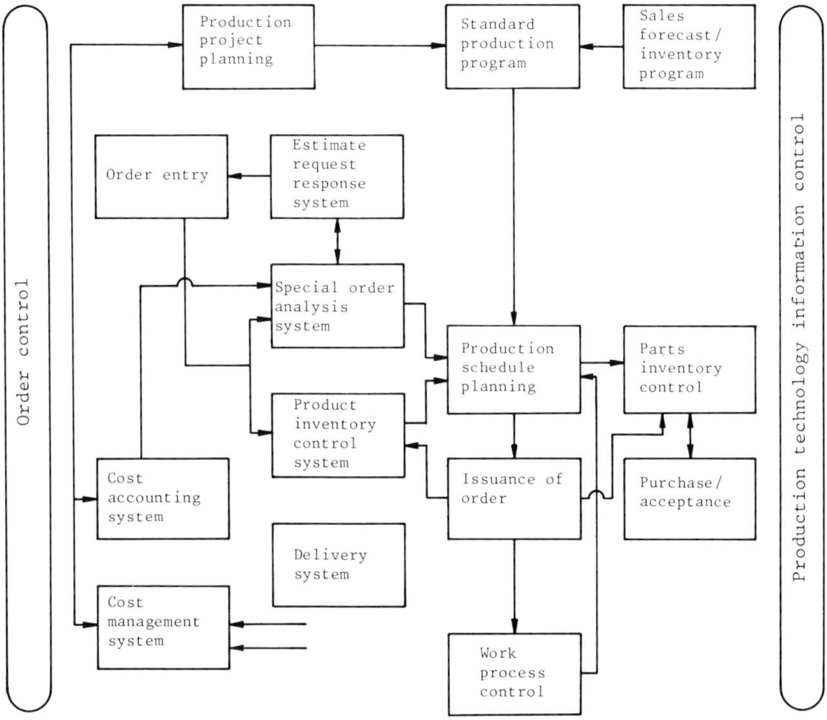

Fig. 14.4 Production information control system.

supply of outside components, purchase of finished products, inventory, sales, money received, personnel salaries, requests for appoval of expenses and the cash system of the accounting sector.

The accounting system deserves special mention. It is so designed that each sector can access the host computer, to codify the data the sector normally inputs for its own business.

For example, in the warehouse sector, when an inventory input is made at the time of sale, this input automatically becomes a record of sales and accounts receivable for the sales sector and an invoice is issued. Then, when an input of authorization of payment by the customer is made, a decrease in accounts receivable and an increase in cash received are automatically recorded. In this way, each subsystem is integrated with the accounting system, leading to paperless, labour-saving accountancy (see Table 14.1).

14.7 Future Tasks

The company's efforts to develop a comprehensive management information

Table 14.1 Major benefits of the management information system

Item	Labour, cost savings (men)	Remarks
Labour-saving in office work		
Marketing/delivery	560	Possible reduction in manpower resulting directly from the introduction of EDPS and necessary manpower that will be needed to do manually the functions enabled by EDPS are conservatively estimated for each business
Home office	70	
Plants/warehouses	180	
Affiliated companies	50	
EDPS-related sectors (including outside agencies)	40	
	820	
Reduction in interest and inventory cost burdens	29 million yen/month	This reduction is the result of improved turnover ratio of current assets. Estimate of accounts receivable is impossible due to variable conditions
Reduction in telephone charges	10 million yen/month 8 million yen/month	This is the outcome of use of electronic mail in the company. This concerns inquiries into inventory and progress of arrangement
Improved productivity per employee. Better delivery schedule management. More complete management materials. Eliminated bad credits. Reduced opportunity costs.		

system have been based on a concept initiated in 1973. Until now these endeavours have been made mainly in a "top–down" manner. All employees have used an on-line terminal in some way, and the system now seems to be firmly established. However, when one looks at each part of the system, one finds that many need upgrading or revision owing to changes in environment and background.

One important factor in the future development of the system will be how far participation of employees in management will proceed. Future improvements will not be achieved unless there are efforts to rationalize plant and office productivity through new ideas supplied by each employee in each workshop. Fortunately, the company expects to develop decentralized processing systems that will enable these new ideas and devices to materialize (in the past, a centralized concept was prevalent).

To sum up, a good management system will be attained when definition of policies by top management ("top–down") and employees' willingness to participate in management ("bottom–up") coincide with each other. Efforts will therefore continue towards system improvement in accordance with this basic philosophy.

15

Integrated Quality Information System in Kakogawa Works

Kobe Steel Ltd, Kobe, Japan

15.1 Introduction

Since the completion of Kakogawa Iron and Steel Works in 1968 the volume and variety of raw materials and products treated have been getting bigger year by year. These materials are passed and treated through many processes. Therefore, the quality and operation data generated during processing are enormous.

Effective use of the data greatly influences quality improvement and cost reduction. The sooner the data can be utilized, the more profit can be obtained. What is essentially required for such a data processing system is easy and quick retrieval and flexible data processing.

Kobe Steel Ltd established the interactive quality and operation data analysis system (KIS: Kakogawa Information System) in April 1981. It is the first system connecting database with APL (A Programming Language) in Japan.

15.2 Background

15.2.1 Requirements Necessitated by the Current Economic Environment

Since the oil crisis in the early 1970s, cost reduction activities in the works

CAD/CAM and MIS in Japan
ISBN 0-12-682580-7

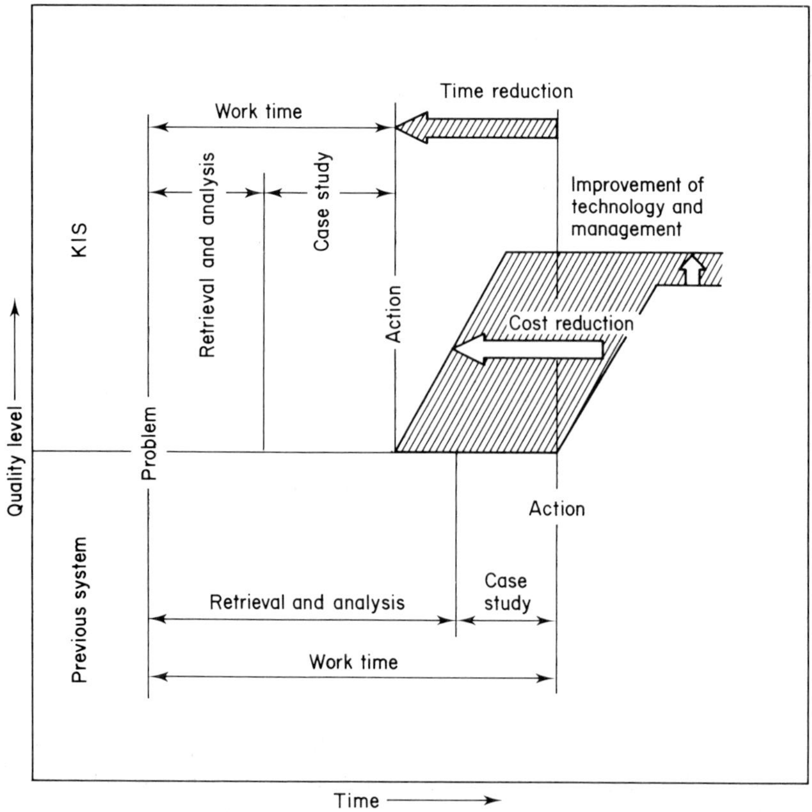

Fig. 15.1 Effects of KIS.

have achieved outstanding results in the area of energy saving, equipment rationalization, operation and yield improvement. It is now becoming increasingly imperative to utilize the data generated in the works for further cost reduction.

In order to reduce costs caused by ineffective quality and process design and reprocessing as well as yield reduction caused by delay in taking action against irregular conditions, it is clear that timely data processing is very important, as shown in Fig. 15.1.

15.2.2 Status of Technical Department

In the middle 1970s unprocessed information often caused significant trouble to the Technical Department. Moreover, the information which they

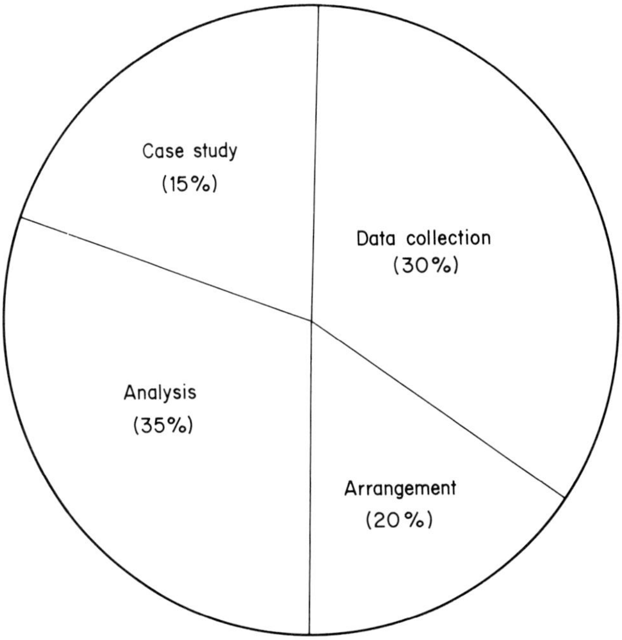

Fig. 15.2 Job analysis of technical department.

received was becoming more complicated. Since June 1977 the jobs in the Technical Department have been analysed, focusing on those involved in searching for irregular conditions of quality and operation.

From the results of this analysis (Fig. 15.2), it is clear that the collection, arrangement and analysis of data occupied 80–90% of staff work time.

15.2.3 Problems of the Previous Technical Analysis System

The previous technical analysis system was not able to solve certain kind of problems quickly enough, because each problem required a new program. If communication between the staff of Technical Department and System Section was not sufficient, completion of the program was delayed even further. Consequently it took a lot of time to solve the problems. The procedure of the previous system is shown in Fig. 15.3.

In addition, the data were neither retrieved quickly enough, nor sufficient in quantity. To solve these problems, the KIS project, staffed by the Technical Department and the System Section, was begun in July 1977.

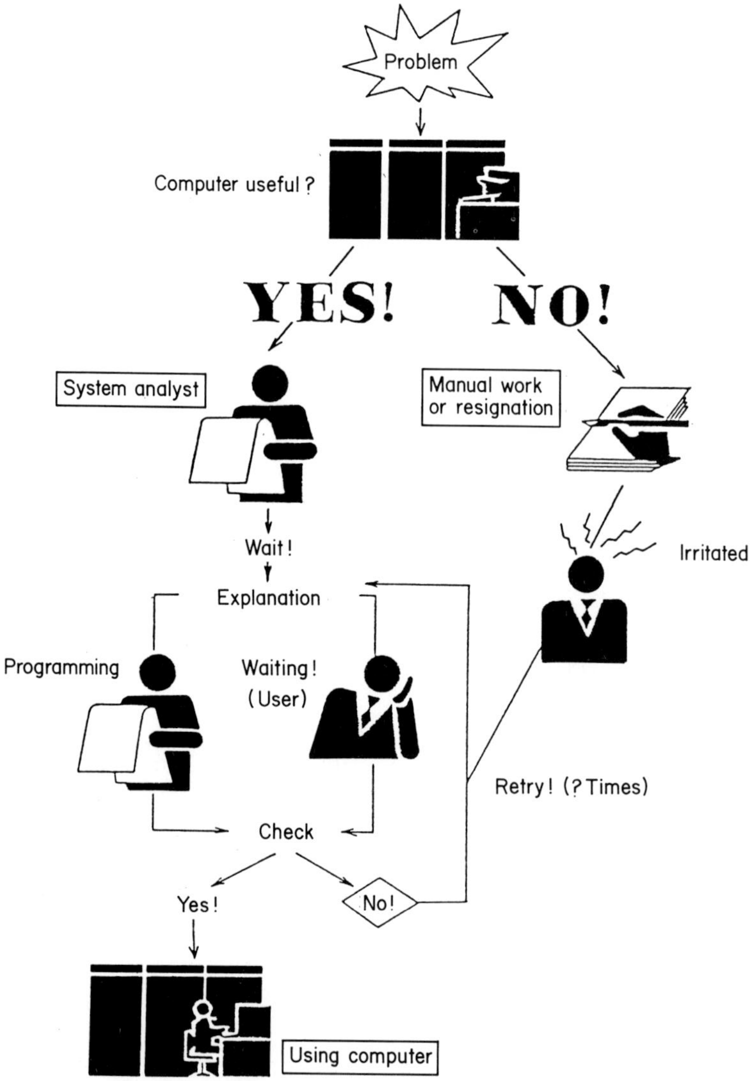

Fig. 15.3 Procedure of previous system.

15.3 The Concept of KIS

15.3.1 Purpose of Development

From an analysis of the work of the Technical Department and the problems of the previous system, achievement of cost reduction by timely data processing was set as the primary goal for KIS, as follows.

(a) reducing defective products, improving operation technology, and optimizing quality and process design by effectively utilizing quality and operation data from all processes of the works;
(b) expanding the productivity of the staff by reducing data processing time.

The organization of KIS project is shown in Fig. 15.4, and the development schedule and tasks in Tables 15.1 and 15.2.

15.3.2 System Features

The following system features were decided in order to achieve the primary goal (Fig. 15.5.).

(1) Data from all works processes must be totally managed and stored on magnetic disk and in the optimum form for quality and operation analysis—adoption of DB (database).
(2) The system must be capable of being used by those without programming ability. Ideas for solution can be applied to problems using the computer directly without the help of System Section—adoption of APL.
(3) Data must be processed interactively in real time and analysed by trial and error method—facilitation of continuous problem solving.
(4) Terminals must be installed close to the staff and capable of use at any time—adoption of open system.
(5) The system must be equipped with wide statistical and analysis functions, such as regression analysis, time-series analysis, scatter diagrams and linear programming—adoption of software packages.

15.4 Outline of KIS

15.4.1 System Configuration

The system configuration is shown in Fig. 15.6 and the system structure in Fig. 15.7.

Quality and operation data are gathered through the production control system and stored in DB. The DB is updated by the DB generation subsystem.

Data can be retrieved and analysed in real time through the retrieval and analysis subsystem. Alternatively, data can be retrieved from DB through the

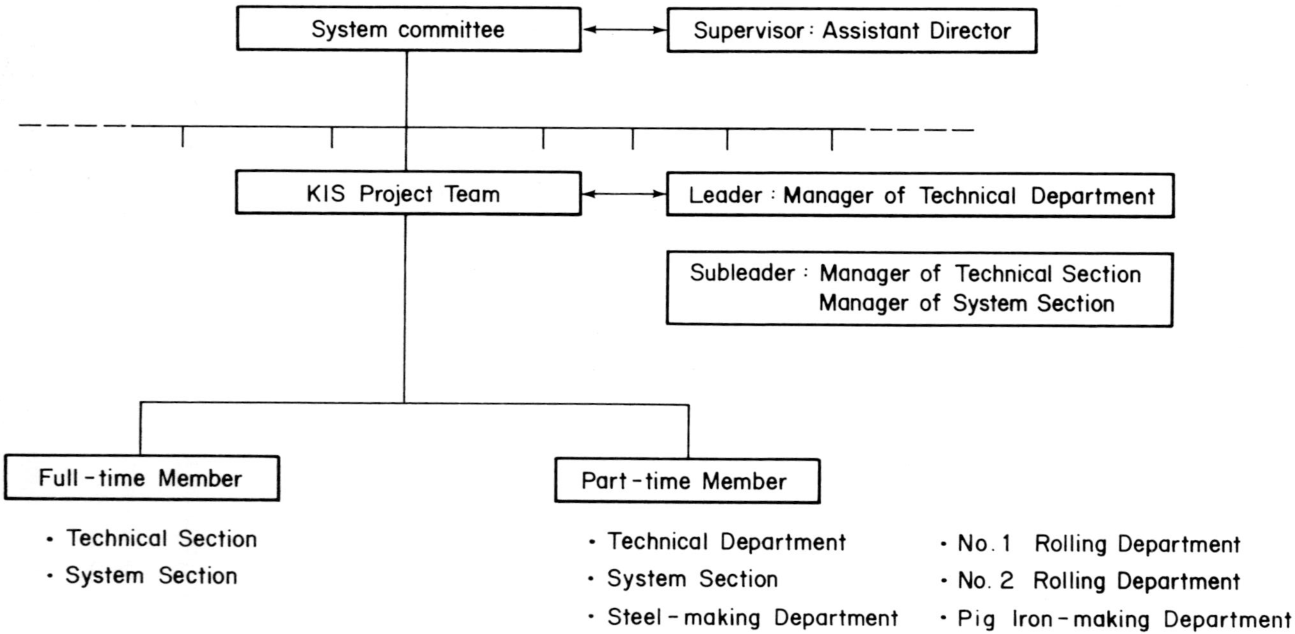

Fig. 15.4 Organization of development KIS.

Table 15.1 Schedule of development KIS

Item	1975	1976	1977	1978	1979	1980	1981
Job analysis of Technical dept.	⟶	⟶					
Analysis of previous system			⟶				
Establishment of system concept			⟶	⟶			
Development of iron and steel making process					Steel making ⟶	Blast furnace ⟶	⟶
Development of wire rod process					⟶	⟶	
Development of plate rolling process						⟶	
Development of hot/cold strip rolling process						⟶	⟶
Cut over							⟶
Development organization	Project team of technical dept.			Project team of system committee			

DB retrieval subsystem and converted to the DI-file (data interface file) for analysis, or data can be accessed through the OCR (optical character reader) and then converted to the DI-file through the DI-file conversion subsystem.

15.4.2 Outline of Subsystem

DB generation subsystem
The quality and operation data are gathered and stored in DB by off-line

Table 15.2 System volume and work load of development

Program steps			Work load (System design–cut over)
APL	PL/I	Total	
2800 (56 000 by PL/I)	59 300	115 300	280 man-month

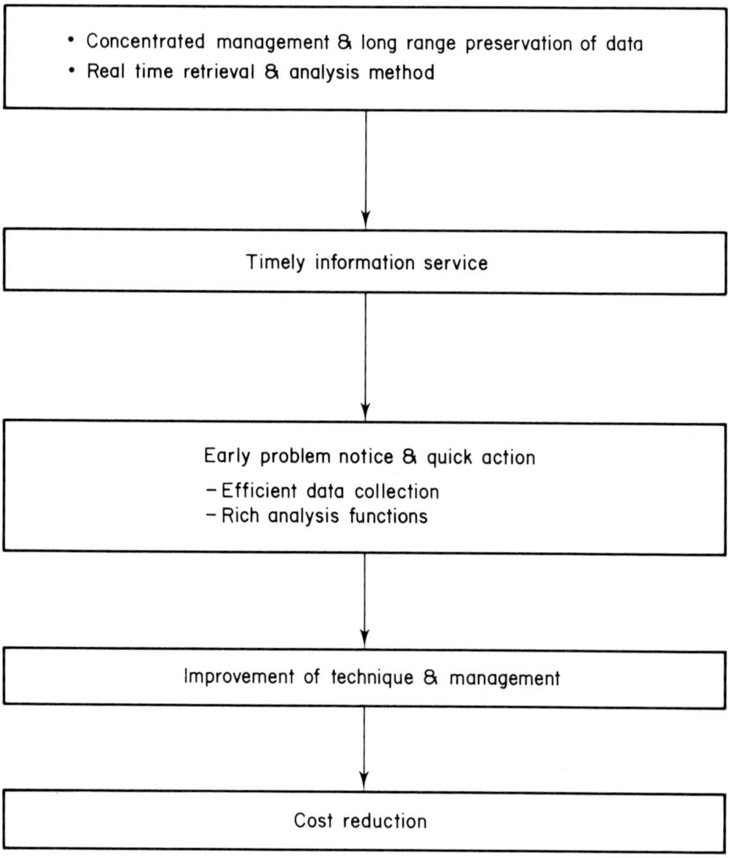

Fig. 15.5 Concept of KIS.

processing. The DB structure has been designed to get highest retrieval performance after consideration of usage patterns.

Database structure of DB generation subsystem is shown in Fig. 15.8.

DI-file generation subsystem.
The resident DI-file is generated once a day from the data most frequently used.

DB retrieval subsystem
The DB retrieval subsystem has been developed in APL and PL/I, in order to solve the restriction that APL cannot directly retrieve DB.

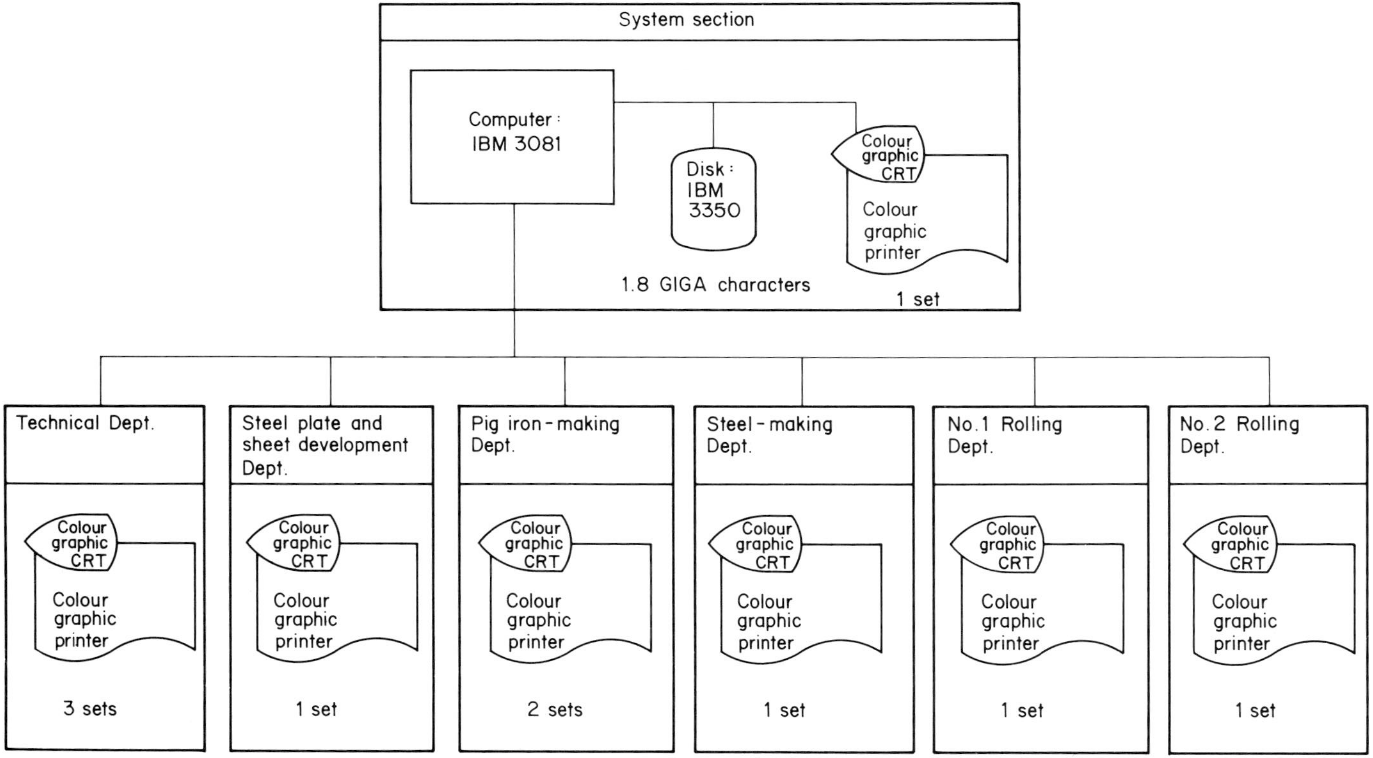

Fig. 15.6 System configuration (CRT = Cathode Ray Tube).

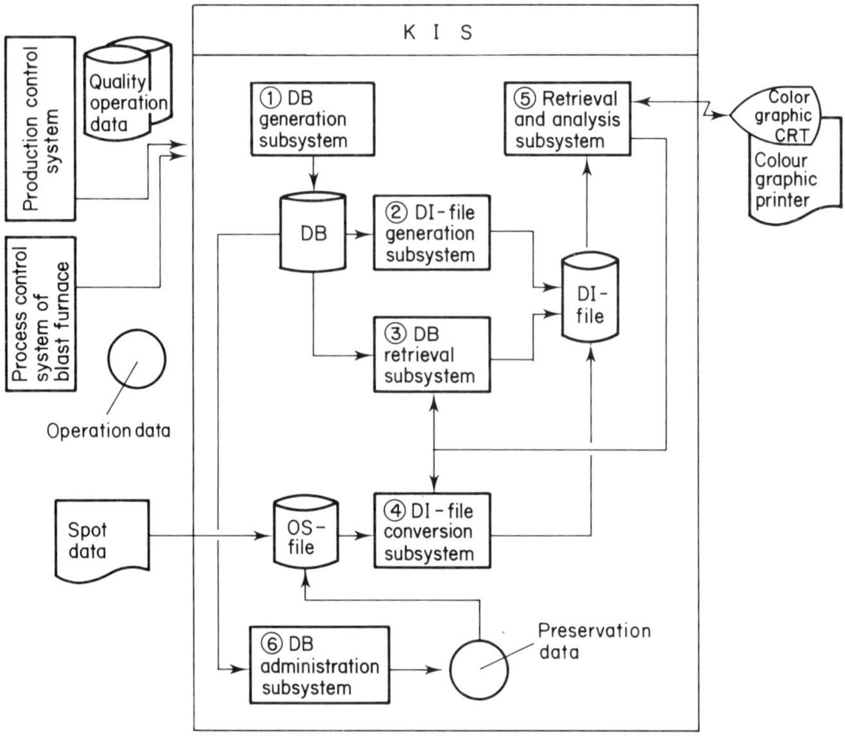

③④⑤ : Directly used by terminal user

Fig. 15.7 System structure.

The dictionary file was developed to minimize the work load of terminal users accessing the DB retrieval subsystem. This file contains all the parameters for the DI-file generation subsystem and the DB retrieval subsystem, such as type, length and position of items in the DB.

When data not in the resident DI-file are needed, the data can be retrieved by on-line batch processing and converted to the DI-file. The operation procedure for DB retrieval subsystem is shown in Fig. 15.9.

DI-file conversion subsystem

The DI-file can be retrieved and analysed directly by APL in order to secure real time data processing. This file is the starting point for all data processing. When data not in the DB are needed, the data are accessed through the OCR and generated to the OS-file. After parameters have been input from the terminals, the information is transferred to the DI-file.

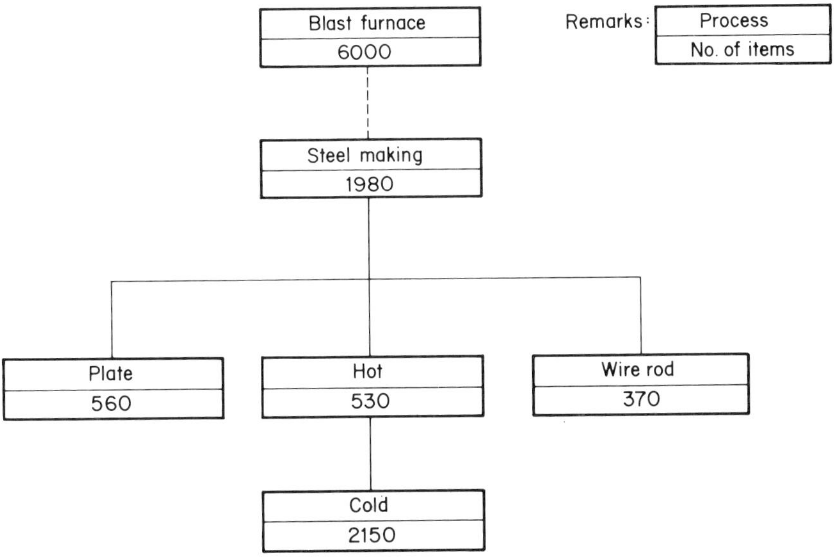

Fig. 15.8 Database structure.

Retrieval and analysis subsystem

The data in the DI-file can be retrieved interactively and analysed in real time by the APL-DI. This subsystem consists of three groups of functions as follows.

(a) Functions of the APL-DI: The data in the DI-file can be retrieved, sorted, counted and analysed (65 types. An example is shown in Fig. 15.10).

(b) Statistical analysis functions: The system is equipped with powerful statistical analysis functions, such as regression analysis, discriminant function analysis and linear programming (14 types. An example is shown in Fig. 15.11).

(c) Graph functions: Graphic analysis is one of the most effective ways of processing data. The system is therefore equipped with many graph functions, such as histograms, distribution charts and time-series graphs (15 types. An example is shown in Fig. 15.12).

DB administration subsystem.

To maintain performance, the DB is regenerated once a month. Data over 6 months old are deleted from DB and stored in magnetic tapes.

```
       BATCH ←——————————————————————————————— START DB RETRIEVAL SUBSYSTEM
****** BATCH KENSAKU START ******
MAN N0. : 11342
BUKA CODE (XX) : 16
SELECT KEY : SYUKKOUBI ←——————————————————————— SPECIFY RETRIEVE KEY ITEM
KEY HANI SITEI (FROM/TO) : 791201/791231 ←————— SPECIFY PERIOD
KEY HANI SITEI (FROM/TO) :
KEY TANPIN SITEI :
MUKESAKI (A,H,B,G,Z) : H
DASSAN (R,C,S,K): C
ZOUREN (0,1,2,3): I
SELECT SEGMENT : CHARGE                         SPECIFY OBJECT SEGMENT
SYOUKAI FILE NAME
SELECT KOUMOKU : 1,7,49,56,64,71,74,77,80,85,88
SELECT KOUMOKU : 103,106,108,110,112,196,197,320 ←—— SPECIFY ITEMS
SELECT KOUMOKU : 1485,1488,1490,1492,1520,1527
SELECT KOUMOKU :
SELECT SEGMENT :
SITEI NAIYOU HYOUJI ? YES(Y), NO(N) : N
****** SYOUKAI FILE SAKUSEI (H1451510) START ******
JOB H1451510 (JOB00324) SUBMITTED
ENTRY (A) H0160 WORK DELETED
****** BATCH KENSAKU END ****** ←——————————————— END DB RETRIEVAL SUBSYSTEM
```

Fig. 15.9 Operation procedure of DB retrieval subsystem.

```
           INQ
FILENAME : CHARGE
SELECTION : CH0049 = [xxxxxxx]
FUNCTION  : STATS  <------------------- SPECIFY STATISTICAL CALCULATION
FIELDS    : CH1485, CH1488, CH1490 <--- SPECIFY ITEMS
PAGE NO. 1

15:14   5/22/81   KIS DATA INTERFACE
 KIS CHARGE : [CH0049 = xxxxxxx : STATS]
    CH1485   CH1488   CH1490

 ZERO VALUES EXCLUDED :

59.00    59.00    59.00    SAMPLE SIZE
54.93    16.86    11.42    AVERAGE
 3.10     1.06      .89    2 × S.E. OF MEAN
11.92     4.05     3.42    STD DEVIATION
75.00    27.00    23.00    MAXIMUM
31.00     7.00     6.00    MINIMUM

 ZERO VALUES INCLUDED :

61.00    61.00    61.00    SAMPLE SIZE
53.13    16.31    11.05    AVERAGE
 3.92     1.28     1.01    2 × S.E. OF MEAN
15.32     5.01    3.94   STD DEVIATION
75.00    27.00    23.00    MAXIMUM
  .00      .00      .00    MINIMUM

 TIMES (SEC) : ELAPSED = 66.7, CPU = .8, KEYING = 57.6
```

Fig. 15.10 Example of statistical calculation.

```
        INQ
FILENAME  : CHARGE
SELECTION : KOUSYU = [xxxxxx]
FUNCTION  : REG <------------------------------------------ SPECIFY LINEAR MULTIPLE REGRESSION
CONTROL FIELDS :
DEPENDENT VARIABLE(S) : CH0116 <--------------------------- SPECITY DEPENDENT VARIABLE
INDEPENDENT VARIABLES : CH1641, CH2104, CH2110, CH2114 <--- SPECIFY INDEPENDENT VARIABLES
 11 : 46     5/22/81     KIS DATA INTERFACE
   KIS CHARGE : KOUSYU = xxxxxxx, REG
   CH0116        CH1641        CH2104        CH2110        CH2114
   (N = 348)
 MEANS
   51.408        149.974       15.3448       4.33908       1.07759
 STANDARD DEVIATIONS
   5.97313       39.8994       94.9352       27.6237       10.4112
 CORRELATION MATRIX
   1              -0.0984282      -0.096824       -0.126196       -0.0590546
   0.0984282       1               0.205629        0.195354        0.148622
   0.096824        0.205629        1               0.873126        0.695563
   0.126196        0.195354        0.873126        1               0.373417
   0.0590546       0.148622        0.695563        0.373417        1
 REGRESSION COEFFICIENTS (B1, B2, .... CONSTANT)
   0.0114934      0.014882      0.0599256      0.0623513      53.2306
 STANDARD ERROR OF COEFFICIENTS
       .008         .012         .032     .057
 T-VALUES
       1.408        1.247        1.885        1.090
 ANALYSIS OF VARIANCE OF REGRESSION
   SOURCE              DF          SS              MS
   REGRESSION           4        3.2339E02      8.0848E01
   ERROR              343        1.2093E04      3.5256E01
   TOTAL              347        1.2416E04
   STANDARD ERROR OF EST = 5.93764
   F VALUE = 2.29319
   MULTIPLE R 2 = 0.0260462
 TIMES (SEC) : ELAPSED = 68.8, CPU = .6, KEYING = 60.4
```

Fig. 15.11 Example of linear multiple regression.

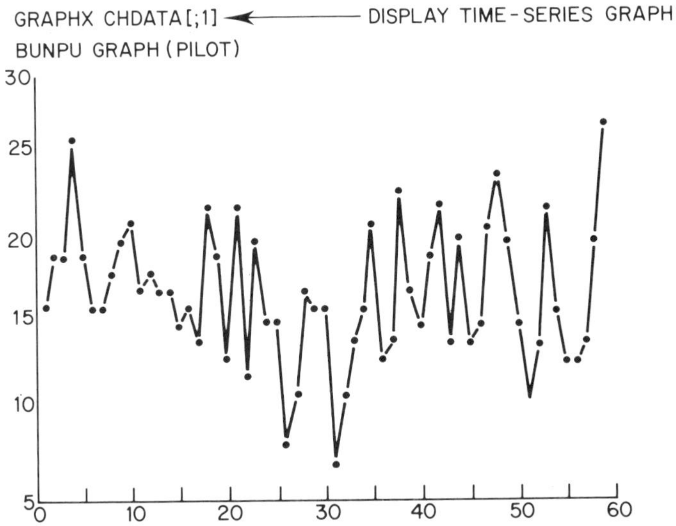

Fig. 15.12 Example of time-series graph.

15.5 Advantages of the KIS System

15.5.1 System effect

(1) Quality and operation data analysis throughout the processes can be easily processed.

(2) Other systems in use adopt different programs for each problem and batch processing. Moreover they use a menu, which is fixed beforehand. Whereas KIS adopted interactive and real time processing which complied with flexible retrieval, statistical analysis functions. Because of these features, KIS is a superior system both in data processing time and flexibility.

(3) The open system, with terminals installed near the staff, has increased the staff performance efficiency.

(4) When data not in DB are required, the data can be accessed and easily processed through the terminals.

15.5.2 System Evaluation

Since its installation in October 1979 the effects of KIS have exceeded expectations. The interactive feature allows data to be processed by the trial and error method, thus facilitating continuous problem solving.

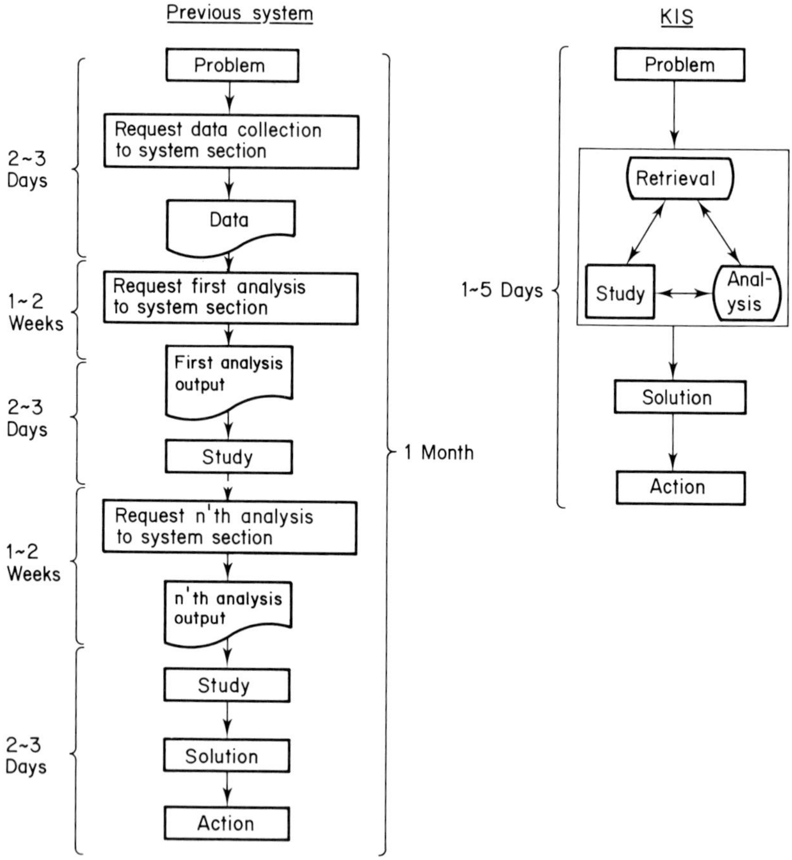

Fig. 15.13 Comparison of work pattern in technical department.

KIS has changed the work pattern of the Technical Department, compared with that of the previous system. This is shown in Fig. 15.13. The work time which was required to obtain solutions is reduced to 1/5–1/15 of the time required by the previous system (Table 15.3). The data analysis by effectively utilizing quality and operation data from all processes of the work, which were impossible before, can now be examined. The results of analysis have been effectively applied to quality and process design and operation conditions (Figs. 15.14 and 15.15).

15.6 Future Developments

KIS is the epoch-making system using DB and APL, and its functions have

Table 15.3 Reduction of work time by KIS

	Days necessary by previous system (rank)			
	1–10	11–20	21–30	Over 30
Process time	9	6	3	4
(a) Previous system (estimated days/case)	3.7	18.3	28.3	60 : 1 case Designated: 3 cases
(b) KIS (days/case)	0.6	3.8	4.7	4.0
(c) Reduction ratio (%) $(\frac{a-b}{a})$	84	79	83	93

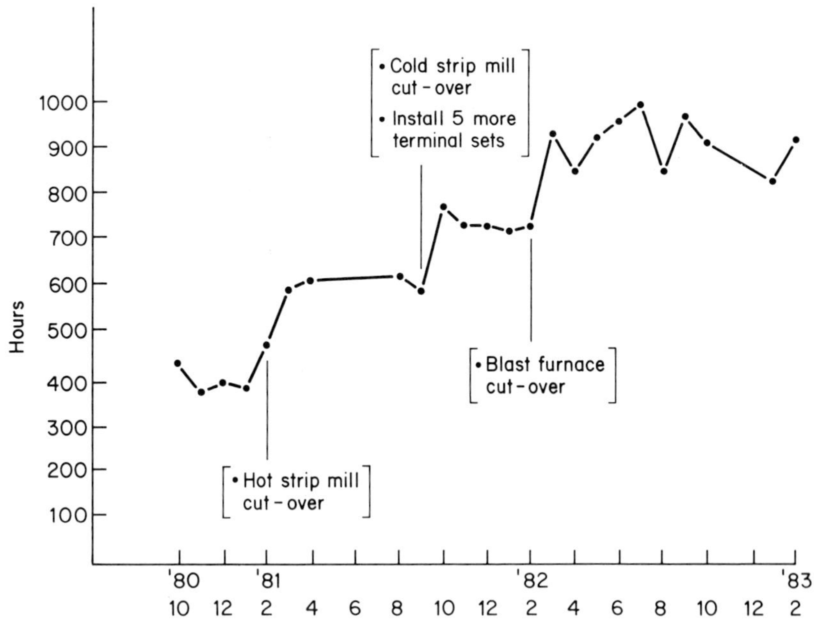

Fig. 15.14 Trend of total usage time.

proved extremely powerful. In order to make the system more complete, however, a few problems remain:

(a) Development of a simulation model to minimize production cost and to optimize quality and process design.

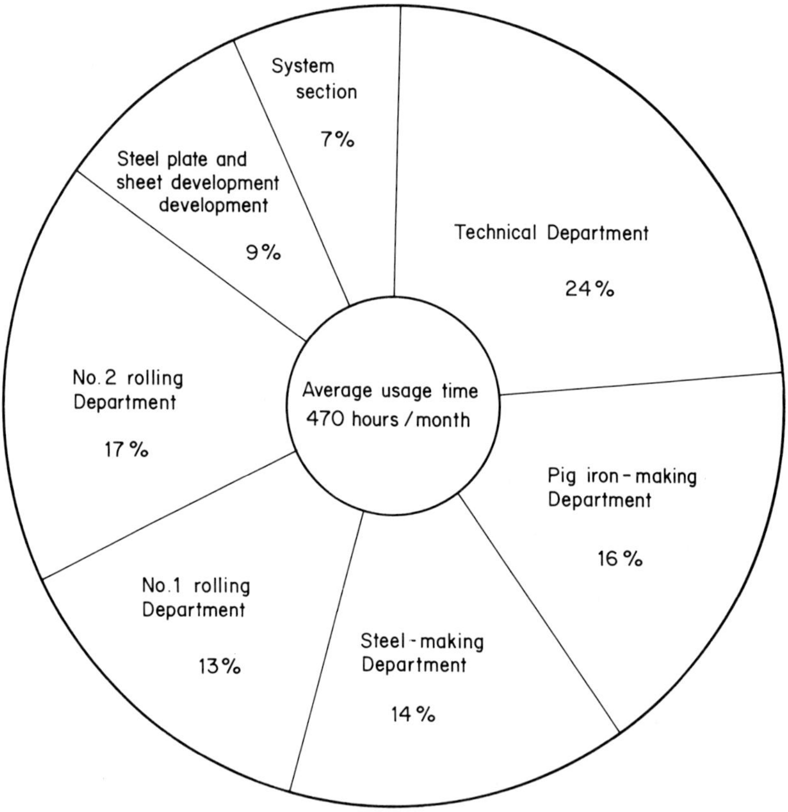

Fig. 15.15 Usage time ratio of department.

(b) Extension of graph functions.
(c) Data precision.

When these problems are solved, KIS will become an even more effective tool for cost reduction, quality improvement and technical development.

16

Information and Control System for Ethylene Plant

Showa Denko KK, Tokyo, Japan

16.1 Introduction

Built in Oita City under the leadership of Showa Denko KK, the Oita Petrochemical Complex commenced commercial operation in April 1969. It consists of one ethylene plant (220 000 tonnes per year), fourteen derivative plants, three power-generating boilers, one air separating plant and other miscellaneous utility supply facilities. The plant layout and materials flow are shown in Figs. 16.1 and 16.2, respectively.

The plant site is 1 640 000 m^2 in area, and the employees of Showa Denko group companies working there number 750. The major raw materials, such as naphtha and fuel oil, are supplied from the adjacent refinery of Kyushu Petroleum Company through underground pipelines.

As the centre of the complex, the ethylene plant produces ethylene, propylene, C_4's, cracked oil, etc., which are supplied to the derivative plants or delivered out of the complex by ship, road tanker and pipeline.

Showa Denko has developed a total control system that accomplishes advanced control of the various phases of the activities of the complex. An Ethylene Plant Information and Control System (EPICS) was specially developed for the ethylene plant.

The Oita petrochemical complex satisfies the criteria used in establishing

CAD/CAM and MIS in Japan
ISBN 0-12-682580-7

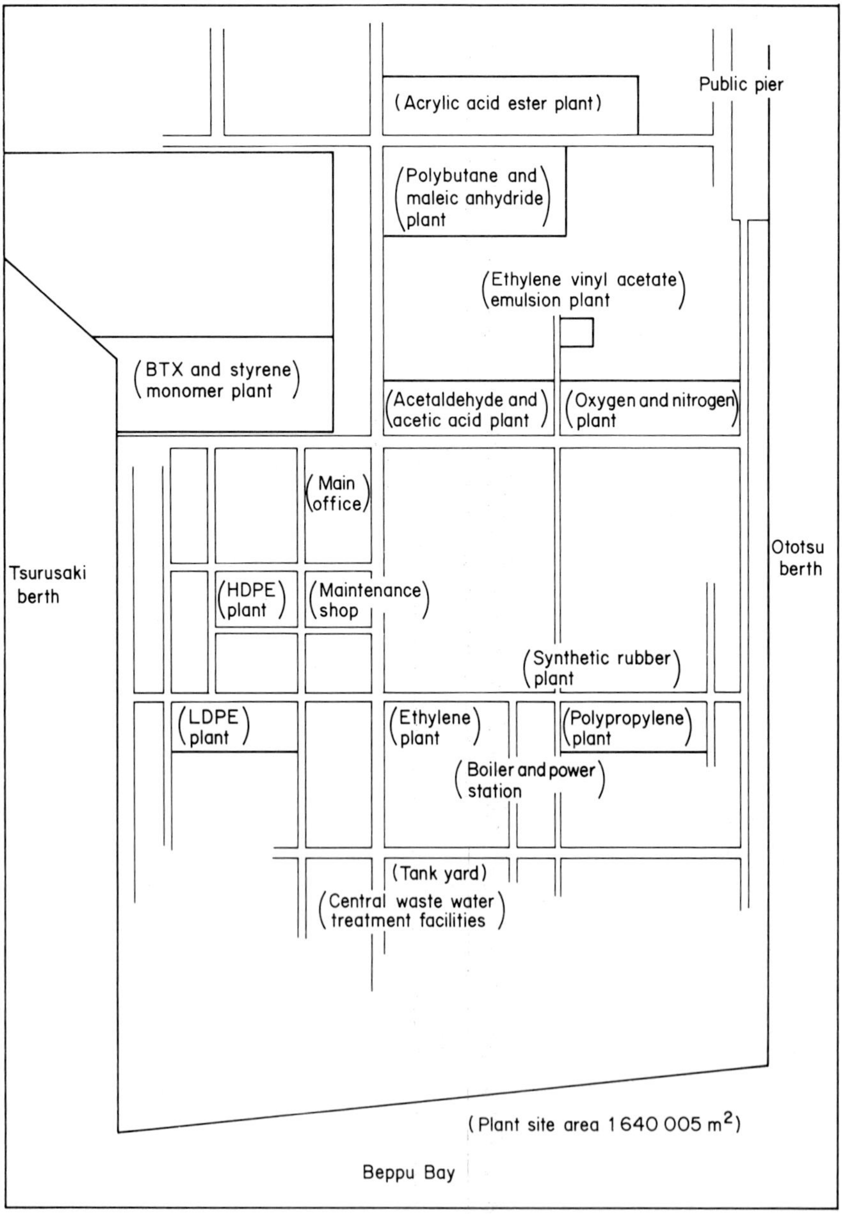

Fig. 16.1 Plant layout of Oita Complex (1969).

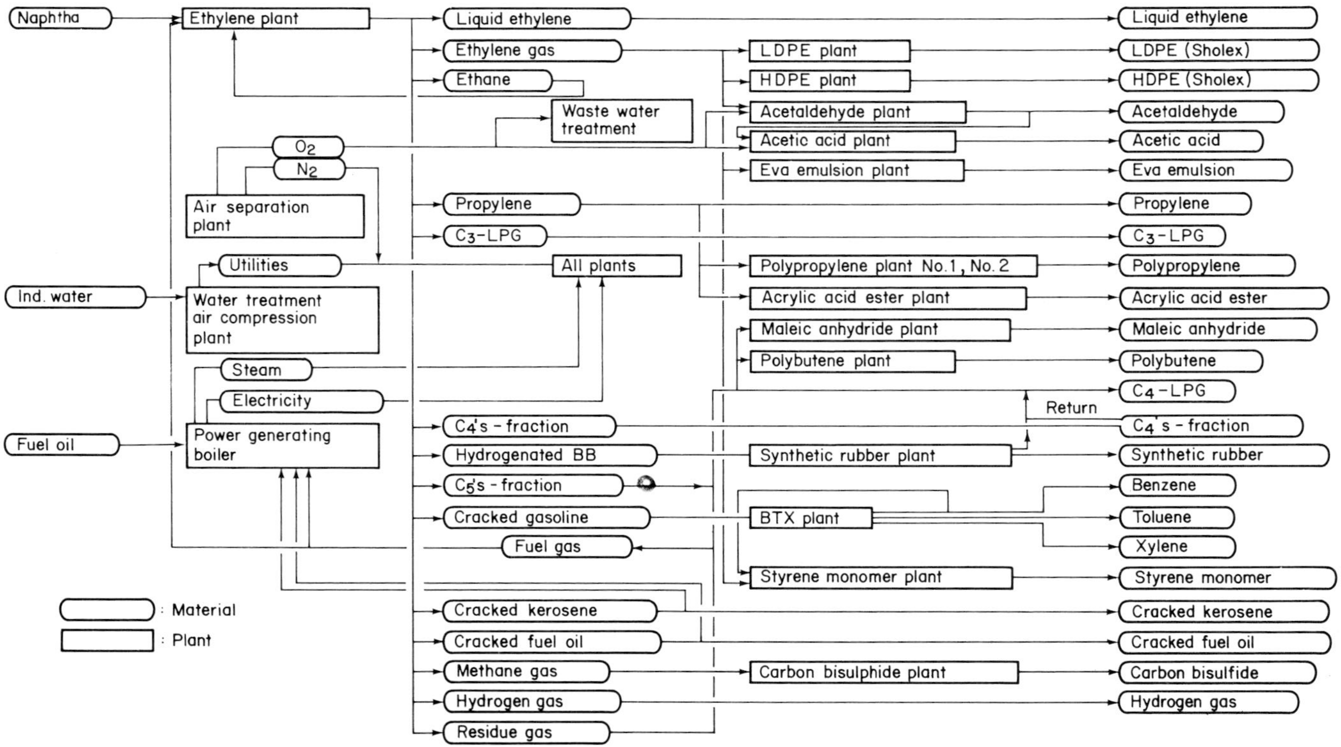

Fig. 16.2 Materials flow in the complex.

technical justification for a computer control system offering potentially large economic benefits from improved control and, in addition, ensuring safety and preventing environmental pollution.

The success of the advanced control system for the complex owes much to the fact that almost all the plant systems, including those for raw materials and utility supply, have been consistently designed and operated under the direction of Showa Denko from the planning stage of the complex, since 1967. A hierarchical computer system, consisting of four process control computers, was installed at the initial plant start-up to facilitate development of the entire computer control system in combination with the conventional analog control system.

Additionally, certain factors made it possible to establish an improved total control system over a period of four years. These included the wide technical experience gained in process computer application with over fifteen computer systems since installation of the initial systems at the Showa Denko subsidiary Tokuyama Petrochemical Company in 1963; the knowledge accumulated on process characteristics, and the availability of improved instruments and reliable on-line process analysers.

16.2 Characteristics of Ethylene Plant

The operations of all the derivative plants are closely related to those of the ethylene plant. The power plants, which supply steam and electricity to the complex, use low sulphur, low nitrogen fuel from the ethylene plant. Accordingly, the complex is strongly influenced by the ethylene plant, the outstanding characteristics of which are presented in Fig. 16.3 and outlined below.

(1) The ethylene plant consists of many items of large-scale equipment (for example, seven cracking furnaces, four compressors and over fifteen distillation towers) which are interrelated by material flow and heat recovery. Consequently, it is very difficult to control their operation satisfactorily using process information (almost 2000 measurements and about 300 control loops), unless a computer system is applied.

(2) So large is the quantity of energy handled (for example, the total cooling duty reaches the order of 100 mm kcal per hour), that even a slight improvement in thermal efficiency brings about significant economic advantages.

(3) Production scheduling and operation of the downstream sectors are affected by unavoidable cyclic changes in cracking load caused by the decoking of the cracking furnaces and quenching boilers.

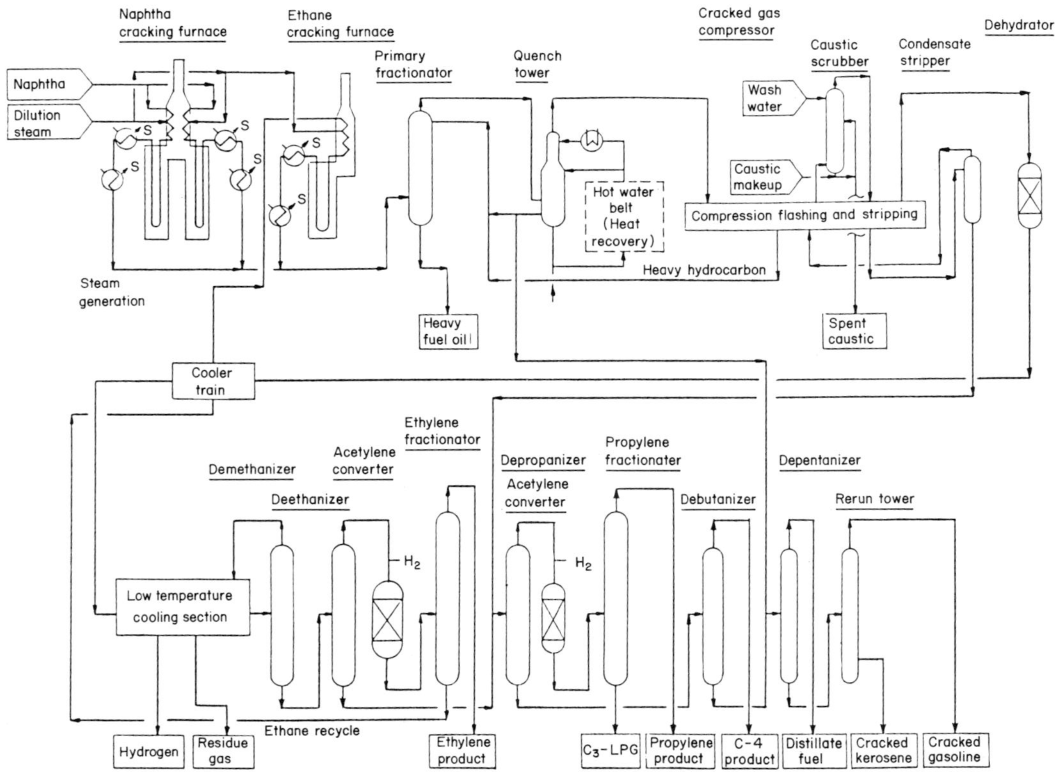

Fig. 16.3 Process flow of ethylene plant.

(4) As shown in Fig. 16.3, many high purity products are simultaneously produced, and the product yield distribution is affected by various factors such as naphtha specification, cracking conditions, operation of the ethane cracking furnace and the product specifications required. This makes it very important to find optimal operating conditions to meet the product demand profile.

In addition to the items mentioned above, there are many other factors that have to be taken into consideration in establishing a control strategy.

Further, the petrochemical complex strongly influences both the life of the people and the economic activities in the surrounding districts. It is, therefore, important for the complex to operate in a steady and safe manner, without causing environmental pollution, and to keep its operating procedures flexible in order to satisfy various community requirements.

16.3 Computer Application

For an ethylene plant having the above-mentioned characteristics, it is necessary to utilize computers by applying their various functions to attain safe and rational economic operation and management. For the purpose of computer application, the ethylene plant, the derivative plants and the power plants were divided into three groups, and four process computers were installed. The configuration of the computer control system is shown in Fig. 16.4.

From the initial start-up stage, an IBM 1800 DACS (32 kiloword core memory, three 512 kiloword disk memory) has been used as the central computer for high level applications. For simple supervisory control services, two FACOM 270-20s have been used, linked with the central computer by communication lines. One of them supports a YODIC 500 for DDC (direct digital control).

This hierarchical computer system was judged in the planning stage to be the most feasible, offering the lowest investment, the widest capability of covering basic applications, easy development of advanced level functions and capability for future expansion.

The basic level functions of the process computers are to accomplish improved operability, to ease supervision and to ensure stable and safe operations. Accordingly, data acquisition and treatment, alarm and sequence monitoring and control were put into operation from the initial start-up.

Advanced level functions are divided into two major categories: planning and control.

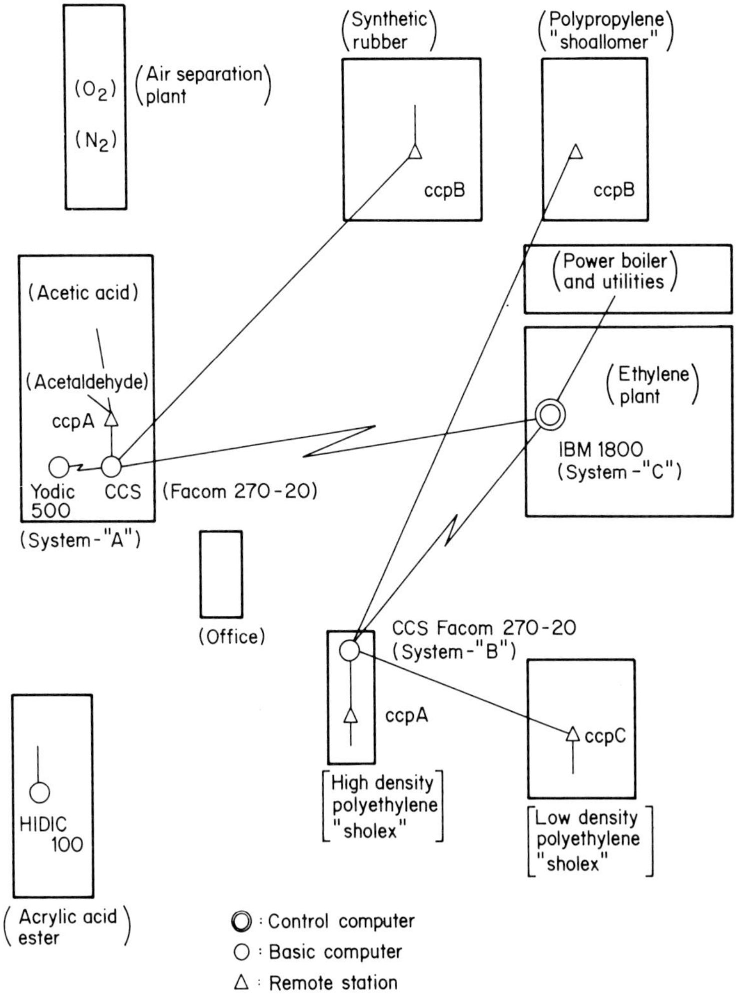

Fig. 16.4 Computer system.

The control level functions are for closed loop execution of optimal process operation based on mathematical optimization taking into account the target indications from the planning system. The targets of the planning level functions are to obtain the most economic production and sales program for the individual plants in the complex.

The planning system is also important from the viewpoint of keeping the plant operating stably and safely, because it minimizes rapid changes in operating conditions caused by changes in production load and product grade.

16.4 Development of EPICS

First, control level application was developed for stable and optimal plant operation to lower the running cost of the plant.

For fundamental process variables, such as temperature, pressure and flow rate, and analysis data obtained by on-line gas-chromatography which was technically incomplete at the time, much effort was put into making them sufficiently reliable for hardware and software. This facilitated analysis and control of the static and dynamic characteristic of the process in operation. Optimal control of the process was applied, first, to the fractionation section and, then, to the compressor section, which accounts for a large part of the running cost.

In order to achieve successful process control downstream of the cracking furnaces, computer control of the cracking furnaces was first made available, based on an analysis of their performance from the initial start-up stage of the plant, by establishing furnace operation technology in addition to developing measuring techniques for the surface temperatures of reactor tubes in the cracking furnaces, and for the analysis of cracked gas.

For planning level functions, an ethylene plant production scheduling program (EPPS) was first developed and applied to keep inventories at optimal levels, and then an ethylene plant optimal planning system for production and sales (EPOPS) was developed. It consisted of process models, including process data collected through computer control and analysis data obtained on naphtha and cracked gas.

Owing to development of the cracking furnace control program and EPOPS, optimal control of the entire process was made possible. By the end of 1972, EPICS had been completed as a total system.

16.5 Outline of EPICS

For an ethylene plant having the characteristics described in Section 16.2, computers can be an important means of raising profitability. The fundamental parts of computer application are as follows:

(a) To produce information to support the development and determining of fundamental strategies for sales, purchase and plant operation.
(b) To develop optimal production plans and determine optimal plant operating conditions.
(c) To carry out optimal operation.
(d) To maintain stable operation (ensuring stability, raising the production rate and maintaining product quality).

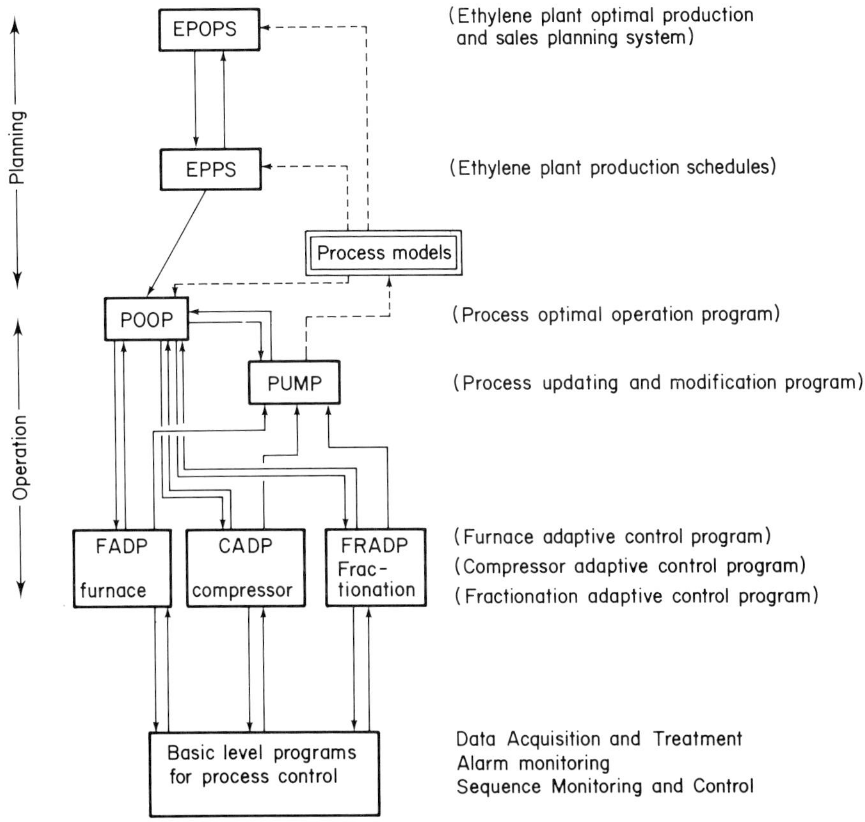

Fig. 16.5 EPICS total control system.

(e) To improve the operability of the plant (minimization of operating manpower and prevention of mis-operation).

The system configuration of EPICS, which is for the above-mentioned purposes, is shown in Fig. 16.5.

The planning level application system for the ethylene plant consists of an optimization program to seek the best production and sales plan for the ethylene plant under such constraints as may be imposed by the ethylene plant itself, the availability of naphtha, the demand and cost trends of various cracked fractions and the need for a stable supply of product to the derivative plants.

Figure 16.6 shows a block diagram of this system, i.e. the ethylene plant optimal planning system for production and sales (EPOPS).

The results of the optimal planning computation are fed to the ethylene

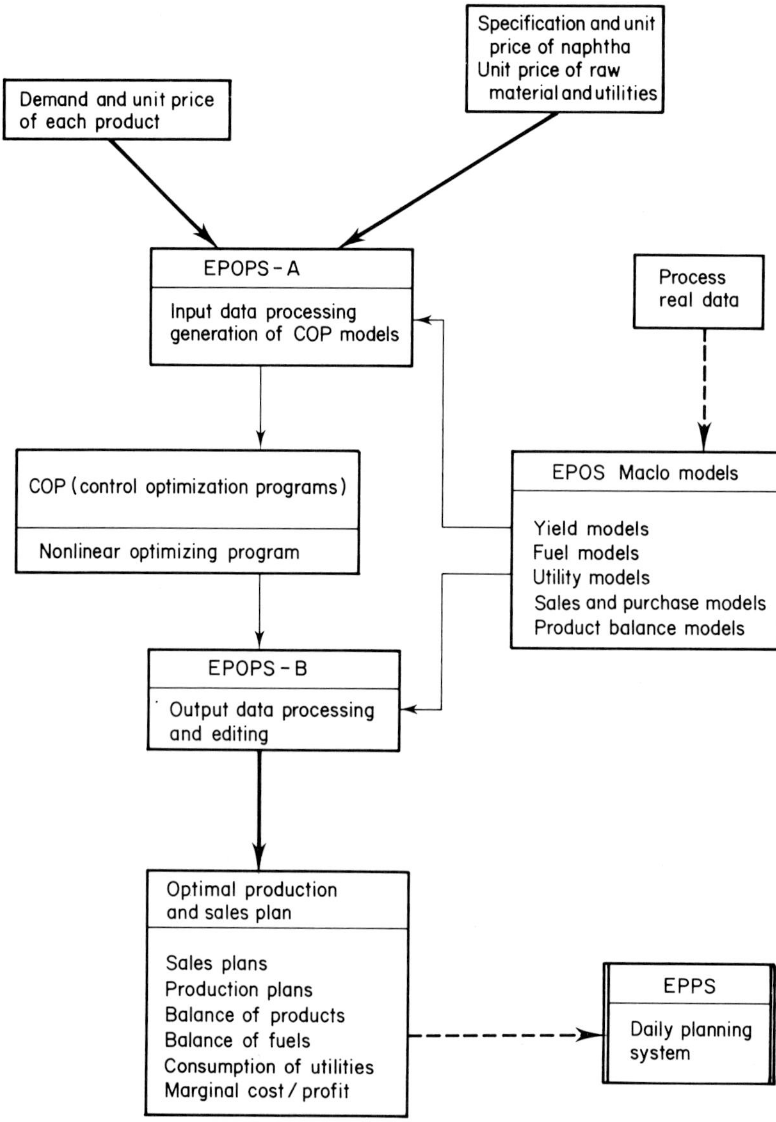

Fig. 16.6 EPOPS block diagram.

plant production scheduling (EPPS) program shown in Fig. 16.7 to determine the optimal daily operating conditions and product shipment schedules. At the same time, this program makes it possible to keep adequate levels of products in inventory, to minimize fluctuations in plant operation and keep a

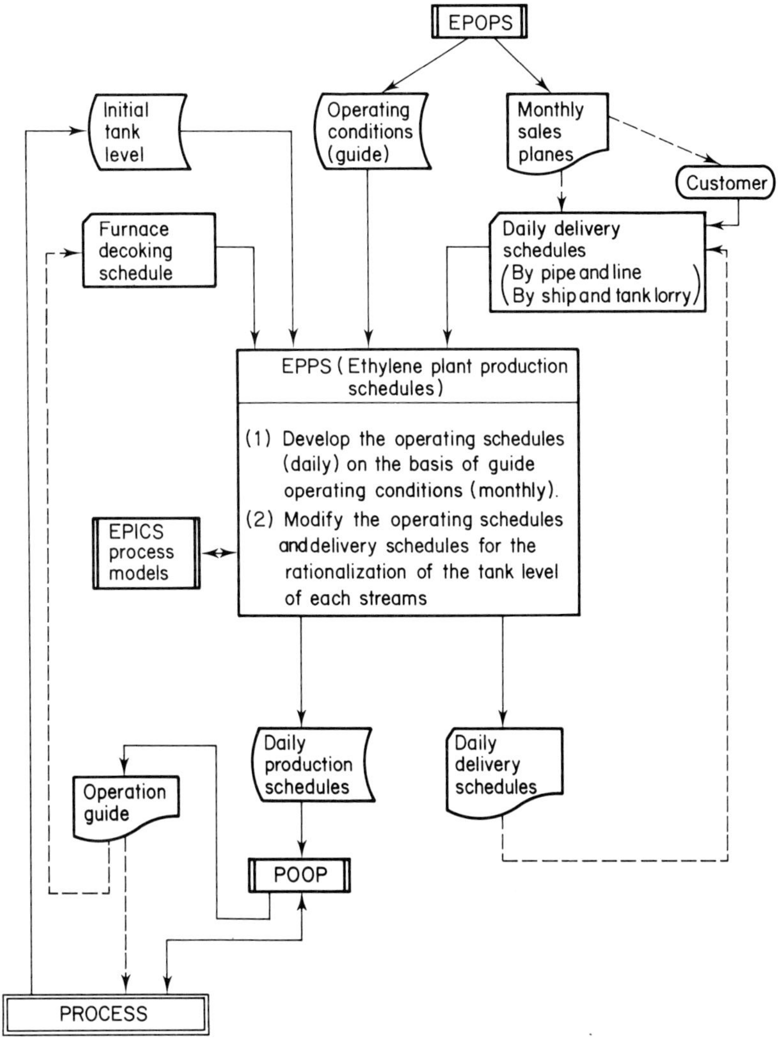

Fig. 16.7 EPPS block diagram.

proper balance of fuel and utilities. These planning systems are normally used on a semi-annual and monthly basis to determine the production and marketing plans for the following year and the daily shipment schedules and operating conditions for the following month.

The mathematical models included in EPOPS and EPPS were developed from operating data and test data obtained from the plant, by applying chemical engineering theories and statistical methods.

The control level application system consists of a process optimal operation program (POOP), a process updating and modification program (PUMP) and individual supervisory control programs such as a furnace adaptive control program (FADP) for the cracking furnace section, a compressor adaptive control program (CADP) for the compression section, and a fractionation adaptive control program (FRADP) for the fractionation section.

POOP determines the targets of optimal control of the plant according to the production plan calculated by EPPS and the process status fed back by PUMP. FADP, CADP and FRADP perform the output from POOP, and, at the same time, carry out their own sectional optimal control.

Showa Denko has developed instruments for measuring the cracked gas composition and the reactor tube surface temperature, both of which are important in determining the operating conditions of the furnaces. Ever since these process variables (which had been difficult to measure up to that time) were connected to the computer, FADP has controlled the furnaces to keep them in safe and steady operation, and contributed to optimal operation and increase in production rate.

CADP controls the compressors to keep them operating at maximum efficiency under variations in cracking condition, gas quantity and cooling water temperature.

In the fractionation section, where cracked gas products are fractionated and purified, disturbance from upstream columns can influence the downstream columns and thereby cause production of off-spec. FRADP has facilitated stable operation and sectional optimization, ensuring product quality and reduction of purification costs despite variations in feed flow rate or composition.

Other programs, which are the basis of the planning and the control systems, fulfil functions such as data acquisition and treatment, alarm monitoring, sequence monitoring and control, etc.

16.6 Effect of EPICS

16.6.1 Economic Effects

EPICS is highly effective in achieving rational management, safe operation and operability. It also leads to optimal production plans, a reduction in variable costs and increase in production rate.

These effects can be classified according to the economic environment into the following patterns.

When there is small demand and adequate supply (that is, where sales are bottlenecked, as in 1972), high profitability is attainable by:

(a) careful management of production, such as effective control of cracking furnace yield in accordance with the trend of product demand, keeping appropriate levels of inventory and avoiding missed sales opportunities;
(b) proper setting of operation conditions or process state, to enable safe and optimal operation and decreasing the consumption of raw materials and utilities like steam and electric power.

On the other hand, if the state of the ethylene market is one in which production is bottlenecked owing to prosperity of the world economy and difficulty of obtaining naphtha (as in 1973) then, in addition to (a) and (b) above, the following are also important strategies:

(c) tracking of maximum plant capacity and detection of any bottlenecks in the process, and the development of immediate measures to raise capacity by improved operation or de-bottlenecking;
(d) executing stable operation and grasping sales opportunities;
(e) effective use of the raw material naphtha by increasing the yield of ethylene and propylene, to raise profitability.

EPICS has made it possible for these various requirements to be met.

16.6.2 Effects on safety of operation

In each petrochemical complex, great care is taken to prevent accidents. EPICS has contributed to the safe operation of the plant in the following ways.

One effect is that, by computer control, EPICS maintains stable operation against changes in process variables, and stable operation is, of course, essential to safety.

Another is that EPICS monitors moment-to-moment plant status. Pre-detection and pre-alarm by the computer of abnormal or out-of-control status enables action to be taken to avoid accidents and to maintain normal status. Monitoring by computer can detect the rate of change in process variables, or abnormal relationships between the variables, and is especially valuable for maintaining the safety of the reactor.

16.7 Conclusions

The quality of management has been improved by application of EPICS through improved control of production and operating conditions.

Ten years have elapsed since EPICS was established in 1972. In the Oita petrochemical complex, the second ethylene plant, which has a capacity of 320 000 tonnes per year, and several new derivative plants were completed and put into commercial operation in 1977. The plant's capacity is now greater than ever. Further, computer application has become increasingly necessary to ensure stable, safe and economic operation.

Both hardware and software have also been markedly improved as improvements have been made in their RAS (reliability, availability and serviceability); they have become relatively cheap; and have become easier to use because of development of man–machine interfaces like video terminals.

It was against this background that a new computer system has been established in the Oita complex and added to the old one; computer applications have also been expanded. For example, automatic and immediate loaddown of the cracking furnaces can be carried out in emergencies by applying DDC (direct digital control), with resulting advantages in both economy and safety. Planning level applications have also been expanded to enable all the plants, including the derivative plants and utility facilities as well as the ethylene plant, to be operated most economically.

17

A Production Planning and Control System Based on Material Requirements Planning

Shinichi Itoh, Associate Senior Vice President and Director, Nippon Electric Co Ltd, Tokyo, Japan

17.1 Introduction

A production planning and control system based on the material requirements planning (MRP) concept has been developed and has been successfully implemented at the Nippon Electric Co (NEC) and many other companies. This chapter discusses both the basic concepts and features of the system which have made it very successful, and its future developments.

17.2 Fundamental Concept and Background for NEC's MRP-based Production Planning and Control System

17.2.1 Background of the System

After the rise in oil prices, NEC and its customers alike were confronted with the problem of adjusting to a low-growth economy and, among many other measures taken, were faced with the need to redesign their production systems. The new systems had to meet various requirements, including shorter lead times and more flexibility, in addition to reduction in production and inventory costs.

CAD/CAM and MIS in Japan
ISBN 0-12-682580-7

The above requirements presented an obvious contradiction to the old systems, which were required to produce more in a shorter lead time while reducing inventory levels. Confronted with this dilemma within the company, NEC did an extensive survey and study to redefine the basic concept for its production planning and control systems.

MRP was known as a concept, but was at that time hardly implemented anywhere in Japan. NEC, after studying the concept carefully, decided to adopt it as the basis for production planning and control, for its own operations and for a package it offered to its computer users. By the time Japanese industry and business began talking about actual implementation of MRP-based systems, NEC already had an MRP-based package to offer.

This early start in the field, and the resulting accumulation of experience in implementing MRP-based systems, are behind the current MRP-based system. It has been successfully adopted and implemented by many users, including major companies in Japan. Its merits include a saving of development resources for equivalent systems, reduction in inventories, a reduction in administrative man-hours, and a shortening of production lead-times, amongst others.

17.2.2 Basic Concepts of the System

The original MRP concept covered a phase of calculating requirements for materials via the bill of materials (B/M), giving orders for final products. Since NEC had a system developed for this phase, it designed a system incorporating further functions required at various stages of production management. The resulting system, based on MRP in its broader sense, has the following characteristic.

(1) *Emphasis on planning.* The days when one could sell whatever NEC made are long gone and the company has to employ limited resources optimally to produce only what can be sold. The system, therefore, emphasized the master production-planning phase, as the following phases are completely dependent on this.

(2) "*Just-in-time*" *concept.* In order to minimize inventories along the production process, the system plans to deliver tens of thousands of items, produced in-house or purchased, as they are required throughout the factory floor. It specifies what actions are to be taken by whom, and at what time, for each of the materials and parts which go into a diversified array of products.

(3) *Dynamically responsive system.* The environment in which the system is put into use incorporates the following conditions. Increasing uncertainty

accompanying demand and the resultant changes in orders, more diversity in customer needs and the ensuing changes in design specifications, aggressive search for new and better parts and materials and the resultant changes in the bill of materials, and unplanned changes in inventory levels due to off-specification items, produced and/or purchased. Thus the system must be able to adjust to the many possible changes both quickly and dynamically.

(4) *The formal system.* Any production operation involves many shops and people, and this often results in what is referred to as "informal systems", where different people have different plans and criteria. As a result, the production system is not coherent and has high levels of work-in-process. The new "formal" system can remedy this situation since the planning and control functions are worked out by one centralized system.

17.2.3 Other Supporting Factors

To back up the MRP concept the company had to have the capability to translate it into an efficient and effective system, both software and hardware. As for the software development, NEC had many years of experience in developing various software products in the production planning and control areas. Drawing on this experience gave it fund of inexpensive, efficient and easy-to-use tools, including database systems, on-line systems and higher level languages, for use in the development of new software products.

Computers and peripherals being some of NEC's major products, it could accommodate the hardware requirements for the new system in terms of cost-effective terminals, high capacity disks, and other hardware to support the new system.

17.2.4 Development of a Package for the System

When NEC decided to develop a package for the system, it defined five standard types of production systems so that the package would be equipped with the flexibility to cope with a variety of production systems. These are shown in Fig. 17.1.

17.3 Outline of the NEC Production Planning and Control System

17.3.1 System Description

The system consists of the following subsystems, and their relationships are depicted in Fig. 17.2.

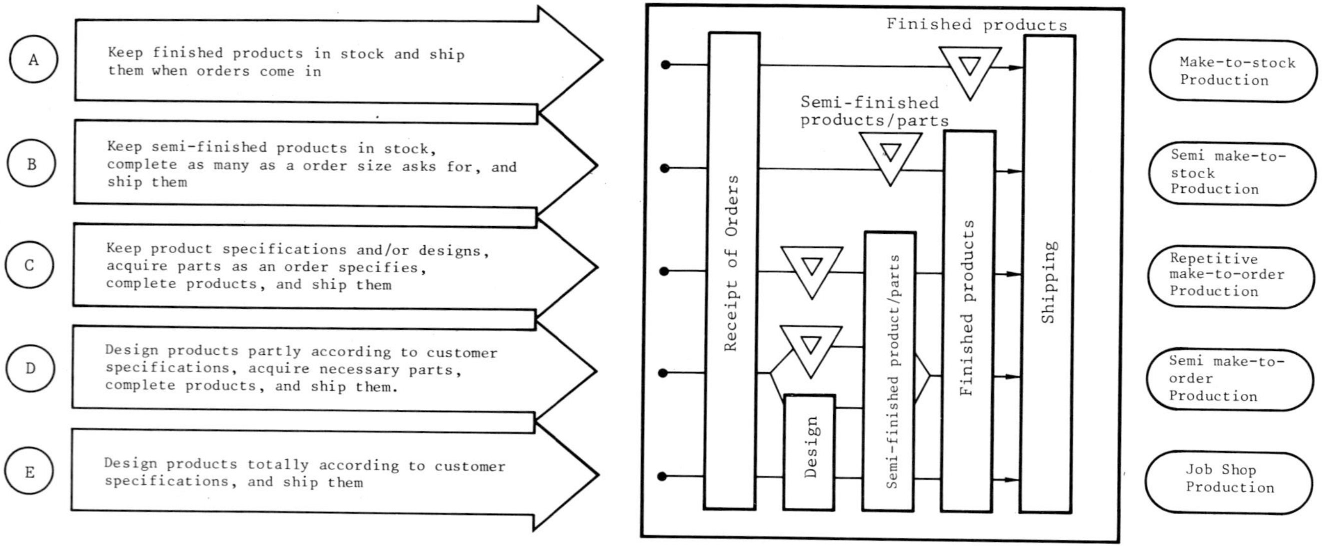

Fig. 17.1 Classification of production types.

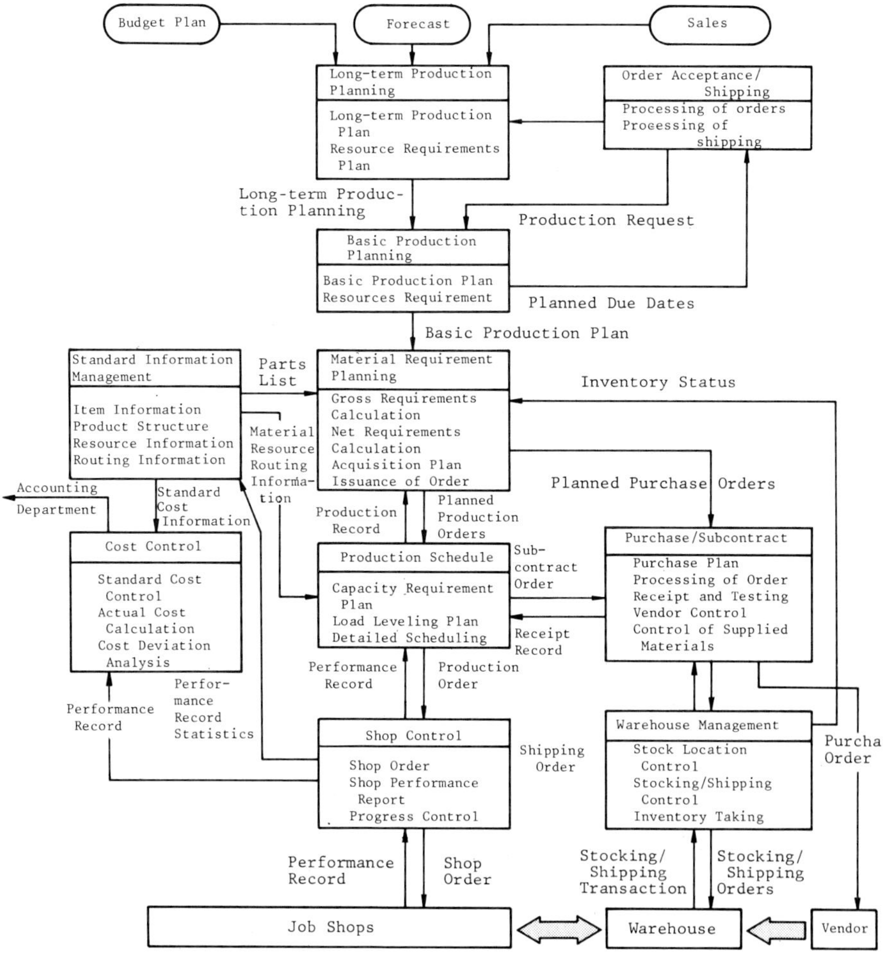

Fig. 17.2 Relationship among subsystems.

(1) *Subsystem for standard data-management*. This is used to add/update the basic information in the database which is the basis for planning and control. Such information includes, for example, item number, bill of materials, resource information and routing information.

(2) *Subsystem for production–sales interface*. This system acts as an interface between the sales system and the production system. Its main function is to transfer pertinent information from one system to the other.

(3) *Subsystem for long-term production planning*. A long-term production

plan is developed by this system. Requirements for resources (human, material, monetary and machinery) are roughly calculated by a long-term production plan which reflects company policies and the future trend of demand. This is then checked for feasibility.

(4) *Subsystem for basic production planning.* This system is used to make the long-term plan concrete by specifying which products to produce, how many, and when. The resource-requirements-planning feature of this system is vital to ensure feasibility.

(5) *Subsystem for material requirements planning.* Based on the basic production plan, this subsystem calculates what products (semi-finished, parts, raw materials) are needed, at what time and in what amounts. It then calculates by what date how many of the parts are required, and plans when to issue orders for them.

(6) *Subsystem for production-scheduling.* Based on the material requirements calculated in the previous subsystem, this subsystem schedules the in-house production to meet those requirements.

The scheduling is done according to the following planning hierarchy:

(a) *Capacity requirements planning.* Work centres are loaded for the planning horizon without capacity constraints.

(b) *Load-levelling.* Loads are smoothed to meet the capacity constraints for a part of the planning horizon.

(c) *Detailed scheduling.* Detailed scheduling of imminent jobs is done for a several-day period for each machine, to optimally utilize its capability.

(7) *Subsystem for shop control.* From the production schedule worked out in the previous subsystem, the subsystem issues shop orders for jobs to be processed within the next few days, and collects status and performance data. This also provides information for job-shop control.

(8) *Subsystem for purchasing orders.* Based on the resource requirements planned by the subsystem for resource requirements planning, items to be purchased from outside are assigned to vendors, and purchase orders are issued. This subsystem also provides and collects data concerning purchasing operations.

(9) *Subsystem for inventory control.* This subsystem calculates inventory levels and keeps track of how much of which item is stored and where, whenever items are received or shipped in response to requests from other systems.

(10) *Subsystem for cost control.* This subsystem calculates the costs of items. It also collects actual costs, and analyses the differences between the two.

17.3.2 Special Features Imbedded in the System

Several features were required to be implemented when NEC developed an application package for the MRP system. They are as follows.

(1) *Blending with conventional Japanese methods.* Traditionally, in the "make-to-order" environment, each part or material is often treated as being part of a specific order. In other words, each part is assigned an order number which indicates to which customer it belongs. NEC has adopted a "bucketless" system (on database requirements) to facilitate this conventional method within the MRP system. This lets both the MRP and conventional methods coexist, and allows the latter to be easily converted into the former. The bucketless system also allows customers' requests for detailed planning to be met.

(2) *Step-by-step implementation of the system.* In introducing the system, there are certain conditions that have to be met, such as preparation of the basic database.

As a result, some users have to introduce the system step-by-step. NEC's system consists of "modules" so that it can be implemented gradually.

(3) *Flexibility in ordering policy.* Besides the standard option of lot-sizing, there are other options available for two other cases. The first case is where customers want to order certain items individually, and the second is where they want to use an "order-level" policy.

(4) *Improvement in system efficiency.* By recalculating material requirements, a new set of plans corresponding to any changes introduced can develop. Since this is time-consuming, the "net change" approach to shorten the process has been adopted.

(5) *Utilization of the database.* In the production planning and control process, there are many people involved at various stages sharing the same set of data; this has to be changed frequently. In this system not only the standard data, but other data (such as order information and status data) are stored in the database, thus ensuring efficient use and integrity of the data.

(6) *On-line and real-time capabilities.* Various types of terminals have been developed to make data collection possible at the point of data generation. Also, on-line capabilities make real-time information retrieval and interactive planning systems possible.

17.4 Implementation

17.4.1 Breakdown of Implementations

The MRP-based package has been implemented by more than 100 companies (although not all the subsystems are implemented by all companies). The following table shows the breakdown of the users.

Breakdown of NEC Production-System Users

Industry	Number of companies
General machinery	20
Electric and mechanical equipment	39
Transport machinery and equipment	15
Precision machinery and medical equipment	8
Chemical industry (medicine, petrochemicals)	5
Rubber products	3
Pottery, leather and wooden products	3
Metal products	5
Textiles	1

The reason why machinery makers are the number one users is that MRP system was modelled after (and has evolved from) the assembly-type industry. Users in the process industry are however gradually increasing, and more users have advanced to the stage of daily MRP.

17.4.2 Cases of Implementation

Two cases are presented from among the users. The first is a manufacturer of industrial machinery with long lead times and many kinds of materials; the second is from the process industries, where it is supposed to be difficult to implement MRP systems.

The industrial machinery manufacturer implemented a full MRP-based system for its assembly-type operations, but its old "make-to-order" system still coexists with the new. It wanted to synchronize production schedules with purchasing schedules, shorten the time spent calculating material requirements for large products of long lead time, gain a better and faster control of progress by direct input from the shop, and promote standardization of data, all by introducing the MRP-based system.

Implementation of the system has resulted in a reduction in its planning cycle from 15 days to 3–4 days, and a reduction of work in process by 30%

(in money). The system covers 95% of the products, and has also improved the service in maintenance parts supply.

The areas of future development include more effective usage of the database, linkage of the system with the CAD system and faster order entry via on-line networks.

The company in the process industry is not an assembly-type operation and not a large company. It has long been engaged in developing an MRP-based system for its production process, which has been undergoing automation over the last ten years. In the current computer system, the factory floor, the control division and the management division are all connected by an on-line network. As a result, the top management can have access to data concerning current performance and predicted trends. Sensor-based terminals keep collecting data real-time, and production performance and progress status are calculated by the system on a daily basis.

For this system, it prepared a B/M specifically for its tree- or network-structured production processes. Machine-loading is done by machine, with a finite capacity using ESD (earliest start date), and scheduling takes into consideration the network structure of the production process.

With the MRP-based system the company switched from its made-to-order operation to the lot-sizing approach of MRP. Monthly load-planning by work centre has enabled the sales department to accept or turn down orders from customers. For orders accepted, delivery dates are given, resulting in a greatly improved utilization of machine capacities.

Since feedback of current status is available, schedules can be adjusted dynamically. When an order is imposed on the current schedule, the system regenerates a new schedule, giving the current status of the other orders.

Employees having access to this integrated computer system, work with much more accurate information and greater confidence, which is reflected in their performances in the data collected by the system. Inventories of raw materials and semi-finished products are reduced.

The current plan to make the system more useful in the future includes extensions to a much higher level of factory automation, distribution of the processing to lessen the burden on the host computer, an integration of the database of the system with the other databases, and eventual development of a computer-integrated management system, embracing other business functions.

17.4.3 Approach to Implementation—Emphasis on the Training and Education of Users

There are two aspects to this: one is in designing MRP-based systems; and the other is in running them. In designing systems, the participation and education of end-users was first requested. Points to emphasize in the process are:

(a) MRP-based systems are not automatic. They are designed to help people in decision-making.
(b) They emphasize planning, particularly at upper levels. They should not be made into a black box by being filled with algorithms and techniques.
(c) Formal systems should replace informal systems.

In running systems, the following points are emphasized:

(1) The production plan developed from the subsystem of basic production planning is the basis for everything else.
(2) The subsystem of basic production planning is run repeatedly until a feasible plan is developed.
(3) Lead times and lot sizes are to be constantly checked and modified, if necessary.
(4) Correct execution of stocking/shipping transactions and correct reporting of unplanned stocking/shipping transactions are critical.
(5) Education in MRP must be constantly continued.
(6) Top management has to be committed, since the introduction of MRP-based systems requires everyone to discard previously-used "informal" systems.
(7) Ultimate responsibility must be taken by people, not the computer.

17.5 Future Directions for the System

Parallel to efforts in developing production planning and control systems, there are three new types of automation effort currently being developed: office automation (OA), engineering automation (EA) and factory automation (FA). These are all computer systems, and closely connected to production planning and control systems.

Their relationships are examined first from the functional, and then from the data processing, viewpoint to assess the future implications for production planning and control systems.

17.5.1 Functions

Linkage with logistics systems

In the system NEC has been concerned with the management of shop operations. Once the system achieves its initial goal, one of the next directions might be to expand the system to include the distribution of final products. This logistics system consists of factories, factory and, distribution warehouses, end-users and the transportation vehicles connecting these nodes.

In addition to the need to optimize physical distribution operations, what is happening in the market must be reflected in what has to be done at the factory. Thus the need to expand the computer system to include this logistics information system.

Linkage with CAD/CAM system

Computer-aided design and computer-aided manufacturing have made remarkable advances in many industries. Many companies are using CAD systems in their design work and some use CAD/CAM systems for process planning.

In the near future many companies will have CAE (computer-aided engineering) systems in which CAD systems will be supported by many analytical tools and simulation features to make the whole system more creative. The specification data from the design phase will be transferred directly and used to carry out the process planning. Automatically generated NC programs will then be sent to the manufacturing line.

Thus developments in the CAD/CAM area will have a direct impact on the production and production planning and control systems. The conventional system uses various standard data concerning product structures and manufacturing processes which are generated manually outside the system. But with CAD/CAM developments and their eventual linkage with the production system, MRP will be directly accessing data generated in the product design and process-planning phases.

Figure 17.3 shows the linkage of a CAD/CAM system to production planning and control. In this framework, not only will the direct flow of data from design and process-planning phases to the database be possible, but also direct control of the product structure and the fabrication sequence by the CAD/CAM system.

Linkage with automated manufacturing processes

Developments with economic microcomputers have enabled the automation of many production processes. Robots, CNC and DNC technologies are utilized by many factories, and the flexible manufacturing system (FMS) has been making inroads into many factories.

Much of this expensive hardware has to be well-planned and controlled for it to be economically justified. This indicates a linkage (one way or another) to the production planning and control system at the level of detailed scheduling. Figure 17.4 shows a schematic view of this linkage.

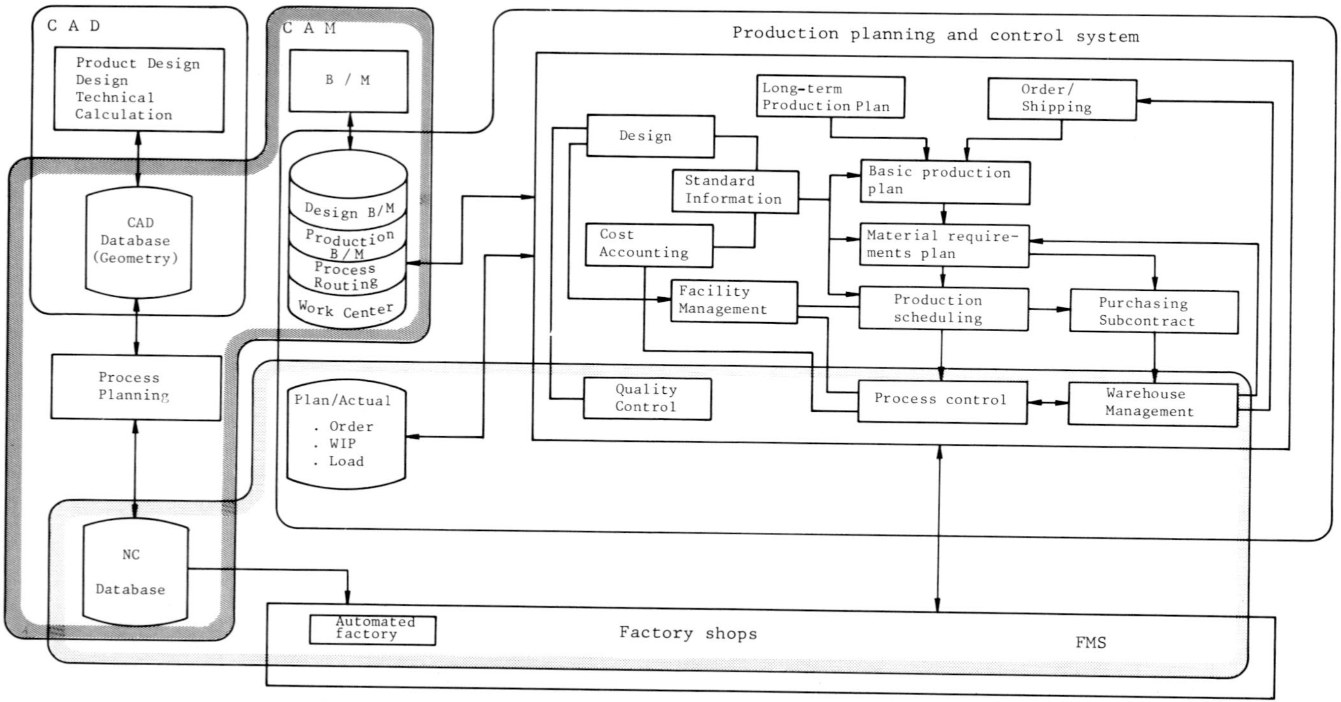

Fig. 17.3 Integration of CAD/CAM and production planning and control.

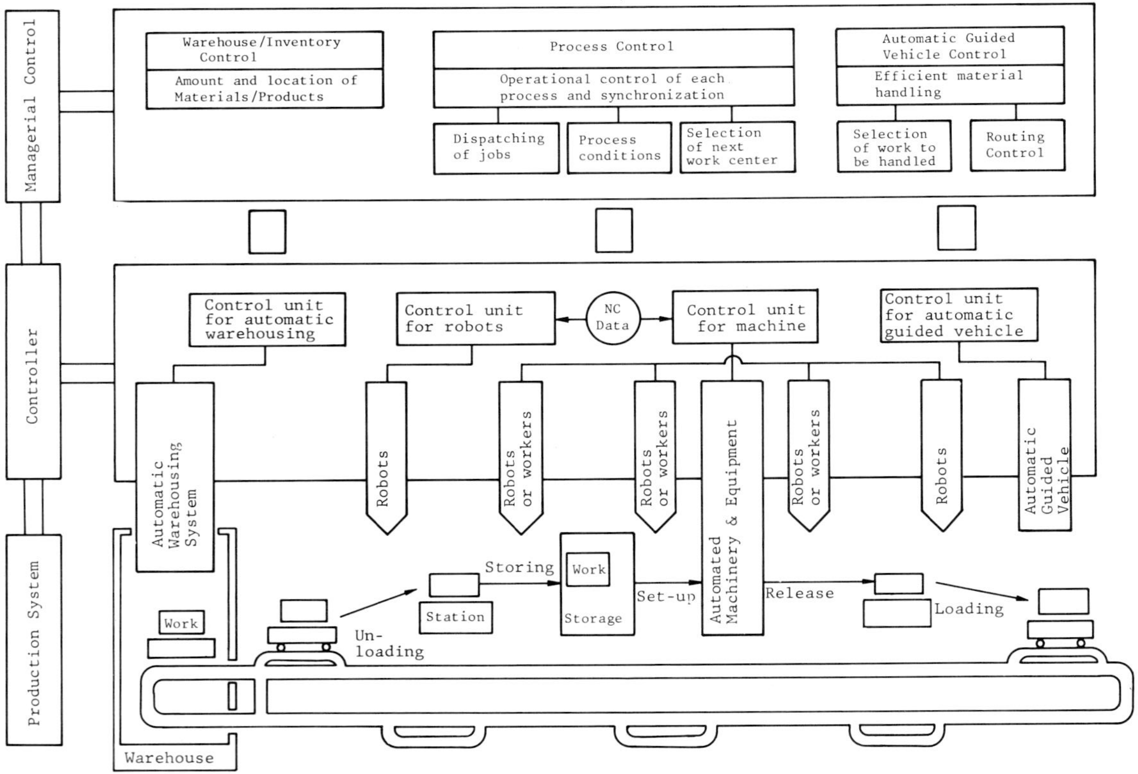

Fig. 17.4 Linkage of production planning and control factory automation.

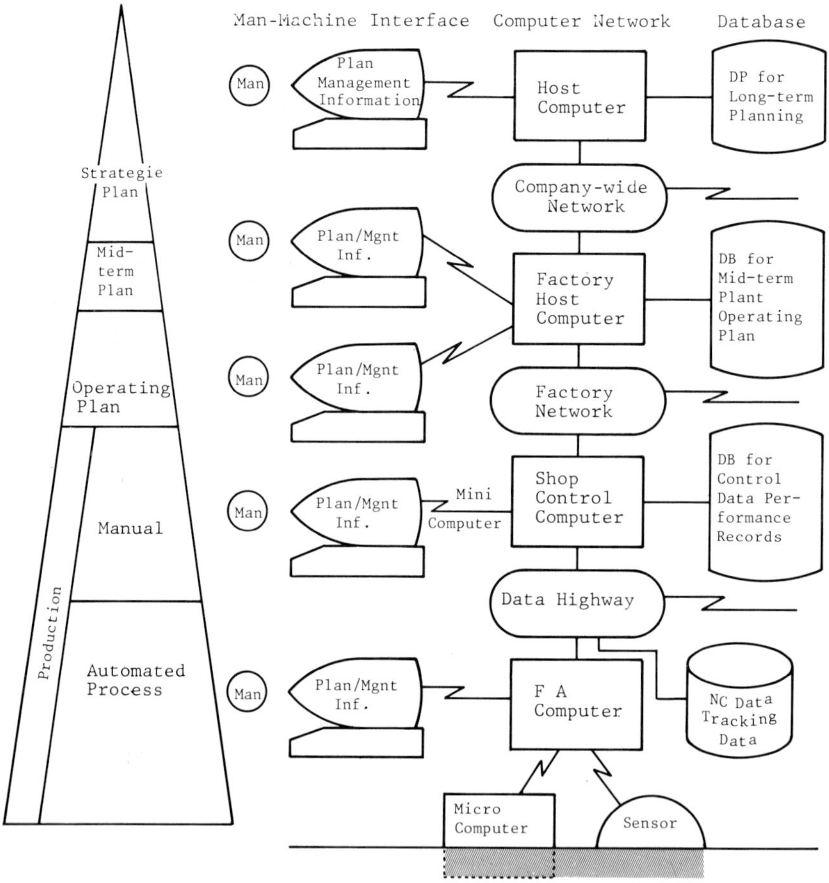

Fig. 17.5 Hierarchy of production system.

17.5.2 EDP System

Progress in database technology

With increasing improvements in cost-performance the databases may be implemented on smaller computers, which will encourage greater use of databases in the distributed processing environment. Coupled with this, advances in the relational database will soon enable personnel unskilled in the EDP system to create and manage their own databases on the factory floor.

Progress in man–machine interface.

Progress in image-processing, voice-processing and graphic technology, in addition to various easy-to-use languages, will make the interface more user-friendly.

Hierarchical structure and distributed processing of the production and control system. The advances mentioned above will have significant implications for the system. Figure 17.5 shows the hierarchy of the planning process against that of future computer systems in the production planning and control area. Strategic planning will be handled by a host computer at corporate headquarters, and the mid-term and short-term production planning will be done by a host computer at the factory. There will be smaller computers in charge of daily production operations, with various automation equipment reporting to it.

The number of levels will of course depend on such factors as the size of the factory operations, and the distribution of processing by function, geographical location and processing load.

18

Total Production Management System

Toshiba Corporation, Tokyo, Japan

18.1 Aims and Background of the System

The Fuchu works is responsible for the industrial and power generation products of Toshiba's heavy electrical division. Increases in the volume of production, and a high proportion of new products, together with requirements for maximum efficiency in resource and energy use, have constantly presented problems for mid-term, as well as current, planning.

FUTURES (FUchu Total User-oriented system with REquest Service) was established as a means of solving the main question in the field of production management, i.e. "What is the most efficient method of developing production activity?" The project has two primary aims:

(1) Establishment of a flexible production management system able to cope with a doubling of current production volume in plants producing a large variety of products in small numbers.
(2) Revisions and improvements in indirectly related activities as a means of radically improving management practices.

Sixteen representatives of middle management (section heads, managers, chiefs) from all sections of the works were assigned to this project, and FUTURES resulted from vigorous discussion between a number of working groups (30 groups, 1000 personnel).

CAD/CAM and MIS in Japan
ISBN 0-12-682580-7

The basic design philosophy of FUTURES is a production management system of method selection via interactive communication between operator and the computer.

Project development was begun in 1973, main system operations began in 1976, on-line system operation in 1977, and the initial monthly target production system commenced in 1979 using the management database.

As a result, the ratio of overhead personnel was reduced from 31% in 1975 to 26% in 1981, and a monthly production of 17 000 000 000 yen was achieved with a 2% increase in personnel.

18.2 Outline of the System

18.2.1 Environment of FUTURES

The Fuchu works is organized into divisions for each product, each division being organized vertically. The information system, however, consists of a horizontal system for design testing, and FUTURES for production management. Operations head offices of the management technical division are linked on-line to the plant to enable transmission of manufacturing specifications.

18.2.2 Characteristics of FUTURES

FUTURES is an on-line interative production management system employing a large-scale computer, its main features being:

(1) All management data for the Fuchu works is collected, merged centrally and stored in a database (FDB: Fuchu Database). The information is divided into three types: basic data, status information and management information.

(2) On-line terminals are placed at a number of locations within the works to permit interactive communication with the computer.

The information used in the Fuchu works is of considerable variety and subject to frequent alterations. In order for it to be of use for management purposes, it must be accurate. The on-line system and FDB are therefore used to ensure that data reflect changes accurately, and that common information is from the same source.

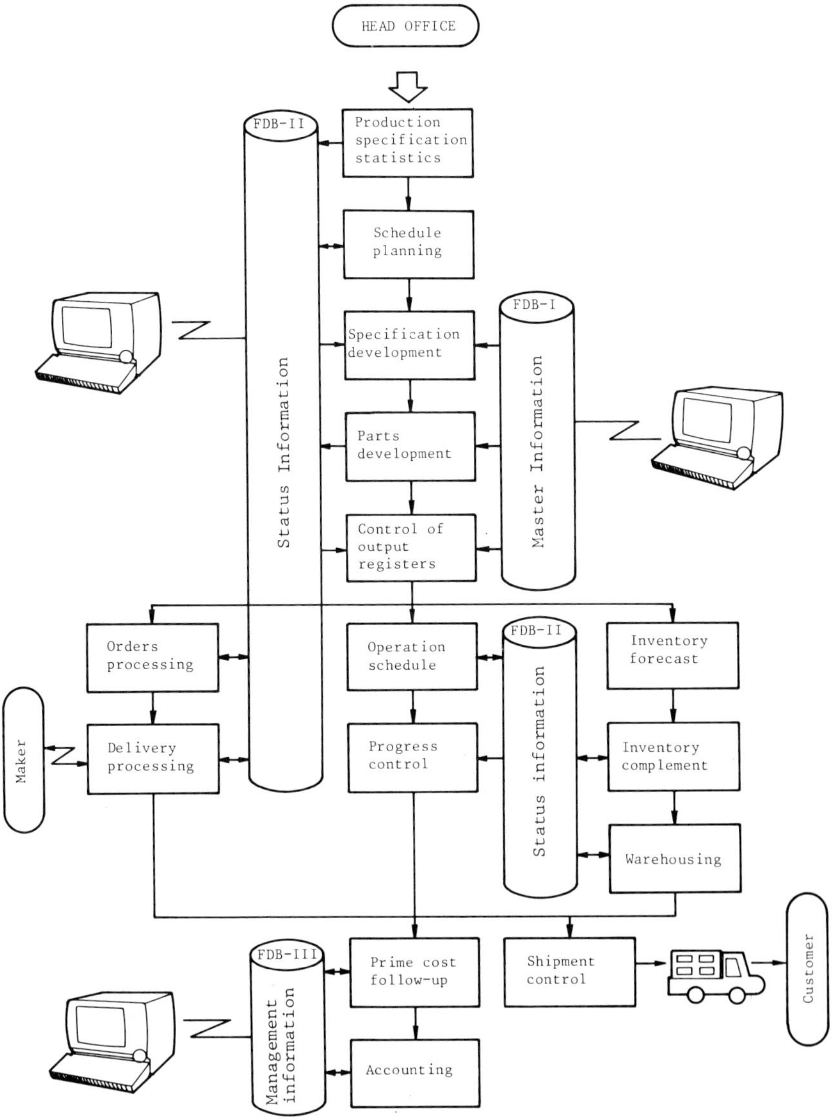

Fig. 18.1 Daily batch process flow.

18.2.3 FUTURES Configuration

FUTURES is composed of four main systems (management, daily schedule, design, material) and sixteen subsystems. These systems mutually interact whilst at the same time maintaining their independence (Fig. 18.1).

(1) The management system collects all information related to costs at the Fuchu works, and utilizes it at the planning execution and completion stages as an aid to works management.
(2) The daily schedule system develops production-related information primarily for daily scheduling. It develops schedules daily in accordance with production specifications, design, materials supply and shipping requirements.
(3) The design system collects all information relating to what is to be produced, and where, and interfaces design and production.
(4) The material system is used to maintain a balance between materials already in stock or procured from outside suppliers, and delivery times and availability of capital.

Information processing is divided into on-line processing, daily batch processing and other batch processing.

18.2.4 FUTURES Scale

This is shown in the table below.

Computer (TOSBAC-5600)	CPUs, 768 kilo-word (3072) kilo-byte)
FDB data items	20 000 000
Total transactions	5 000 000 per month
Terminals (both design and production management)	300
Distributed processing computers	Nine
Total monthly operating of terminals	20 000 hours
Use of information retrieval program	23 000 times per month
On-line menus	135
Programs	3600

Computer configuration is shown in Fig. 18.2

18.3 Characteristics of the System

The following describes three representative features of the system.

18.3.1 Output of Drawings

Products manufactured at the Fuchu works cover a wide range, from compo-

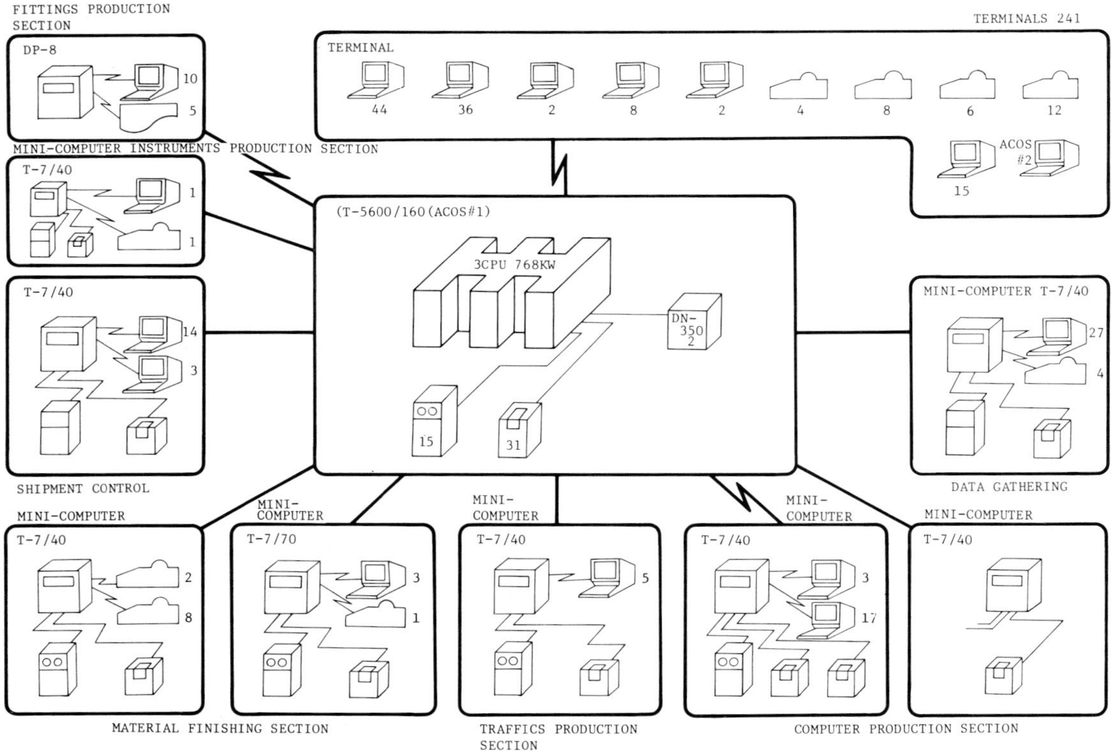

Fig. 18.2 Computer configuration (for production management).

nents for system control equipment to the equipment itself. Previous design involved separate preparation for the production of each model. This involved input to a manufacturing order list (MOL) table to simplify all aspects, from preparation of drawings to the supply of parts and assembly. A switch board list (SBL) is output for equipment, and parts tables are produced for fittings. SBL functions both as a table for collection of parts (pre-assembled electrical relays and meters etc.) and as a means to instruct the division responsible for production (fittings production section, external procurement section, material finishing section).

The parts tables function as lists of parts used in manufacture and are used by the fittings plant and materials finishing section.

18.3.1.1 Specification development

Standardized products may be developed to customer specifications. This involves the selection of existing drawings and production of an MOL. This system permits a reduction in the labour requirements involved in product design. It also reduces the input and specification stages for models produced in large numbers.

18.3.1.2 Preparation and Supply of Drawings

Formalized rating conversion
Formalized ratings are converted to drawing numbers to simplify selection of supplies and parts.

Revisions
When standard parts are revised or discarded, the DB information is checked and automatically updated to prevent mistakes and product defects.

Non-standard rejects
The use of standard parts permits improvements in design efficiency and productivity, and prevents the use of parts not conforming to plant specified standards.

Use of the FDB
The development of design, production, manufacturing and cost information permits the automatic issue of accurate production instructions.

Batch supply
Batch supply is used wherever possible as a means of restricting increase in stock with increases in production.

Control of output
Over-production in response to supply alterations is prevented.

Distribution of supply instruction slips
Slips are distributed to the required sections in accordance with the production plant instructions developed from the FDB.

18.3.2 Production Management System

Products manufactured at the Fuchu works are system products. Thus, the number of production processes differs with each model, the number of processes involved is large, and a large number of components are linked together. The order situation, the progressing of supplies and components, and job status, from inquiries to completion and shipping, is recorded in the database (FDB II). This then allows one to search from all angles, and produce a system for production management (Fig. 18.2).

Daily schedule planning
As specifications become more clearly defined through the process of inquiry, ordering, pre-design and design, large, medium and small daily schedules are developed.

Rapid response to inquiry and ordering
Sections subject to supervised production processes are given management priority, and on-line processing is used to permit an understanding of commitments, and consequently provide both management and the customer with a delivery date.

Visible production scheduling
Scheduled dates for the start and completion of approximately thirty tasks (specifications, development drawings, manufacturing drawings etc.) are set automatically for each ordered product. Information on the start and completion of these processes is input on-line to provide progress information.

Collection and management of supplies and parts using on-line operation
Approximately 3 million parts per month are used, and on-line operation is employed to provide information on progress and to search to monitor the situation for each ordered product. The system also provides information on assembly instructions and abnormal values etc.

The primary production and supply searches are as follows:
Production progress (all processes between ordering and completion), design

processes (progress for all design), drawing output (checks on design drawing output), collection of supplies (proportion supplied from stock), progress for supplies and parts, external procurement (situation of orders despatched to suppliers).

18.3.3 Presentation of Management Information

Status information changing daily is summarized and the management information to date is presented to support management activity.

Budget simulation
Trial calculations of such management details as capital and manufacturing profit and loss are used to determine budgets at the production planning stage.

Pre-cost management
Costs may be determined beforehand, and management informed of estimated totals for standard parts at the design stage.

Management of cost/price progress
All costs incurred up to the previous day are totalled in individual cost/price units, and profit and loss information is available on-line.

Monthly analysis
Data on design, daily schedules, capital cost and price, stock levels and customer management are output to plotters and presented at terminals. This information is used for works management meetings.

Various types of management information are stored in the management databases. This information is retrieved and is an important aid to works management under changing conditions.

18.4 Features of FUTURES Database

The use of the FDB permits efficient processing of large amounts of information, as well as the ability to cope with frequent system changes. Development involved the analysis of the production management characteristics of the works and the development of a database suited to the system used.

FDB configuration
The frequency of use of the information held in each system, similarities in

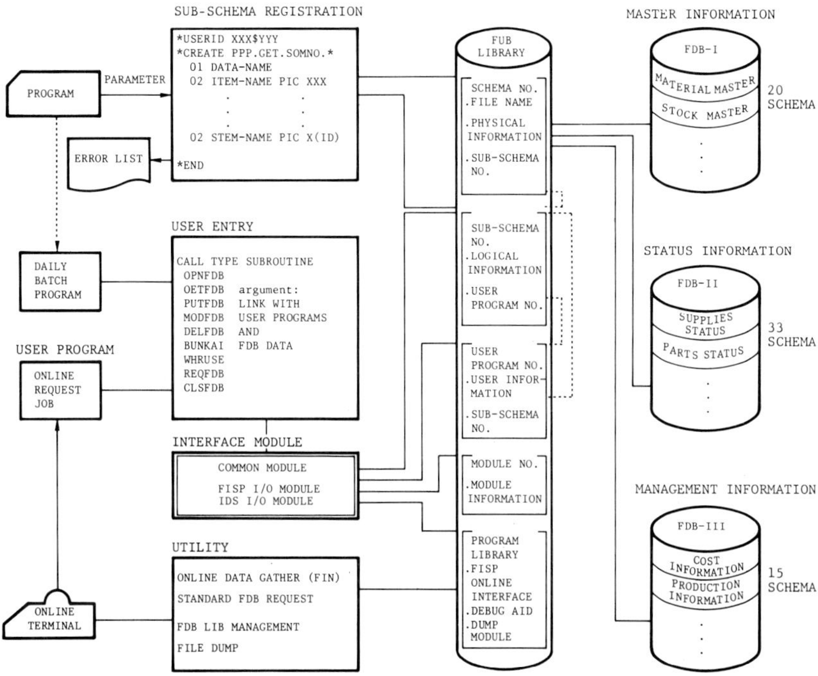

Fig. 18.3 FDB structure.

characteristics and search effectiveness were adjusted for 53 schemas. In order to provide more diversified search facilities, supplementary schemas are also provided (Fig. 18.3).

FDB interface development

An I/O module able to search efficiently and to alter large amounts of information, as well as perform random search operations, was developed. Search and update is performed only via a common interface which may be used for a wide variety of purposes.

On-line and batch processing

Both on-line and batch users may use the FDB simultaneously without interrupting on-line search, even during FDB updating.

In addition to the above, many utilities are available (e.g. complete schema update, simplified on-line search and automatic recovery, setting of update debug files) both for program development and improved efficiency of operation.

The functions above provide rapid response to system changes, and 21 schemas have been added to the FDB, and 12 schemas changed, each being completed in a few hours. The disk use rate is as high as 80%, with an average search time of 8 milliseconds.

18.5 Benefits of FUTURES

The benefit of FUTURES is shown by the fact that production has increased 2.5 times without any increase in the number of personnel not directly involved in production. This increase has been achieved in the face of considerable technological change, and the accompanying diversification of the production system. The primary contributing factors to this have been:

(a) a reduction in the number of supply documents (manufacturing instructions, order forms, etc.);
(b) an understanding of the on-line progress of 3 million parts;
(c) elimination of handwritten documents and management material;
(d) rapid action resulting from the availability of on-line management information.

Production	2.5
General personnel not directly involved in production	0.9
Inventory turnover	1.5
Average process time	0.9
Number of documents	0.9

FUTURES is composed of a range of subsystems developed over a considerable period of time through co-operation between users and system engineers. The main factors in the success of this system are as follows:

(a) a user-oriented development system;
(b) the use of the accumulated knowledge and experience of all personnel in the Fuchu works;
(c) PR and familiarization activity, from the planning stage to actual introduction;
(d) increase in equipment in accordance with the degree of familiarization.

The following points should be noted from the technical point of view.

(1) Entry of large amounts of information into databases and the maintenance of on-line response speed.
(2) Simplification to control input of large amounts of data.

(3) Introduction of partially distributed processing.
(4) Inclusion of an evaluation system for self-diagnosis.

The rationalization of works management, and the increase in the speed of information processing brought about by the introduction of this system, have permitted a monthly production of 17 000 000 000 yen in the midst of the changing demand structure of the 1980s.

19

A Production and Sales Management System on Three-phase Induction Motor Production

Toshio Doshida, Manager, Small Rotating Machine Department, Mie Works, and Kazunobu Maruyama, Manager, Development Department, Heavy Apparatus, Standard Products Division, Toshiba Coporation, Mie, Japan

19.1 Introduction

Toshiba Corporation was awarded the 1974 Ishikawa prize for the development of a production and sales management system in the field of three-phase induction motor production.

During the late 1960s the Japanese economy grew rapidly. The annual production quantity of three-phase induction motors increased tenfold over 15 years. The growth of the "non-standard" motors was especially remarkable. These are designed and manufactured to customers' orders. Non-standard motors are sometimes called non-stock motors or custom-made motors.

The annual production of "non-standard" motors has exceeded that of "standard" motors since 1970. Standard motors are offered in standard ratings with standard operating characteristics for general purpose use. They are manufactured using a mass production system, stocked and shipped off the shelf as orders are received. Standard motors are called stock motors.

In order to improve the productivity of non-standard motors, a production and sales management system was introduced. Its aims included:

(a) standardization of anticipated variations to non-stock motors;
(b) a parts-oriented production system;
(c) a computer-aided production system from order receipt to shipping.

CAD/CAM and MIS in Japan
ISBN 0-12-682580-7

Since receiving the 1974 Ishikawa prize, Toshiba has developed and improved the system further as described below.

(1) On-line telecomputer system for standard motor production.
(2) A production system similar to the Toyota Kanban production system (see Section 19.3.1).
(3) A computer-aided shop management system.
(4) Advanced computer-aided design.
(5) Flexible automated production system.

19.2 The Production and Sales Management System

19.2.1 Three-phase Motor Market in Japan

Figure 19.1 shows the annual production of three-phase induction motors (not larger than 15 kW) in Japan over the period 1958–1982.

The reasons for the remarkable growth in non-standard motors are:

(1) Equipment manufacturers who purchase motors in large quantity require purpose-built motors which are more suitable for their particular applications.
(2) In the USA, the NEMA (National Electrical Manufacturers Association) standard for motors covers special purpose, as well as general purpose, motors. The JIS (Japanese Industrial Standards) and JEM (Japanese Electrical Manufacturers) standards, on the other hand, do not apply to special motors.

19.2.2 The Production and Sales Management System

The growth of the market for non-standard motors caused an increase in the variety of products.

There are about 150 standard motors and roughly 4500 non-standard motors. On a production basis there are approximately half a million models, if one includes modifications such as different voltages, special shaft extensions, different classes of insulation, and so on.

Non-standard motors are designed and manufactured for particular applications to meet customers' requirements, while the standard motors are manufactured in a large quantity using a mass production system.

The productivity, not only in terms of manufacturing efficiency, but also the efficiency of indirect jobs such as the preparation of quotations, order processing, design, and production control, of the non-standard motors is

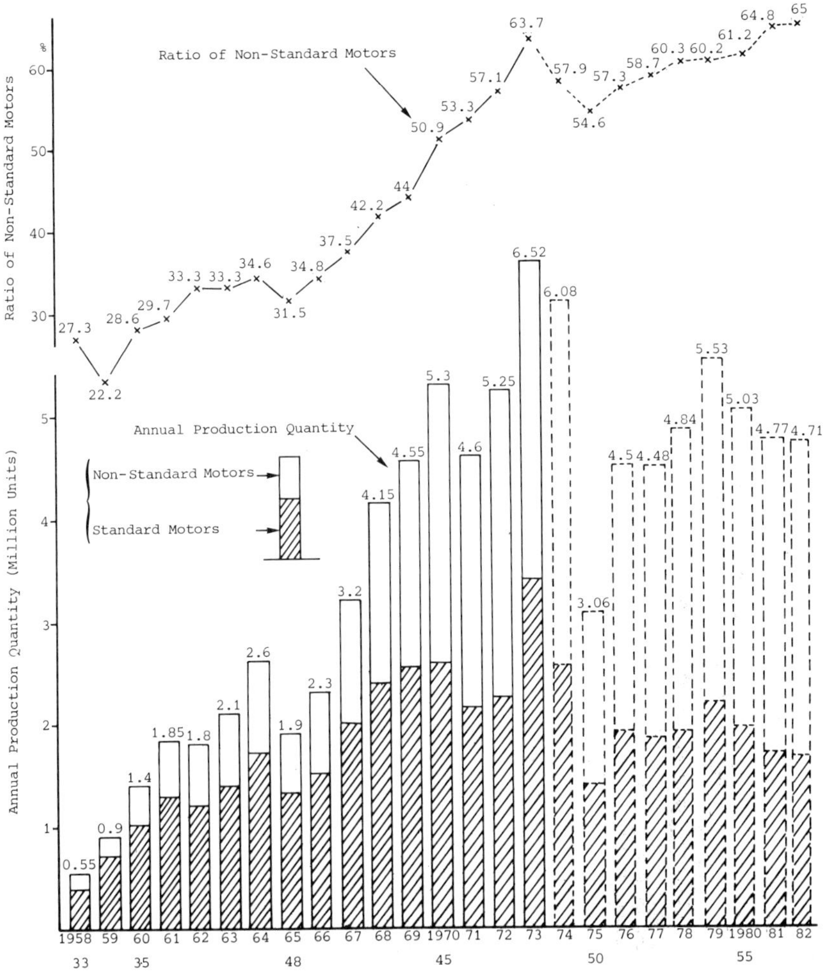

Fig. 19.1 Annual production of standard and non-standard motors.

relatively low compared with that for standard motors. It is therefore necessary to improve the productivity of non-standard motors and so offer a quick service to customers. It was under these circumstances that the production and sales management system for three-phase induction motor production was developed in 1974.

The system shown in Fig. 19.4 is called the computer-aided production and sales management system (CAPS). The system consists of the following three parts:

(1) Order system with comprehensive catalogue.

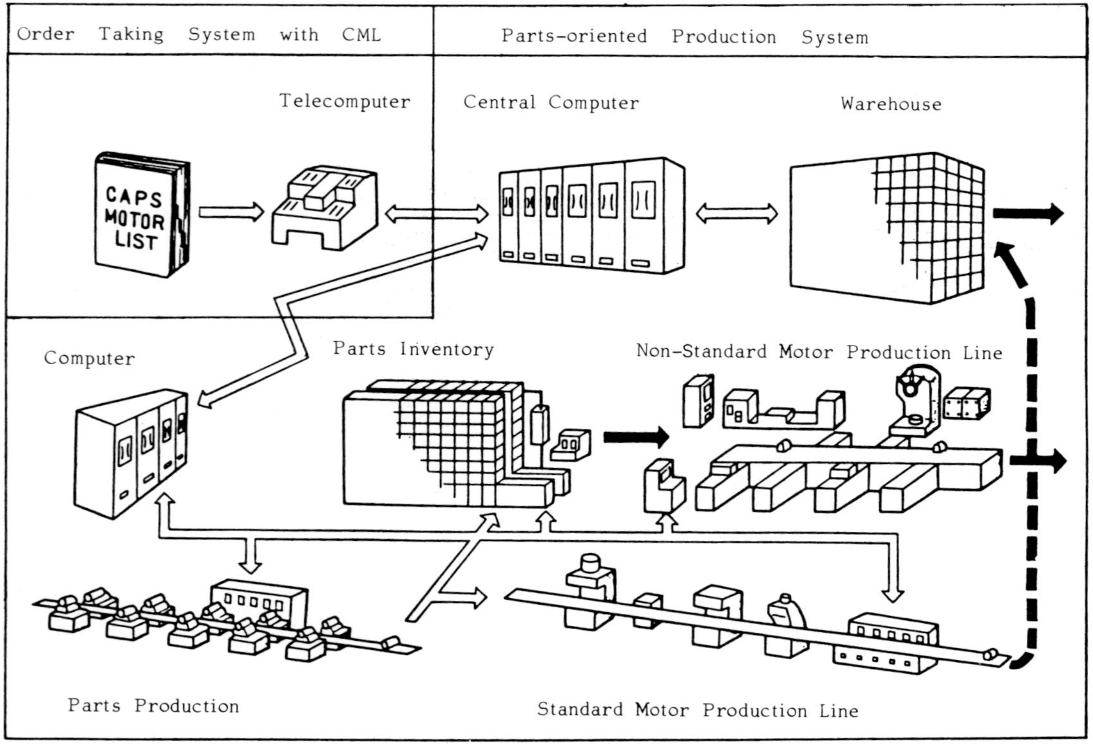

Fig. 19.2 Production and sales management system.

(2) Parts-oriented production system.
(3) Computer-aided production control system.

The order receipt system

In order to improve the low productivity of the non-standard motors and offer better service to customers, Toshiba has developed an order receipt system with comprehensive catalogue facilities (CAPS motor list).

Conventional catalogues are usually lists of standardized motors classified by either type or construction, so that customers have little choice in selecting the correct motors for their special requirements.

After analysing past requirements and specifications for non-standard motors, Toshiba published the CAPS motor list (CML). By utilizing CML, it is possible to offer a better service in selecting suitable motors to meet customers' requirements, and to quote price and delivery times without the time lag necessitated by special engineering.

CML covers about 5000 models and half a million modifications, and consists of the following:

(a) Flow charts for selection. There are check lists to aid selection of the proper motors.
(b) Standardized modifications. Every anticipated modification is predesigned, standardized and coded.
(c) List of models. Five thousand standardized models are listed and coded.
(d) List of modifications. Half a million modifications are included with codes.
(e) List of ratings.
(f) Dimensional drawings.
(g) Price list.
(h) List of delivery times.

In order to avoid inflexibility due to standardization, continuous updating is undertaken to meet new customer needs.

A variation check list is filled in for every inquiry for special motors not already held in the CAPS list. A CAPS committee, composed of salesmen, application engineers and design engineers, review the variation check lists, analyse market trends, and discuss the planning of new products.

The CML is published and copies are distributed to Toshiba salesmen and to Toshiba distributors each year.

Parts-oriented production system

Non-standard motors are custom made on receipt of order, together with any electrical and/or mechanical requirements. The productivity of non-standard motors is therefore lower than standard motors because of the necessity to produce models in smaller quantities.

In order to improve the productivity of non-standard motors, Toshiba has adopted a parts-oriented production system. Parts are standardized after analysing past order for non-standard motors. Special motors are predesigned by using standardized parts as far as possible.

In order to minimize material set up and handling time, the layout of machines for manufacturing parts is arranged according to the kind of parts, frame size, and production quantity.

The main standardized parts are manufactured in a large quantity monthly, similar parts being produced using the group technology method.

The sequence of machine loading is scheduled to minimize set up time.

Parts manufactured in this way are used partly for manufacturing standard motors, the balance being transferred to the parts inventory operated for the manufacture of non-standard motors.

The stock of each part is determined by considering its anticipated usage, its hold-up (raw material, semifinished, finished, or sub-assembled parts), standardization level (standardized or special parts), and the lead time for its production. The difference between anticipated and actual usage of each part is fed back to revise the stock quantity, and stock level. This is the function of the computer-aided production control system which is described more fully below.

On receiving each order for non-standard motors (parts-oriented production system), a bill of materials is prepared using the computer. This integrates the machining of stocked semi-finished parts used in different orders with a certain amount of batch processing. On the assembly and testing line, a production sequence of orders is arranged by frame size, output kW, pole, voltage and mounting.

Thus, the set up time at each production stage is reduced, productivity of non-standard motors is improved and production cost is accordingly reduced.

Computer-aided production control system

A computer-aided production control system has been established so that the order receipt system, the CML and the parts-oriented production system are combined.

All models, ratings, modifications, dimensions, prices and deliveries are coded, input and transferred via the private telecomputing line.

Business transactions (including order processing, non-stock design, production scheduling and production preparation) are all processed by the computer. Immediate and accurate information processing is thereby assured, and indirect labour saved.

(a) Fully automated business transaction: special motor designs and bills of materials are completed by the computer. Firm delivery information is output by the computer after stocked parts are assigned.
(b) Accuracy of information: machining and assembly instructions are produced using the group technology method, and stocked parts are automatically replenished on a basis of actual usage.

19.3 Further Developments of the System

Since Toshiba was awarded the Ishikawa prize in 1974, the situation has changed in the following ways:

(1) The Japanese economy has decelerated following the oil crisis.
(2) The growth in total annual production of three-phase induction motors has decreased accordingly (Fig. 19.1).
(3) The ratio of the non-standard motors to the standard motors has increased.
(4) More improvement in productivity of non-standard motors has been necessary.
(5) Information technology has developed rapidly over the past ten years.

The production and sales management system for three-phase induction motor production has been developed further to meet the above changes.

Figure 19.3 shows the new production and sales management system, which has the following features.

(a) Distributors, sales offices and the factory are connected through an on-line telecomputer network. Through this network variations in actual sales can be relayed to production with minimum time delay.
(b) A rational production system similar to the Toyota Kanban Production system has been introduced, and has eliminated waste in manufacturing.
(c) A computer-aided shop management system has also been introduced. Production scheduling, progress control and shop management can be controlled through terminals which are installed at the foremen's desks and connected to the central computer.
(d) Interactive computer-aided design (I-CAD) have been applied to special designs.
(e) Computerized numerical control (CNC) machines, machine centres (MC) and robots are connected to, and controlled by, the central computer.

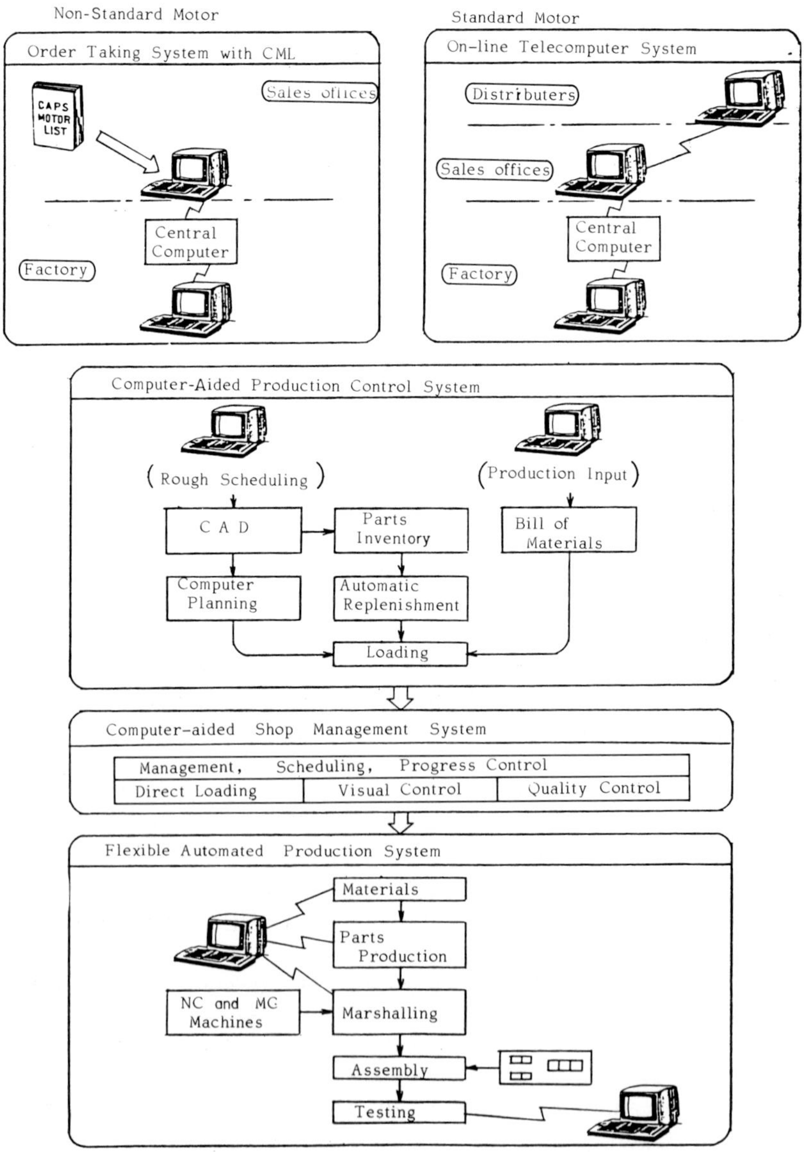

Fig. 19.3 New production and sales management system.

19.3.1 Rational Production System (UFO production system)

According to a work study by industrial engineers, only 30 to 40% of total working hours is spent in manufacturing, 20 to 30% is used in auxiliary services, the balance being idle time.

To improve productivity, it is necessary to eliminate this wasted time. There are several kinds of waste, including the manufacture of too many parts or products, overstocking, the manufacture of defective products, idle time, and so on.

In order to eliminate waste, workers must be encouraged to follow standard procedure, to regulate the pace of work, and to have areas clearly defined.

A rational production system has been introduced to pursue manufacture on thorough elimination of this waste. The system, originally called the "Kanban system", Toshiba has renamed the "UFO production system" (UFO stands for a User-Oriented Flexible production system). It has three main features.

Kanban production system

Under a conventional production system, mass production is pursued to obtain higher productivity. Parts are produced as scheduled, even though machines for the next process may be inoperative. Consequently, the parts produced are overstocked.

There are many processes in manufacturing that cannot operate as scheduled because of (for example) changes in market demand, machine troubles, shortage of raw materials, and so on. The rescheduling frequently required results in chaos and overstocking.

With the Kanban production system each process picks up the necessary parts, in the correct quantity at the right time from the preceding process, which then manufactures the picked up parts, in the required quantity.

Since the assembly line is the final process, production instructions are given only to the assembly line, the required parts in quantity, and at the necessary time being supplied from the previous process.

The system can also reduce management time, since the Kanban system is used to relay instructions on the necessary requirements to the front-end processes.

Stop system

To improve productivity by eliminating waste, visual control is used. Every worker on the assembly line has a "stop" switch in front of him. When he does not have a necessary part, finds a defective part, or when he cannot finish his job within his working area, he must stop the line by pushing the switch. Every machine is also equipped with an automatic stop mechanism, so when a wrong part is loaded, a part is installed in the wrong position for machining, or when something wrong happens, the machine is automatically stopped.

Since it is obvious that production falls when a line or a machine is stopped, the purpose of the automatic stop system is to call the attention of the managers to the line or the machine and solve the problem immediately.

Well-balanced production

Generally speaking, production fluctuates because of changes in the market, machine troubles and shortage of materials. The more production fluctuates, the more waste occurs.

Production at every stage should therefore be helped by elimination of waste and prevention of overstocking between processes, and less handling of material.

Monthly production of standard motors is divided into three lots manufactured every ten days. Thus sales variations which are monitored through the on-line system can easily be adjusted three times a month.

19.3.2 The New Computer-aided Production Control System

As previously mentioned, Toshiba introduced the order receipt system with the CAPS motor list to improve the low productivity of non-standard motors. The system has since been improved, and Fig. 19.4 shows the new computer-aided production control system. The system is connected to computer-aided design (CAD), numerical control (NC) machines and direct NC (DNC) machines. All orders are processed and classified using the computer. Outputting bills of materials, loading and production control are done automatically. New designs, which are approximately 5% of all orders, are handled by modifying data held in the computer.

Special designs are done using the interactive computer-aided design (I-CAD) system shown in Fig. 19.5. Standard drawings, together with drawing standards are held in memory. New drawings are generated by inputting dimensions through the host computer and parameters through the terminals, or by combining several conventional drawings. Data on dimensions are transmitted directly for input to DNC machines.

In addition to CAPS for non-standard motors, Toshiba has established an on-line telecomputer system. Distributors, sales offices and the factory are connected through the telecomputer. Distributors input their sales activities daily, so the production of standard motors can be planned on distributors' actual sales.

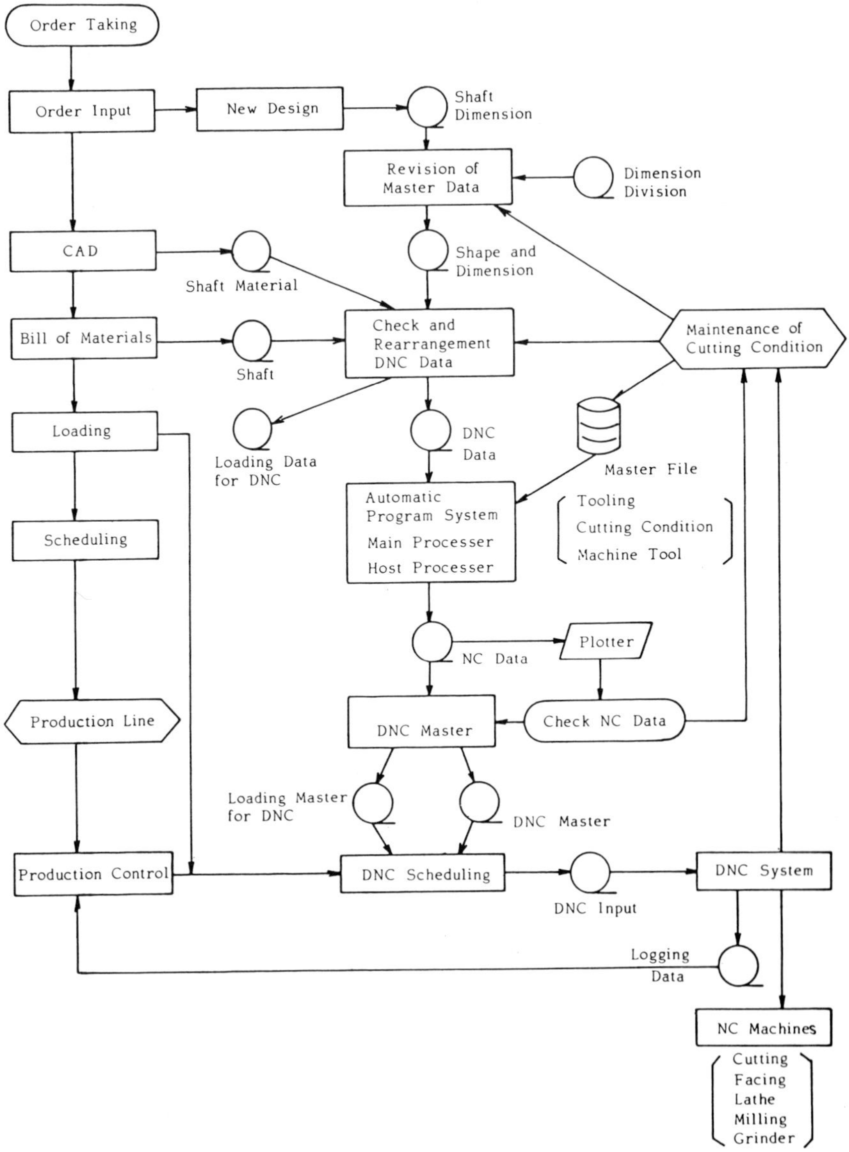

Fig. 19.4 Computer-aided production control system.

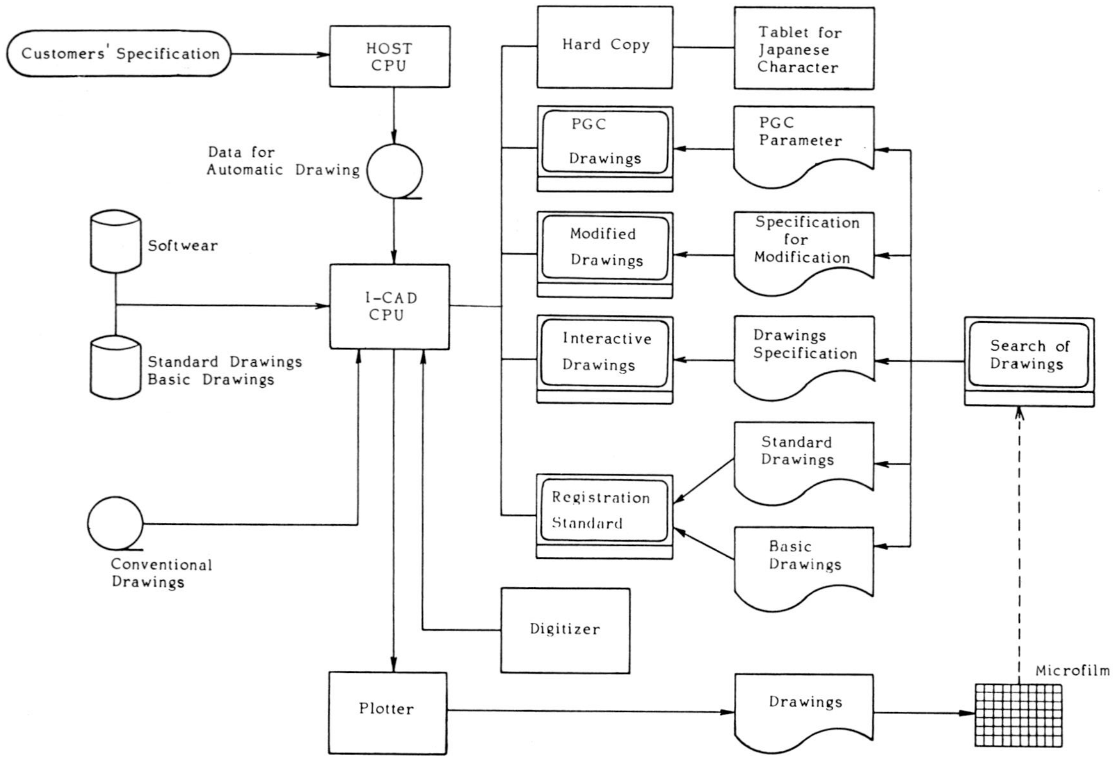

Fig. 19.5 Interactive computer-aided design system.
PGC = Parametric Geometrical Construction

The new order receipt system for non-standard motors is integrated with the on-line telecomputer system for standard motors to form the new computer-aided production control system.

19.3.3 Computer-aided Shop Management System

Toshiba has introduced a computerized shop management system. Every foreman has a terminal which is connected to the central mini-computer. This system leads to flexible manufacturing by eliminating information delays.

The system has the following seven functions:

(1) Job instruction and progress control.
(2) Production planning.
(3) Material inventory control.
(4) Shop management.
(5) Quality control.
(6) Generation of machining instructions.
(7) Despatching.

The last three of these warrant some amplification.

Computer-aided testing has been introduced for quality control. The measured values of all test items are compared with the control range and displayed graphically. They are also analysed statistically, and the necessary materials to replace faulty components are ordered.

In order to improve the productivity of non-standard motors, a flexible automated production system (FAPS) has been introduced. The system consists of numerically controlled (NC) machines, machining centres (MC), direct NC (DNC) machines, computerized NC (CNC) machines, robots and FAPSCON, the Toshiba system for factory automation. This system is connected to the computer-aided shop management system, so that information for machining is fed to FAPSCON together with the job instructions.

After completion of an order, invoices are ouput automatically, and the despatch information is transferred to the production and sales management system.

20

Flexible Manufacturing System (FMS)

Takeshi Nojiri, Executive Director, and Yoshitsugu Odagiri, General Manager, Mikawa-Mizuho Plants, and Toshiaki Ishihara, Manager, Production Engineering Department, Mikawa-Mizuho Plants, Minolta Camera Co Ltd, Osaka, Japan

20.1 Introduction

The production of quality opto-electronic business machines such as plain paper copiers (PPC) for the general office equipment market worldwide encompasses two areas of major engineering challenge.

PPC themselves are technically complex, uniting in one product the most recent developments in the three diverse technologies of precision machinery, electronics and opto-electronics. Acceleration of the rate of development in all three fields is reflected in a highly competitive market, where models frequently change to incorporate the newest advances. This puts pressure on the manufacturing line not only for sophistication and flexibility in basic processing, but for stringent and reliable quality controls that can identify, isolate and correct defects as the assembly process goes on, no matter how often technological advance in product design necessitates changes in the components being assembled.

Because the market is highly competitive and global, speed of response to actual demand is crucial. In a sense, PPCs are becoming consumer items and therefore their manufacturing processes are vulnerable to the whims of a capricious market, as well as to the constant competitive struggle to introduce

CAD/CAM and MIS in Japan
ISBN 0-12-682580-7

functions/innovations that will make each manufacturer's PPC more attractive than the competitors'. This difficulty is further compounded by the fact that any basic-model PPC produced for the world market may need to meet more than fifteen sets of variations in manufacturing specifications, to match local market conditions in the various regions where they will be sold. OEM manufacture can raise the total of per-model variants to as many as twenty. Using Minolta's product family of PPCs as an example, the three mainstream model lines now in production, plus two more due to go into full-scale production this year, mean that there could theoretically be in any one month demand for a total of seventy-five machine variations to be assembled on the same lines, some of which will be ordered in lots of less than 100 each. When it is considered that Mizuho plant's total PPC production is close to 20 000 units per month, the magnitude of the twin problems of product changeover, and the need for maximum speed in answering market demand with production of the right quantity of the model variant in demand, can perhaps be appreciated.

20.2 The Approach

Minolta Camera Co's Business Equipment Operation have developed an advanced technology for PPCs called the Micro-Toning system. Minolta's Micro Toning copy quality, plus the large number of new functional capabilities constantly being introduced into Minolta machines, has rewarded the company with a fine product reputation among customers and a significant share of the global market. In the middle 1970s the company, judging the general market for PPC copiers would grow rapidly for at least the next decade, decided to invest in sufficient manufacturing capability to keep Minolta abreast of any future expansion of the PPC market, anywhere in the world. It was decided to expand the capabilities of the Mikawa Plant, where all of Minolta's opto-electronic business products were then produced, by erecting an entirely new factory, the Mizuho Plant, in the immediate neighbourhood of Mikawa, near Toyohashi City. Construction of the buildings began in 1978, and at the same time Minolta engineers began to consider the basic design of an FMS (Flexible Manufacturing System) that would give the Mizuho Plant nearly total Factory Automation (FA). The resultant system went on-stream in April 1980, and was subsequently judged so impressive a technical accomplishment that it earned the 12th Ishikawa Prize in November 1981.

20.2.1 The Basic Factors

In the strict sense of the term, Minolta's Mizuho Plant is an assembly operation only. More than 95% of the 1000-plus parts that make up the average Minolta PPC are supplied by a network of more than 300 subcontractors, most of whose factories are located nearby and who make deliveries by truck. Coordination of ordering—particularly given the pressures of shortening lead-time response to the demands of the market—would obviously entail an extension of FMS far beyond the assembly lines, and even beyond the Mizuho Plant itself. Minolta would have to totally reorganize its supplier order and delivery system into an organic whole, to achieve the level of time-response needed. In other words, the company would need to engineer an FMS that would embrace the entire Minolta PPC production.

Quality control is of prime concern in the production of a complex, high-technology product like a PPC. Minolta introduced the concept and management systems of total quality control (TQC) into its PPC production operations in 1977 with great success; an overriding consideration being that incorporation into the FMS of TQC attitudes should be assured. In a sense, TQC had to be designed into the system itself, because the goal of maximum speed in production meant that quality inspection and defection could only be undertaken by the assembly line workers themselves; unit production would take place so quickly that reliance upon full-scale downstream inspection to assure quality would result in bottlenecks of defective products at the end of the line, with difficulties in attempting to trace the source of the defects back upstream.

The necessity for speedy changeover of models or model variants with minimum down-time on the assembly line was, as noted, a major concern of the FMS designers. Assurance of maximum flexibility and maximum reliability—including a carryover of all QC standards and guarantees to whatever model was coming on line—would reach far back upstream in the manufacturing process; indeed, not only to the manufacturing operations of parts suppliers, but even to Minolta's initial ordering of the parts themselves.

Total FA, of course, implies a very complex and sophisticated capability for system and data management. Minolta's FMS designers decided early on that the entire system should, in so far as possible, be regulated by a single computer mainframe system, no matter how costly the job of designing specific, and as yet unfamiliar, software. In fact, the very complexity of the task itself encouraged the designers from the start to focus on undertaking a total data-processing integration of every factory-related function that could be automated, from reception of product orders and specifications from the sales division to accounting and disbursement parts procured from suppliers.

In systematizing the above factors, speed of production and of market response was the general guideline. Every step of the manufacturing process was to be reprogrammed from its previous internal logic in so far as possible, and re-engineered as a part interlocking with those steps before and after it, and made subordinate to the logic of the whole system of PPC manufacturing at Minolta. Yet care had to be taken that the overall organization retained sufficient flexibility at the "pressure points" to avoid rigidity.

20.2.2 The System Concept

Minolta's engineers conceived their broadest FMS goals as the realization of a production system designed, just as a Minolta product is, from the standpoint of the users (in this case, assemblers and supervisors); and creation of a plant environment centred around the workers, i.e. one in which assembly-line workers would not be alienated from—or lose control over—automation of their work, but would willingly integrate themselves with it and take personal control of it at the work-station level. Data automation would control the FMS, but human beings must, as the final guarantee of QC, control the work.

In laying out their concept, the engineers decided:

(a) first priority must be given to product quality;
(b) FMS must include a system to minimize parts inventory;
(c) workers and automation must be harmoniously integrated.

These points merit some further comment.

The complexity of PPC machinery and the high levels of competition in the market enforce this priority on quality in the products. Minolta has had exceptionally successful results with its management policy of encouraging factory workers to make the most of their own capabilities, and to assume personal responsibility for high levels of performance in their work. The company had therefore come to believe that a large part of its success in maintaining product quality was due to this human factor. In a sense, Minolta's FMS also incorporates this policy. The engineers adopted a free-flow conveyor system, in which workers are able to take as long as necessary with each unit, until they are satisfied that their work is acceptable. A parts request system based on each assembly station and controlled by the workers reflects this design, as do "Q-UP Signals" and "Q-UP Cards", which will be explained later.

Japanese manufacturing industries have done pioneering work in cost-saving through reduction of the parts inventory, and Minolta engineers incorporated this consideration at every phase of manufacturing, from delivery of the parts to parts racking and assembly-site supply. Efforts were concentrated on developing control software for delivery schedules for parts and materials, organized on a scale of precision heretofore unknown in the industry; quantity and type of container, and simplification of receipt inspection and processing would also be major considerations in the parts-supply system. Efforts would also focus on extending FMS adaptability to embrace the concept of totally automated parts racks. In addition, the assembly line would be not a unified line, but one comprising several self-contained divisions, to each of which the FMS would be applied individually. Within the FMS, the work done by human beings and that done by machines would be clearly divided, and at the contact point between both the basic policy that each worker through direct commands operates that part of the computer system which affects his work would apply. The designers sought an organic harmony, placing esteem for human values on the same level as the goal of improving plant productivity.

Finally, it should be noted that as goals and concepts became integrated into system designs, Minolta's ongoing TQC activities at the Mikawa Plant provided a valuable input of ideas on how to systematize QC itself within the FMS: regular Quality Control Circle (QCC) meetings between line assembly workers and foremen produced the "QC-UP Card" and "Q-UP Signals" explained below, and offered valuable ideas for the formation of the parts-ordering system and incorporation of QC feedback to parts manufacturers themselves, which latter is now an integral part of the system.

20.2.3 The Targets

The emerging concept of the Mizuho Plant FMS soon began to point to two crucial areas where innovation would be imperative—namely, the development of software for production control through data automation, and a completely new parts handling system, capable of being totally computer controlled.

Specifically, the complete FMS would have to embrace:

(1) Modularization of product design.
(2) Development of an assembly system which allows each worker to take as much time as needed to complete his tasks to his own satisfaction.
(3) A three-dimensional factory layout and subdivision of organization, with parts, products and processes moving vertically as well as horizontally.
(4) The promotion of Minolta's "control by sight" concept, which relies on prompt communication between workers and supervisors to avoid production jams and to spot QC problems quickly.
(5) Reduction in the number of PPCs that constitute a production planning unit.

(6) Shortening of production ordering cycles, and greater facility in splitting deliveries of parts to the assembly line.
(7) Types of product packing and handling appropriate for small-lot production, and establishment of standard-quantity production lots.
(8) Reduction in parts and products inventories, including total elimination of plant-site warehouses.
(9) Linkage of data processing between the production system computer and the main office computer, for verifying deliveries of parts and processing orders and accounts.
(10) Full automation of all in-plant movement of parts and products.
(11) Full automation of parts ordering, to be integrated as closely as possible with model switchovers on the assembly lines.
(12) Quality control and delivery date control over parts supplied by subcontractors.
(13) Smoothness of man-machine interface at all levels, and pleasantness of the work environment.
(14) Adoption of an assembly conveyor system mechanism flexible enough to handle sudden changes in products being assembled.
(15) Capability of adaptation to further automation of assembly processes.

To achieve (9) especially, the production control system which resulted would have to interface its materials handling functions with data processed by the main-office computer. In fact, it was judged that an organic union of office and control systems would be crucial to improvement of small-lot production efficiency, by making possible the automation of production planning itself and through allowing control over all changes necessary to switchovers in models and handling. This conceptual interface between the two computer systems is illustrated by the flow chart of information and materials handling below:

(a) detailed production instructions based on production plans;
(b) acceptance and confirmation of data concerning ordered parts;
(c) schedule instructions for a limited number of off-line operations;
(d) basic data concerning automatic parts recording;
(e) master batch processing of parts orders to subcontractors;
(f) verification of on-schedule delivery of parts from each supplier;
(g) data on parts delivered to assembly line work stations.

The minimum lot size of products manufactured is the basic quantifier/index unit; production management of large-lot production models is then calculated through integral multiplication of these minimum lot sizes. In this way, by processing data in terms of the basic unit-lots to be produced *and* the time-series of production, all schedules within the factory are determined by computer. The system concept developed, then, is a "central nervous system" consisting of an efficient FMS designed for multi-variety, small-quantity production of high-technology products.

20.2.4 Implementation

Minolta engineers developed the business software—ordering systems, accounting, etc.—but decided to seek a constructor capable of supplying both hardware and production control system software as a package. Sanki Industrial Co Ltd was chosen as the general contractor for the installation; a team of seventeen engineers from Sanki and Minolta, aided by an independent software development enterprise under contract to Sanki, drew up the plans in the following format. Sanki and Minolta together developed the fully articulated system design; all hardware was ordered and erected by Sanki; all software for the business management side came from Minolta, while all software for system control was Sanki's responsibility.

The Mikawa Plant's installed central computer system served as the head-office data processing unit; for the FMS control, a Toshiba 7/40 CPU was selected, with several sets of I/O and processing peripherals custom-tailored to the Mizuho Plant's design. Total installation of all computer, assembly and materials-handling hardware at the plant site took only three and a half months; operational start-up followed forty-five days after completion of installation.

20.3 The System

20.3.1 Assembly Line

Manufacturing operations take up four floors of the Mizuho Plant: at ground and second-floor levels are receiving bays for deliveries of components from subcontractors; automated rack and conveyor storage and parts handling machinery; and a final crating and shipping bay for packaged final products. There are two independent assembly lines on each of the two assembly floors above, with a total of 200 assembly work stations. Each of the four lines is 250 m long, and each is linked to the automated parts-supply system. The assembly lines are designed to allow adjustment to minor variations in the time taken for each assembly stage by each worker; adjustments for major variations in the time required for each step of assembly when widely different models/variations are being assembled simultaneously; and to minimize downtime in model changeovers (some lines may change products assembled twice a day) due to mechanical failure or maintenance.

As noted above, Minolta set finished-product quality as the end goal of the FMS; it is therefore appropriate to examine here the method of incorporating TQC activities into the new assembly line system.

Each worker is given sufficient time to complete his stage of the assembly task properly; he does not pass on a product until he is satisfied it meets all quality criteria he is competent to judge.

A push-button system called "Q-UP Signal" allows any worker to flash a red light on a panel over his sector of the assembly line. The light is a signal that he has either encountered an insoluble defect problem in a product, or that the amount of time consumed in resolving one or more quality problems on products arriving at his station is causing him to fall too far behind the line's overall pace. In either case, a supervisor responds to the light at once, summoning whatever help is needed.

A "Q-UP Card" system has been effected, in which a small QC "chart" is kept for each individual worker. On this chart the rate of defects encountered by the individual assembly worker is plotted monthly in the form of a graph, so that both the workers and supervisors can tell at a glance at which station quality problems are most frequently arising. These cards are displayed on a bulletin board next to the line for all to see, and if the QC problems are more in the nature of a worker's unfamiliarity with his task than with parts or machine failures, this becomes quickly evident to both the worker and the supervisor. It is the supervisor's responsibility to monitor each worker's chart, consult him on the types of problems he is encountering, and make certain that he receives training in any aspects of his job with which he is unfamiliar. A worker prides himself on the public display of a personal graph that has a downward slope, indicating a decrease in the number of QC problems generated at his station.

The QC Circle activities common to almost all major Japanese manufacturers, including frequent after-work meetings to discuss and analyse recurrent problems and an open worker-suggestion system, are continued within the context of the FMS.

20.3.2 Materials Handling System

From the outset materials handling was recognized as the single most important area, after the assembly line itself, where bottlenecks could defeat the overall FMS aims of flexibility and speed. Parts-processing itself had to be unified and Minolta's Mizuho Plant FMS therefore embraces an entire parts-supply "sub-universe" of its own: a closed-circle system of order, supply, racking, lining, installation and re-order which is integrated into the production plan in all phases.

(A) The process, though essentially cyclic, will be described from the furthest-upstream point; that of order to, and delivery from, subcontractors. With a few exceptions for parts which themselves have a supplier hierarchy and so require more lead-time, all orders are confirmed with subcontractors two months in advance of delivery, through issuance of a voucher ("denpyo") processed and printed by the Mikawa Plant head-office computer in accordance with the confirmed production plan. (Japanese subcontractors have a blanket supply agreement with their customer which covers a broad spectrum of parts; formal contracting and other documentation are therefore held to a minimum.) The supplier makes deliveries to the Mizuho Plant by truck, as often as twice a day. The voucher for each order of parts is delivered by hand with each shipment to a Minolta inspector at the reception bay. After visual confirmation of the number of units received against the number ordered, the inspector inserts a plastic card with the supplier's ID encoded into a special I/O keyboard, then taps out a four-digit code representing the voucher number. The FMS computer flashes an electronic duplication of all entries on the voucher, so the inspector can check the voucher against the computer's data record, as well as against a printout of all orders expected that day. If the information tallies, the inspector enters an "OK" signal. The computer then uses the confirmed data for two purposes: to log the parts as received, placing them into inventory for automated supply to the assembly line and then passing confirmation to the head-office computer for accounting purposes; and to identify the position in the racks where the individual shipment will be stored on "hold" for the assembly line. The inspector then has only one more responsibility: to shunt the delivery on to the proper conveyor system for transfer to the racks. At this point, the FMS system takes total control of the parts inventory.

(B) The key to the success of FMS in the materials handling system is proper reception of the parts. In matching the FMS parts-feed system to the needs of the assembly line, Minolta's engineers created three categories of parts for the purposes of containerized handling, racking and assembly-line supply (see paragraphs). The time criterion was less than one and a half days for all parts handling, from reception at the receiving bay to installation on the assembly line. In actual practice, however, many containers of parts are delivered to the assembly line just two hours after they are received. There are no warehouses, as such, at the Mizuho Plant.

(B.1) Parts designated in the rank (A) include products that are few in type, but relatively bulky and heavy. As such, they represent large units and require fast handling, and the highest level of automation. A standard, high-impact plastic handling case about twice the size of a desktop PPC was designed for all A-rank parts handling. This case exactly fits all conveyors and holding banks controlled by the computer.

(B.2) B-rank parts are small direct delivery parts, which require handling containers only half the size needed for A-rank parts. They are received and processed in exactly the same way as A-rank parts, but because their assembly-line cycles are slower per container than for A-rank parts (one B-rank container may hold up to 200 identical parts; an A-rank container as few as ten) they have a completely separate rack and conveyor-feed system of their own.

(B.3) C-rank parts are small parts with little practical value for automation investment. They are stored in electrically operated revolving shelves as off-line equipment, with location and inventory control conducted separately. Generally, these parts are received, stored and forwarded to the line in B-rank parts containers, though at a much slower rate than with either A- or B-rank parts.

Empty parts containers of both ranks are recycled back to the receiving bay, where the subcontractor's trucks load them for return to the suppliers' plants.

When the inspector clears each container of parts on the receiving bay, it enters a conveyor system under FMS computer control. The containers are conveyed via automatic elevators and conveyors to a series of racks two stories high, to await transfer to the assembly lines above. Like a switch-engine in a train marshalling yard, the FMS racks each container on the appropriate track, and then calls it out again when it is needed on the assembly line, to form a single integrated "train" of containers of parts moving to the assembly line in exactly the sequence needed—no matter what model, variant, or lot is being produced. Figure 20.1 demonstrates the relevant consumption-time cycles in terms of containers fed to the assembly line by the automated parts-handling system.

(C) Although the FMS is capable of supplying all parts in exact sequence to the assembly line itself, the concept has been designed to work in reverse, the assembly line worker being the governing entity in actually bringing parts to his work station. In practice, this means that no parts are sent from the rack to the assembly line until they are *called for*, container by container, by each worker at each assembly line station. This is effected through a "request button" control panel at each such station.

The request button panel has eight simple, numbered buttons.

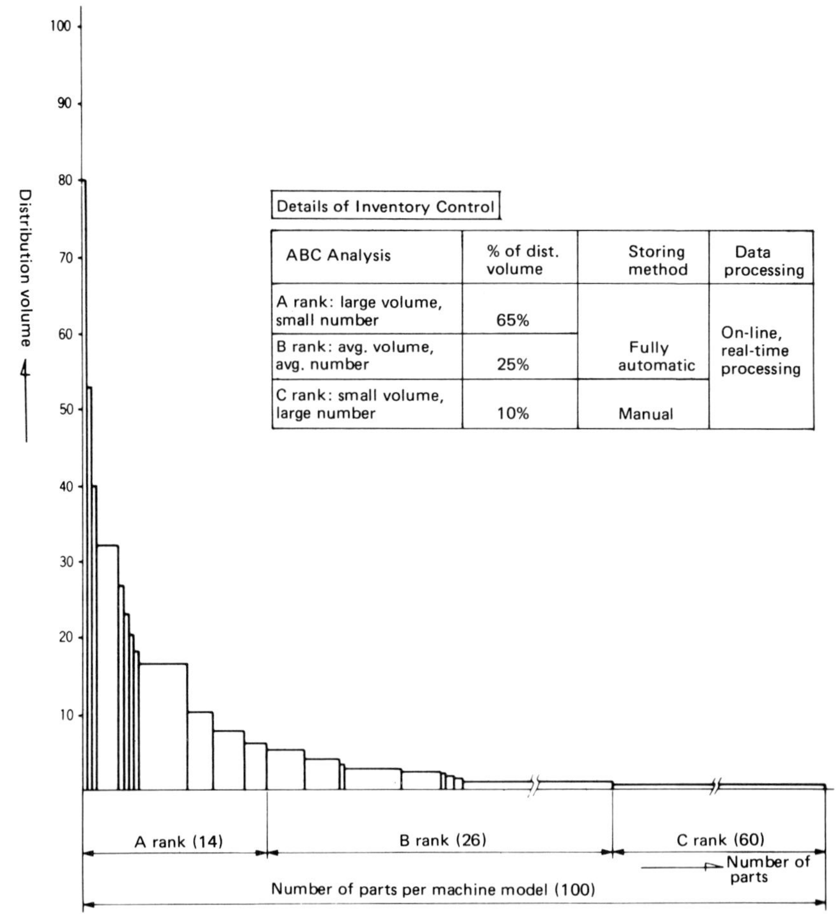

Details of Inventory Control

ABC Analysis	% of dist. volume	Storing method	Data processing
A rank: large volume, small number	65%	Fully automatic	On-line, real-time processing
B rank: avg. volume, avg. number	25%		
C rank: small volume, large number	10%	Manual	

Fig. 20.1 Distribution analysis and inventory control.

Above each panel is a card, with a colour photograph, a name and one or more part numbers for each button number. When a worker needs a new container of parts, he simply consults the card and pushes the button for what he wants. This signals the FMS computer to pull the correct container off the rack, load it on the conveyor and guide it to the station where it is needed. The container of parts arrives by elevator/conveyor within seconds, and is already sequenced against the production plan: if the production model is being changed over on the line, the FMS will deliver the correct variant of the same part for the model that is being changed. For example, instead of 110-volt, 60-cycle motors for the US model, it will deliver 100-volt, 50-

cycle motors for the same model being made for the Tokyo market, exactly as the last US model passes down the assembly line station and the first Tokyo model arrives. The photo, name and numbers are a back-up for verification and each worker is required to check them visually against the first part he removes from a container for installation, any time there is a model change.

This request system, which in effect allows each worker to operate that part of the FMS which directly affects work on his station, has two important effects:

(1) It allows each worker to take whatever time is necessary to complete correctly the task at his station, without parts backing up on the supply conveyor behind him.
(2) It gives each worker a sense of ultimate control over all processes taking place at his station. Rather than feeling he is being pressed into an essentially automated assembly system as an adjunct to the machine, the worker "summons" the machinery to his bidding, and thus maintains his personal sense of identity as master of his craft.

Figure 20.2 shows a simplified chart of the entire FMS concept as applied at the Minolta Mikawa and Mizuho Plants.

20.4 Results

FMS went into full-scale operation at the Mizuho Plant in 1980. The first model to be transferred from production at the Mikawa Plant was the popular Minolta Model EP 310 desktop PPC, of which about a quarter of a million units were produced before it was phased out for a newer model. Since then, other models have been transferred to the system. Each has enjoyed relatively long production runs, and then been phased out for new generation copying machines. There are now six basic models under production in the Mizuho Plant's FMS, and two more were scheduled to be transferred from Mikawa within 1984.

Whilst most system and software "debugging" was completed within the first year of startup of the FMS, it is Minolta's practice to begin commercial production runs for a new model on the older Mikawa Line, and then take it over to Mizuho after quality criteria are fully identified and established. Once transferred, the product is produced only at Mizuho, and Mikawa's lines are used to begin experimental and early commercial production of still newer products, which in their turn will be transferred to Mizuho.

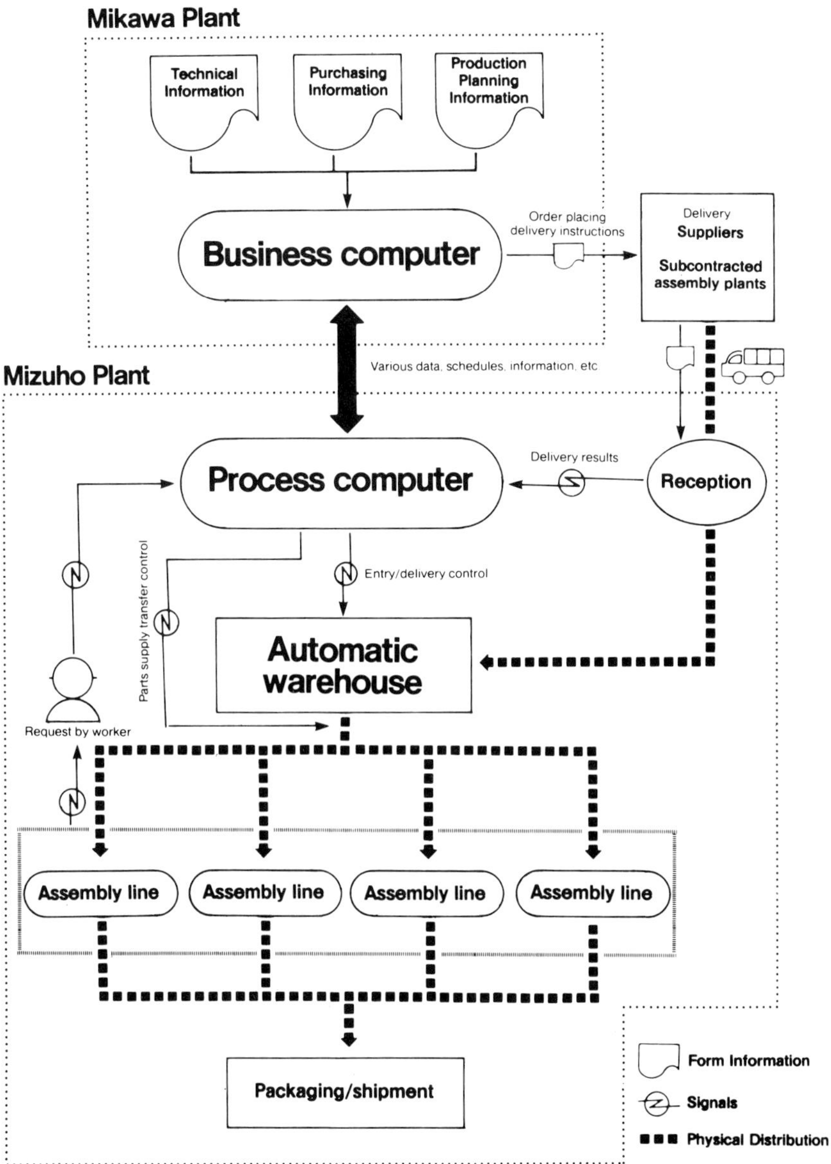

Fig. 20.2 Information and physical distribution at Minolta BE Centre Mizuho Plant.

Quantitatively, the following table summarizes the gains made in production over the old Mikawa system by the new Mizuho FMS.

Output	96% increase
Labour productivity	56% increase
Parts turnover time	93% decrease (within factory)
	21% decrease (in supplier "pipeline")
Processing defects	61% decrease

The smallest feasible production run for any single model is now 50 units; changeover from one model to another on the assembly line can now take as little as two hours.

In terms of speed of reaction to market demand, the final fixing of production schedules is now down from three months under the old system to two months under FMS.

Another important offshoot of the FMS integration of supplier operations with assembly operations has been an increased feedback of TQC information to suppliers, and a subsequent lowering in the defect rate among assemblies and parts supplied: under FMS the supplied-parts defect rate is 500 parts per million.

Difficult to quantify, but equally significant, is the overall effectiveness of the entire FMS managerial system, from order acceptance through production planning, subcontracting and management–labour relations, to finished-product shipping.

20.5 The Future

Whilst FMS was designed to accommodate future automation of the assembly processes, the system is "human oriented". Only minor off-line processes, mainly concerned with inspection and sub-assembly, have actually been automated to the point where no human operators are required. It is not foreseen at present that robotization will affect the assembly line itself in the medium-term future. There are, however, two principal areas in which further extensions of the FMS might increase speed and responsiveness still further.

20.5.1 Integration of the Ordering System

Under the present system, "forecast orders" are received at the Mikawa Plant several months before projected delivery time, and these provide the basis for preliminary production planning. FMS cannot, however, set the final production plan and place parts orders until "confirmed orders" are received, which is now about two months before finishing date. If a way could be found to grade preliminary orders reliably in terms of likelihood of final confirmation, before the two-month deadline, that deadline in turn could be moved closer to the finishing date—meaning that full flexibility to respond to market demand could be cut to as little as two weeks between final receipt of fixed orders and completion date. Discussions are now underway within Minolta to consider ways of classifying preliminary orders in terms of degree of certainty, and integration of ordering itself into the FMS.

20.5.2 Mechanization of Certain Testing and Calibration Procedures

In the Mizuho Plant, certain adjustments and calibration checks can only be conducted when the entire unit has been completed. Minolta engineers are testing various sensors and other devices that could be integrated into the process control system, for quality testing at the time the component concerned is installed in the unit—thereby allowing adjustment or replacement of a defective part at the work station itself rather than in the finished product. Such a device would, for example, examine copy resolution through an opto-electronic sensor, automatically feed data to the computer on-line for comparison with target performance standards, and signal the same assembly line worker installing and testing the component as to what adjustment is necessary.

This QC application of Mizuho Plant could, if fully realized, eliminate the need for most of the final ten workers on each assembly line, which are now primarily concerned with inspection, measurement and adjustment of finished products.

20.6 Conclusion

The Mizuho Plant FMS constitutes one of the most advanced integrated FA achievements in the field of precision machinery manufacture.

21

A Special Steel Bar Production System

Norimasa Kamii, Director, Plant Engineering Division, Engineering Business Operation Bureau; Ryoichi Mohri, General Manager, Equipment Department Muroran Works; Kohji Inazaki, Manager, Electrical and Process Automation Section, Muroran Works and Osamu Ishihara, Manager, Plant Engineering and Technology Bureau, Nippon Steel Corporation, Tokyo, Japan

21.1 Introduction

Special steel bars are used extensively in the fabrication of industrial machinery, and for important automotive parts. Despite this, because of the demand structure peculiar to special steel bars, they have to be produced in small lots in a wide variety of types, thus making it difficult to rationalize production. As a result of research and development since 1971, however, a fully computerized bar mill based on a new concept of special steel bar production successfully replaced the old facilities at Muroran Works in 1974. Improvements have since been made to both the computer system and to the equipment and, as a result, the mill now produces quality bar with the highest operating efficiency ever achieved in Japan.

The production system herein introduced is a total computer control system for the production of special steel bars realized for the first time in the world by combining mechanical automation and computerization. With the advent of the computer control system, builders of industrial machinery, and

CAD/CAM and MIS in Japan
ISBN 0-12-682580-7

the automobile industry in particular, now have access to special steel bars of improved quality and lower price.

21.2 Conventional System

Previously, it was recognized that the quality assurance of special steel bars was impracticable in the manufacturing stages without a subsequent finishing process such as inspection and conditioning because:

(a) surface defects of the starting material (i.e. billet) cannot be completely removed;
(b) reheating and rolling processes inevitably entail decarburization and surface defects;
(c) special steel bars are unavoidably bent during cooling and, therefore, require a straightening step;
(d) it is physically prohibitive to accurately track on-line every piece of bar produced.

Because of these constraints, the following have been commonly practised.

Billets are subjected to magnetic-particle inspection, and then conditioned for the removal of surface defects. Throughout this stage, they are handled by a crane and billets undergo crane handling as often as ten times. To add to the inefficiency so caused, the billet inspection facilities are usually situated in the blooming mill, not in the bar mill.

Moreover, the bar obtained after a set of processes such as reheating, rolling and shearing or sawing is then sent to the finishing stage for size inspection, defect inspection and removal, straightening, cutting, assignment to respective orders, sorting, counting and labelling—a process requiring several weeks in all. All these operations require transfer of bars by crane, the frequency of crane handling being as high as sixteen times per bar. Each of these requires laborious operations such as temporary bundling and unpacking. With the monthly production of special steel bar currently running at 1 million pieces per mill, it is not surprising that mis-identification can occur.

The main problems in conventional practice can be summarized as follows:

(1) The mill design and arrangement are based on engineering techniques, operational practices and a quality control system, all of which rely entirely upon the bar finishing department.
(2) Difficulties encountered in tracking for quality assurance a huge quan-

tity of bar products in the context of the productivity constraining factors of small lot size, and diversity in product types and sizes.

(3) A gap between the users' demand for special steel bars of varied dimensions, and the standards of automation technology.

These difficulties have made the special steel bar department far less productive and far more energy-consuming than other steel departments.

21.3 New Production System

In planning the replacement of the old bar mill by an entirely new type of special steel bar mill to cope with the increasing requirement for special steel bars for automotive and industrial machinery, the basic object was to develop a system that would solve the aforementioned three problems and enable high-efficiency production to be achieved without the finishing stage.

To achieve this, a variety of technologies were combined, and extensive R & D effort was exerted. The result of this effort is shown in Fig. 21.1 which compares the new system with the previous production system, whilst the overall process flow of the new production system is indicated in Fig. 21.2.

The main features of the new production system include:

(a) a quality control system that enables simplification of the finishing process by building in quality during the production stage;
(b) a tracking system that covers not only billets, but also as many as one million pieces of bar per month;
(c) a quality assurance system based on the on-line gathering and feed forward of quality data;
(d) an automatic operation system that offers optimum efficiency and yield.

A more precise description of these features is given hereunder.

21.3.1 Quality Control System

The first consideration for rationalizing the finishing operation is to ensure such high quality in the billet as to permit shipping direct from the line. To achieve this, the following improvements have been made to the quality control system:

(1) A billet conditioning line is installed in the bar mill so that integrated quality control may be assured from billets to finished bars. The condi-

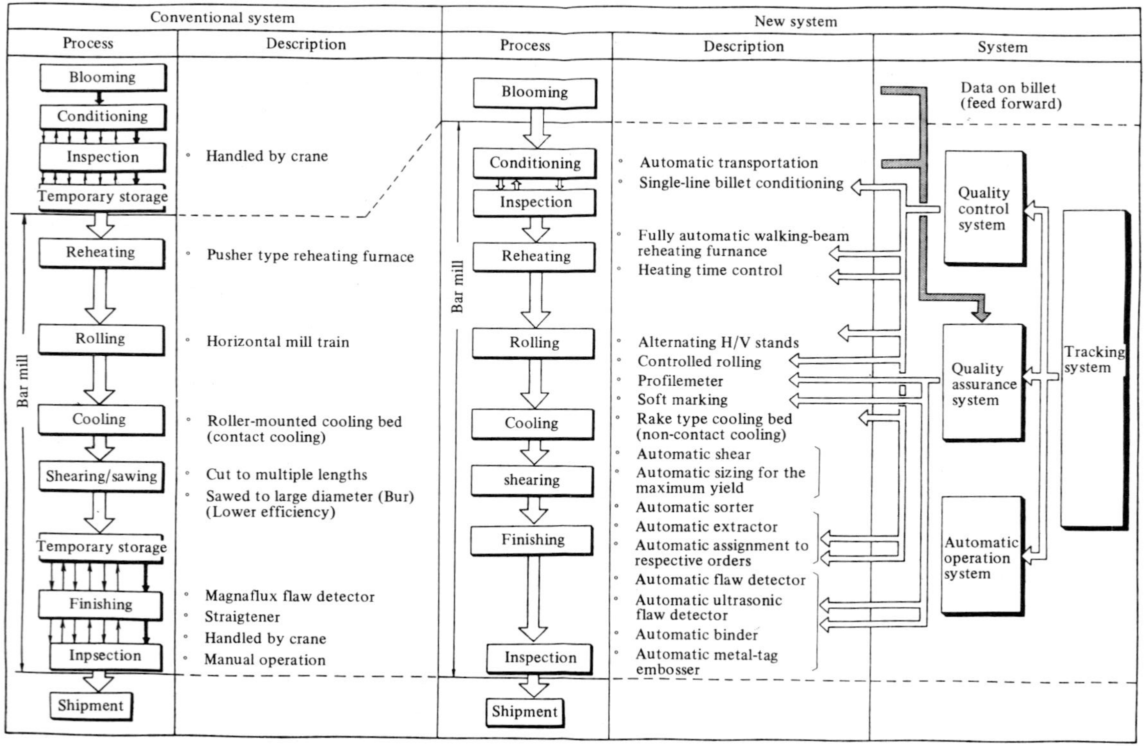

Fig. 21.1 Comparison of new and conventional special steel bar production system.

Billet Conditioning Line

Shot Blast	: 1 strand feed
Magnaflux Flaw Detector	: Dc 6000A, 40 sec/cycle
Magnaflux Flaw Detector	: Dc 3000A, Max 30 M/Min.
#1 ~ #6 Grinder	: 75kW Pressure: 700kg (Max.) Grind stone circumferential speed: 4800 M/Min.
Re-conditioning Grinder	: 37kW pressure: 400kg (Max.) Grind stone circumferential speed: 3800 M/Min.
Billet Car	: 4 wheels 120 M/Min. 13 Ton Max

PRODUCTS
Round bar 22 ~ 100mm ϕ
Deformed bar 22 ~ 51mm ϕ
Square bar 45 ~ 65mm

Capacity
Actual 35,000 Tons per Month
Future 60,000 Tons per Month

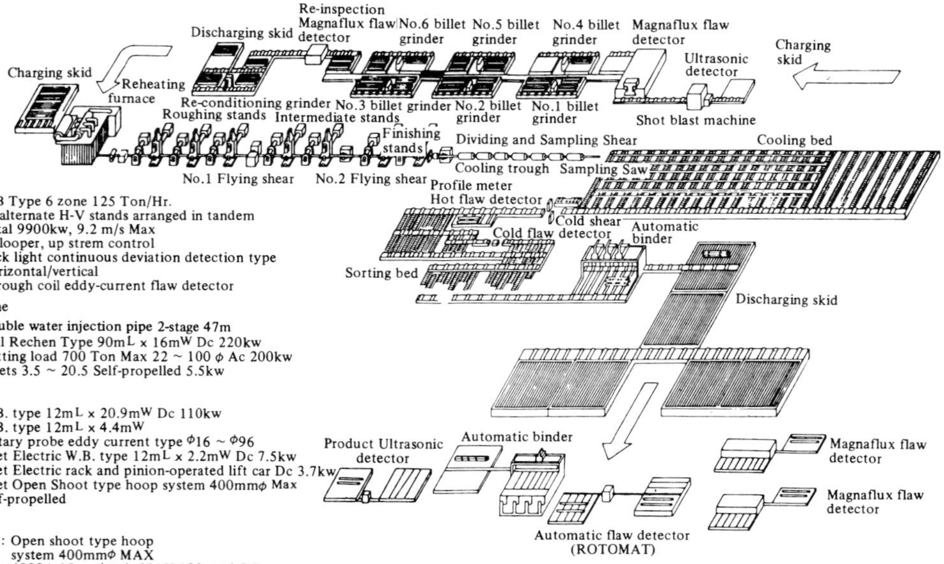

Mill Line

Re-heating Furnace	: W-B Type 6 zone 125 Ton/Hr.
Rolling Mill	: 14 alternate H-V stands arranged in tandem Total 9900kw, 9.2 m/s Max up looper, up strem control
Profile Meter	: Back light continuous deviation detection type Horizontal/vertical
Hot Flaw Detector	: Through coil eddy-current flaw detector

Cooling and Cold Shear Line

Cooling Trough	: Double water injection pipe 2-stage 47m
Cooling Bed	: Full Rechen Type 90m^L x 16m^W Dc 220kw
Cold Shear	: Cutting load 700 Ton Max 22 ~ 100 ϕ Ac 200kw
Shear Gauge	: 2 Sets 3.5 ~ 20.5 Self-propelled 5.5kw

Finishing Line

1st Inspection Bed	: W.B. type 12m^L x 20.9m^W Dc 110kw
2nd Inspection Bed	: W.B. type 12m^L x 4.4m^W
Cold Flaw Detector	: Rotary probe eddy current type Φ16 ~ Φ96
Sorting Bed	: 2 set Electric W.B. type 12m^L x 2.2m^W Dc 7.5kw
Separating Car	: 2 set Electric rack and pinion-operated lift car Dc 3.7kw
Binding Machine	: 4 set Open Shoot type hoop system 400mmϕ Max Self-propelled

Off-Line Finishing Line

Binding Machine	: Open shoot type hoop system 400mmΦ MAX
Magnaflux Flaw Detector	: 4000A 15 sec/cycle MAX 100mmΦ 8.0m
Magnaflux Flaw Detector	: 4000A 15 sec/cycle MAX 75mmΦ 6.0m
Automatic Flaw Detector	: ROTOMAT AC 6043
Product Ultrasonic Detector	: Jet water and spiral scanning type

Fig. 21.2 Layout and equipment specification.

tioning line includes a duplex surface defect tester, an ultrasonic flaw detector and seven grinders.

(2) A walking beam type reheating furnace which is capable of rapid heating and causes fewer surface defects is installed, and simultaneously decarburization is prevented by controlling the heating time.

(3) An alternating horizontal-vertical mill configuration instead of the conventional twisted arrangement; a straight passline; and tension-free control are introduced for improved dimensional accuracy and reduced occurrence of surface defects.

(4) Adoption of a notched walking beam cooling bed in place of the conventional roller-mounted cooling bed makes the straightening step unnecessary.

(5) Adoption of a grooved shear knife eliminates the need for end conditioning.

(6) Computer-aided automatic equipment and automatic control of billet and bar temperatures assure stable quality.

These improvements have virtually eliminated the necessity for such conditioning as rehearsing and straightening which had previously accounted for a large proportion of the finishing process, and have thus made it possible to ship out the product bars straight from the finishing line.

21.3.2 Tracking System

Bars were prevously cooled on a roller type cooling bed and sheared by a flat shear knife whilst in contact with each other. Sheared products were transferred on flat rollers and temporarily bundled and stacked before they were sent to the finishing yard, where they were united, sorted into orders, counted and finally bundled.

This practice of separating such closely packed bars for subsequent processes is, however, not only detrimental to high-speed and high-reliability processing, but is also impracticable because of the size-mix. Moreover, such a practice as identifying and tracking a huge quantity of bars one by one makes computer control of the overall process unrealistic.

Hence, a new concept was introduced to overcome this difficulty, which was to keep products apart from each other right through to the final process.

As seen in Fig. 21.3, the cooling bed is of a grooved walking beam type. All the roller tables after the discharging end of the cooling bed are grooved so that bars will not contact each other during transfer, and the shear knife is also grooved in the same arrangement. After shearing, a grooved table shifts

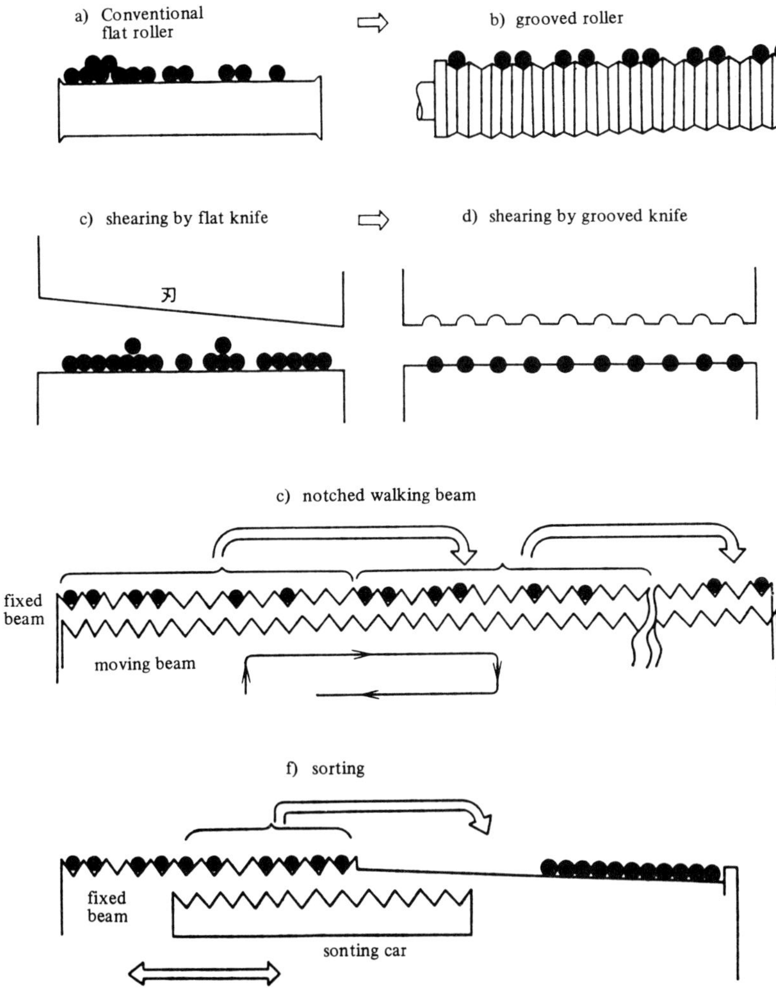

Fig. 21.3 Separate transfer system.

sixteen sheared pieces at a time, and the sorting bed is of the same construction. On the sorting bed, sheared bars are checked for arrangement, and the ordered number of bars are taken up by a raked fork. During this lifting operation the bars are counted. Separate bar lifting makes counting very reliable. After being counted, the bars come into contact with each other for the first time, the ordered numbers are automatically bundled, and a metal tag is attached to each bundle.

This is the first attempt to transfer the bar products separately by means of the grooved roller table and grooved beam throughout the process up to the sorting bed.

In addition to these improvements in mechanism, the tracking technology has also been developed.

A large number of bars are in process from the cold shearing area to the sorting beds. In tracking the sheared products in a unit, a sixteen-groove bar array (sixteen bits) is used as a source for detecting tracking errors. The line is provided at five points (i.e. the branching, rejoining and diverting points) with array detectors based on linear photo-arrays, to correlate every cycle with computer-tracked sixteen-bit array information (Fig. 21.4). In case of

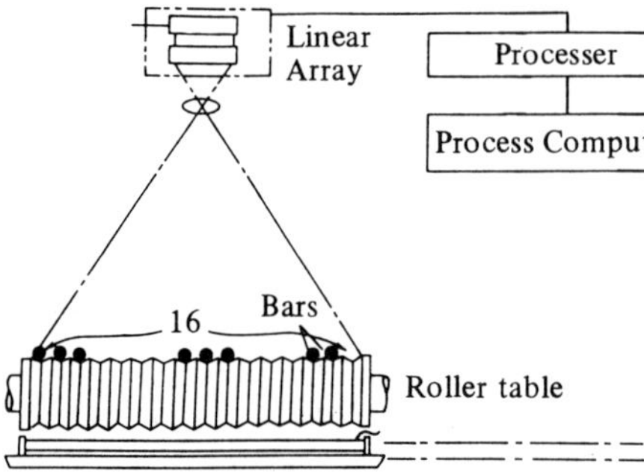

Fig. 21.4 Array sensor.

disagreement, the computer discontinues the sequence until the operator confirms the products concerned and inputs array data to the computer to resume the sequence. At the extracting station, the bars, minus those removed, are compared and checked with the pre-extracting information. Thus tracking errors can be immediately detected and the areas involved shut down to minimize damage and promote recovery through early detection.

Further, a brush switch as shown in Fig. 21.5 is used at the transfer between the walking beam and the table rollers to detect the passing bars ranging from 10 kg to 700 kg in weight.

The finishing line is automatically stopped if the computer malfunctions,

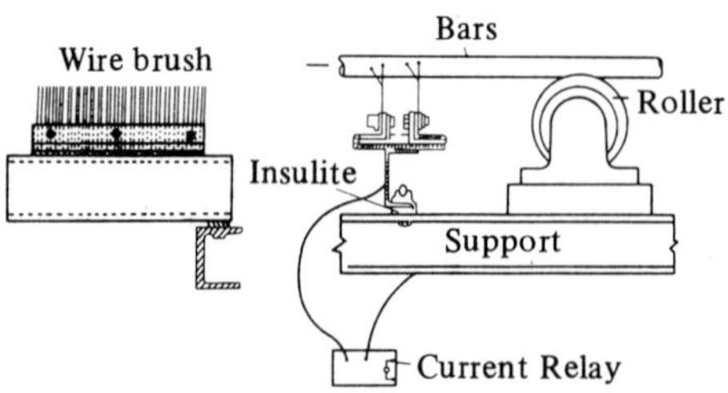

Fig. 21.5 Brush type sensor.

thereby minimizing modification to the tracking system after the computer resumes service. All tracking information in the entire complex is displayed on a CRT to help operators monitor and collate tracking. In March 1975 a high-density colour CRT with a light pen was introduced, allowing tracking modifications to be made using the light pen to indicate the spot directly on the screen.

Thus, a combination of computer technology and engineering improvements leading to rapid and trouble-free handling and tracking of a large number of bars, for the first time in the world has enabled on-line control of all processes from the reheating furnace through to the finishing line and has eliminated the necessity for temporary storage before finishing—a production step which previously required several weeks.

21.3.3 Quality Assurance System

The new special steel bar production system is incomplete without the quality assurance system. Consisting essentially of a soft marking system and a defective bar extraction system, the quality assurance is designed to track and feed forward the quality-related data on surface defects, dimensions, etc. in 10 cm increments of a bar using an eddy-current hot flaw detector and profile meter (Fig. 21.6) installed on the rolling line. The tracking data thus obtained are processed by the computer into a defect map which indicates the relative positions of particular defects. Based on this defect map, defective bars are diverted at the inspection stage from the main line to the extraction line by an automatic extractor (Fig. 21.7) and subjected to cold precision inspection, where defective bars are rejected and conforming bars are returned to their original lots. The latter feature, i.e. the extraction and return line, has made it easier to perform sampling tests to ensure the quality of each lot.

It seems very likely that the soft marking system, unlike the previous system in which marks were put directly on the bars themselves, will be used in the quality assurance of many other products.

21.3.4 Automatic Operation System

The computer-controlled automatic operation system is designed to offer optimum efficiency and yield, and has the following features as seen from the system configuration shown in Fig. 21.8.

Integration of billet conditioning facilities into a single line

Since all billet handling facilities including cars for receiving and sending bil-

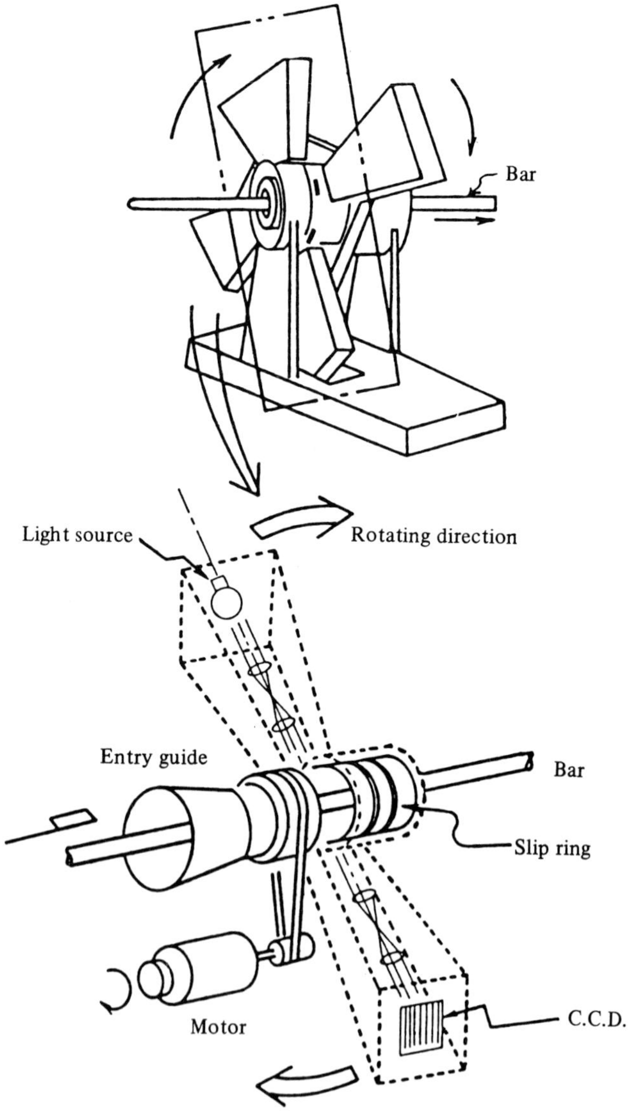

Fig. 21.6 Profile meter.

lets are computer controlled, crane handling is required only at the entry and delivery ends, and the need for temporary billet storage is eliminated.

Fully automatic operation of reheating furnace
The automatic reheating furnace control system consists essentially of a set

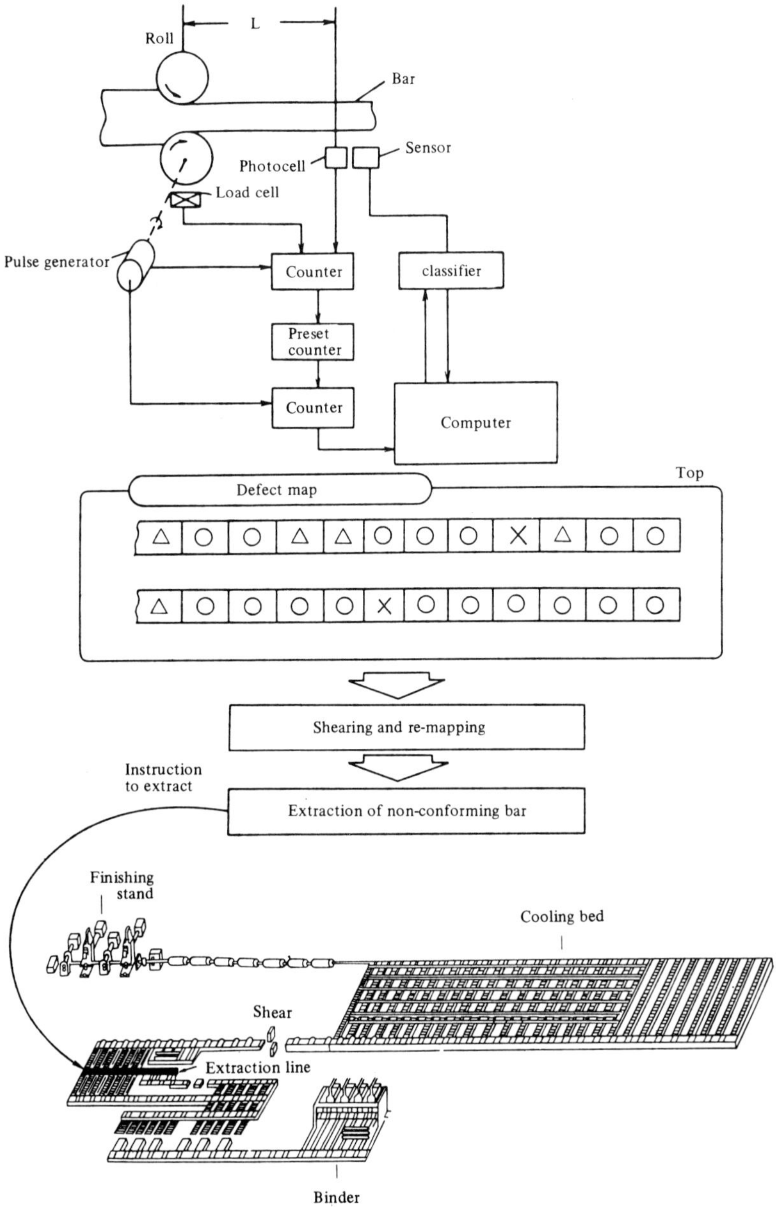

Fig. 21.7 Quality assurance system combining soft making system and non-conforming bar extraction system.

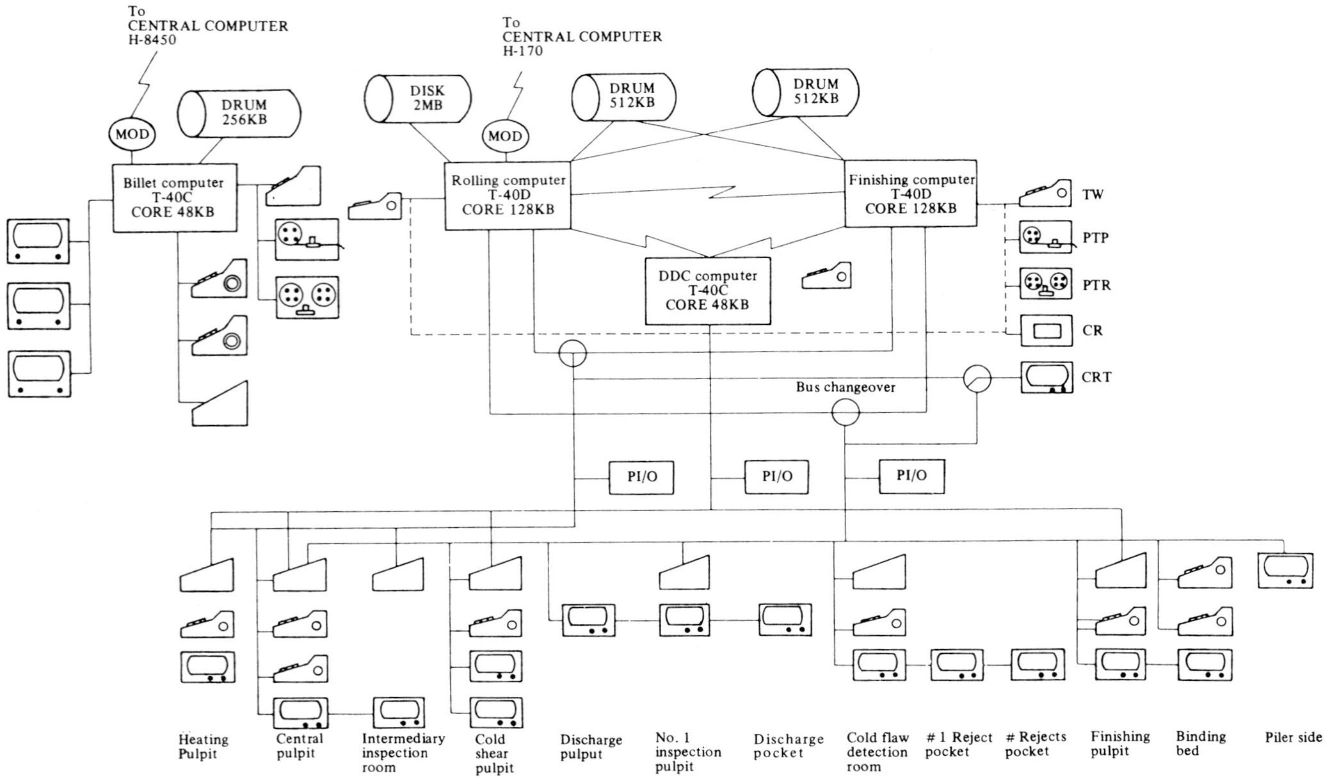

Fig. 21.8 Computer system configuration.

of gamma-ray sensors which detect the position of each billet in the reheating furnace, and a pusher stroke detector which detects the displacement of each billet in the furnace, based upon which the subsequent walking beam stroke and pusher stroke are calculated. This technique is already patented and licensing rights have been granted to other steel companies.

Automatic control of shear

With this system, up to sixteen pieces of bar can be sheared at a time to their desired length by raising them from the roller table by individual lifters, a shearing technique previously unknown in the world steel industry but one promising a great contribution to yield improvement (Fig. 21.9).

Computer-controlled finishing line

All finishing operations including bar handling, extraction, sorting, counting, binding, assignment to orders, output data print-out, etc. are computer controlled. Such an extensive introduction of computer-control technology has

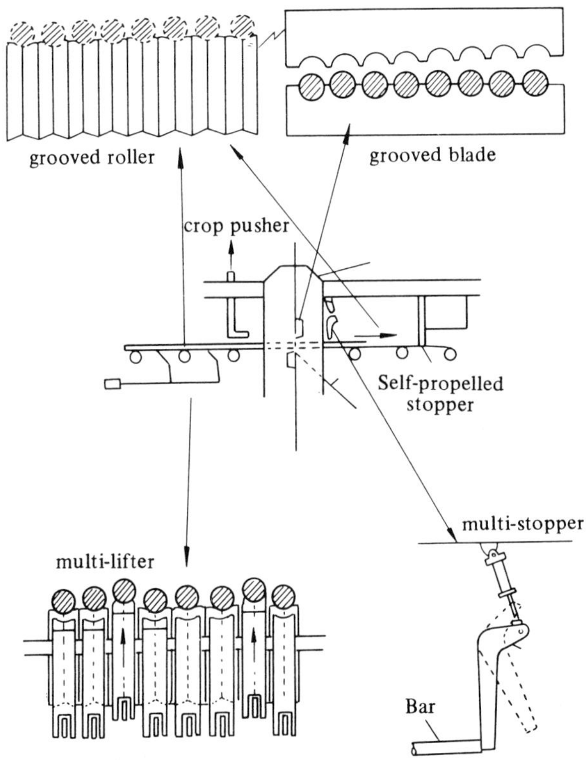

Fig. 21.9 Apparatus of cold shear.

resulted in improved efficiency and higher yield from both the productive and production-supporting capacities. In consequence, the original goal of a high-efficiency system permitting mass production of high-quality special steel bars without the finishing step has successfully been attained.

21.4 Performance

The present system has been operating satisfactorily since start-up with the initial aims fulfilled. The following gives a view of the benefits that have been obtained from the use of the system.

Productivity

The special steel bar mill based on the new production system has registered remarkable improvements over the previous one: about double the monthly production and four times the productivity. Productivity, in particular, outperforms that of other steel companies (Fig. 21.10).

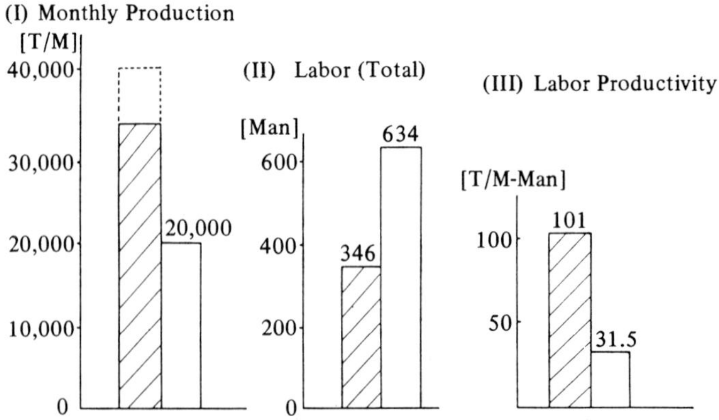

Fig. 21.10 Comparison of productivity.

Production control

The reduced stock volume detained along the production line from the billet conditioning to the bar finishing processes has shortened the manufacturing period from steel tapping to product shipping from one month to a week; a benefit that has brought about not only prompt delivery, but improved efficiency of inventory control and improved yield.

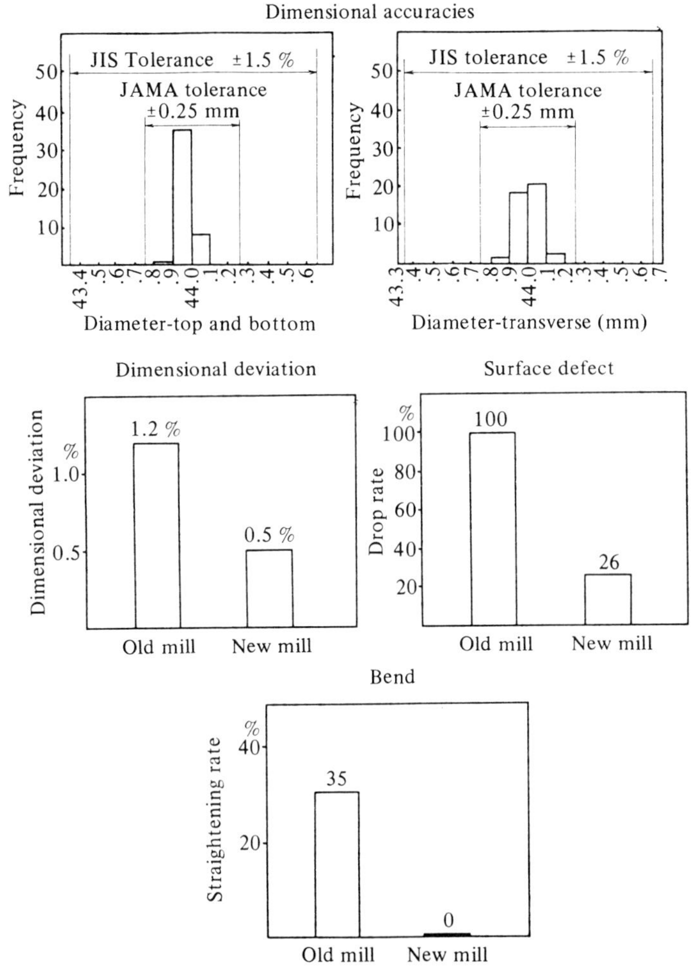

Fig. 21.11 Comparison of product quality.

(*a*) *Quality*

With substantial improvements in dimensional accuracy, bend and surface defects, the most severe quality requirements can be met without reprocessing (Fig. 21.11).

(*b*) *Mis-identification*

Combination of the computer-controlled tracking system with the simplified flow of materials has totally eliminated the mis-identification of finished bars which was previously encountered.

(c) New products

The metallurgical and quality control techniques developed through R & D on the new production system have resulted in the successful development of new grades of steel that require no heat treatment in the secondary processing, thereby contributing largely to the reduction in costs.

Equipment efficiency and optimized yield

Computer-controlled shearing has made the yield of first-choice bars at the Muroran mill unrivalled in Japan, and the automated reheating furnace has slashed fuel consumption per tonne of billet (Fig. 21.12).

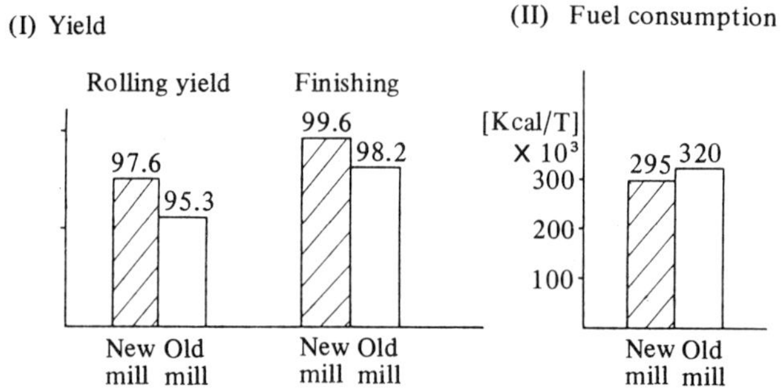

Fig. 21.12 Comparison of operational performance.

21.5 Conclusion

Unlike conventional special steel bar mills, the fully computerized bar rolling mill discussed above is in a twist-free configuration; all lines are continuously integrated in a straight line. However, with regard to labour saving, the equipment before and after the rolling mills, in particular the special steel bar mills, is the least advanced. These are the areas where automation is difficult to apply, but where its potential effects are great. The on-line systems at Muroran, which have resulted in quality improvements and reduced labour costs, have therefore been attracting more and more attention from other steel companies. Secondary benefits in the form of upgraded but less costly bars are also of immeasurable value to consumers.

Acknowledgement

The authors offer grateful thanks to Tokyo Shibaura Electric Co Ltd, Hitachi Shipbuilding and Engineering Co Ltd, and Hitachi Ltd, who co-operated with them in developing this system.

22

Production Control System for a Large Shape Mill

T. Toyoda, General Manager, System Planning Department, Kashima Steel Works; H. Tokuyama, Assistant General Manager, Control and OR Technology Department, Head Office; K. Saito, Assistant General Manager, System Planning Department, Kashima Steel Works and E. Kawamura, Assistant Manager, System Planning Department, Kashima Steel Works, Sumitomo Metal Industries Ltd, Ibaraki, Japan

22.1 Introduction

The Large Shape Mill of Kashima Steel Works has been operating since July 1975. Since the production of large steel shapes was a new undertaking, vigorous research—not only into production technology and plant engineering—but also into a new method of production control had been conducted preceding its foundation. As a result, a new advanced on-line production control system integrating process and business computers has been developed, taking advantage of the experience gained in the development of TOPS

CAD/CAM and MIS in Japan
ISBN 0-12-682580-7

(Total On-line Production Control Systems of Kashima). This system takes over the total management of planning, control and tracing of the entire process from order entry to shipment.

Owing to the development of this system the quality, the productivity, production cost and the delivery times are as good as any in Japan.

22.2 Basic Philosophy of System Design

22.2.1 Basic Target

This system is designed recognizing the following fundamental requirements, which are common to all systems in TOPS:

(a) to produce a large variety of products to small orders with high-speed, high capacity equipment efficiently and at low cost;
(b) to ensure short-term deliveries of high quality products which meet exacting customer requirements;
(c) to limit the labour requirements to an absolute minimum.

With a belief that these problems could be best solved by employing a computer as a central nervous system for all production control functions, development was started with the following four targets:

(1) Higher production control efficiency.
(2) Cost minimization.
(3) Improvement of customer service.
(4) Operation with minimum labour.

22.2.2 Basic Design Philosophy

The characteristics of the production of large steel shapes are as follows.

(a) The number of material grades is comparatively small, but there are a variety of sectional dimensions, for each of which roll exchange is required.
(b) Orders are specified by material grade, dimensions and the number of pieces; products must be produced in small amounts but in large variety.

Hence, in designing an appropriate production control system, the following basic design philosophy is applied.

(1) Adopting a unique lot production control method (where orders and materials are controlled lot by lot before a rolling process) so that flexibility of material supply with the variation of order demand is increased and disturbance of production minimized.
(2) Building upon the on-line system for complete piece tracking, to encompass on-line instruction of operation and process control from material reheating to product shipment.
(3) Fully automating the operation of all the rolling and finishing processes by process computers including intensified control functions for in-built quality during the production stages.
(4) On-line assorting of products by the business computer using operations research techniques so that the allotment of materials to orders becomes optimal.
(5) Introducing automation equipment linked with computers for saving labour.
(6) Controlling quality in all areas be they design, building-in, assurance or analysis.

22.2.3 TOPS

As shown in Fig. 22.1, the functions of TOPS are organized into a five-level hierarchical model, and assigned to business computers and process computers, or to batch processes and on-line processes, as appropriate.

The integration of on-line and batch data processing in one computer, which allows access to the master files from both processes, was completed for the first time in this system, and is a more efficient and rational method for solving the problem of data transfer and synchronization between two computers.

22.3 Principal Facilities

The specifications of principal facilities and their layout with the arrangement of system terminals are shown in Table 22.1 and Fig. 22.2, respectively. The characteristics of these facilities can be summarized as follows:

(1) All facilities from the loading skid (just before the reheating furnace) to the end of the finishing line are fully connected by roller conveyors. This allows fully automatic operation and on-line process management by computers, and contributes to large savings of labour.
(2) For building-in quality a walking beam reheating furnace, open yoke

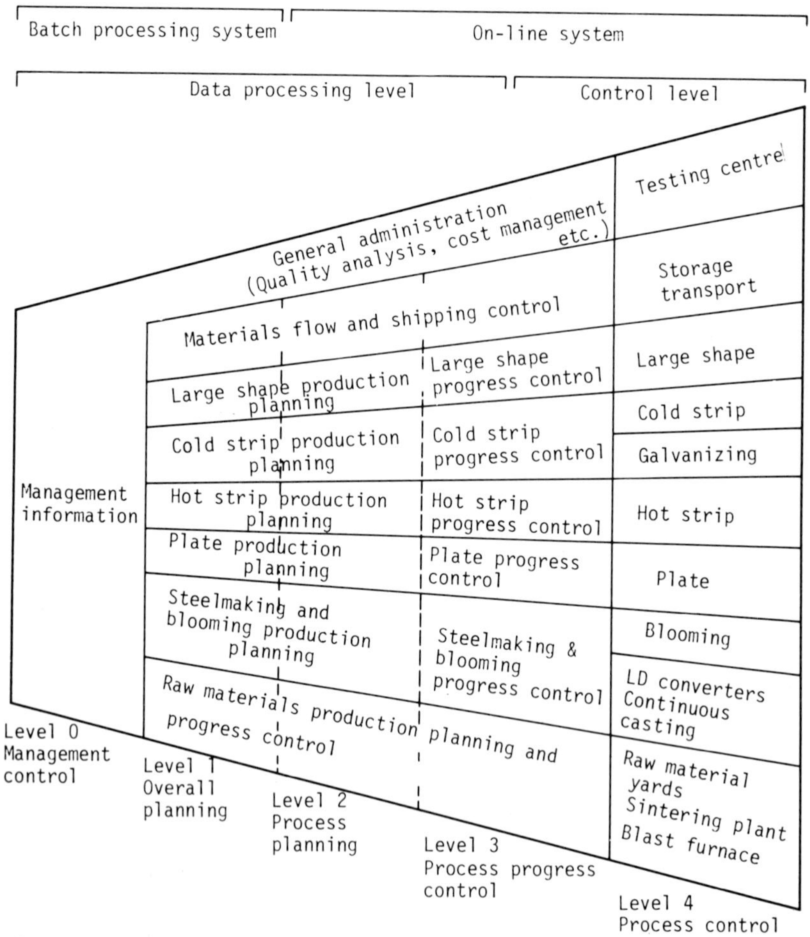

Fig. 22.1 Functions of TOPS (Total On-line Production Control System of Kashima).

universal mills, an electric web guide, and so on are introduced. In addition, a roll stand exchange method is adopted for reducing loss time for roll exchanges.

22.4 Lot Production Control Method

The lot production control method had been established based on the characteristics of large shapes.

Materials of the same standard size (which may provide optimum steel-

Table 22.1 Principal equipment specifications

Equipment	Type	Number of units
Reheating furnace	6-zone, continuous walking beam	1
Breakdown mill	Reversing 2-high	1
Universal roughing mill	Universal	1
Universal finishing mill	Universal	1
Hot saw	Horizontal	2
Cooling bed	Walking beam	2
Roller straightener	Variable roller pitch	1
Labelling machine	Automatic	1
Piler	Automatic	3
Gag press	Horizontal oil pressure and universal oil pressure	Each 1
Cold saw	Vertical	1

making and blooming yields and efficiencies) are grouped in melting furnace units. Then, for improved material handling efficiency and rationalizing process control in the rolling stage, materials are grouped by product size and specification rather than by individual order, and a rolling and sawing plan is formed for each such group. The actual assignment of products to orders is carried out after the product inspection stage.

22.5 System Configuration

As shown in Fig. 22.3, the system actually consists of two major systems: production information control and process control.

For production information control covering the entire process from order entry to product shipment, a large computer is used on which both real-time and batch transactions are processed. To satisfy the diverse demands imposed on the system (such as quick-response, high performance, single master-file and special terminal control) the company's own database/data communication control program (KICS: Kashima Information Control System) has been developed.

Two process control computers are linked and cover all the rolling and finishing lines. They are coupled with four DDC (direct digital control) computers to achieve overall automation of the production lines.

Back-up of the process computers is achieved by automatic switching; namely, when one of the computers fails, the other takes care of the minimum control functions necessary for the rolling and finishing lines. For the

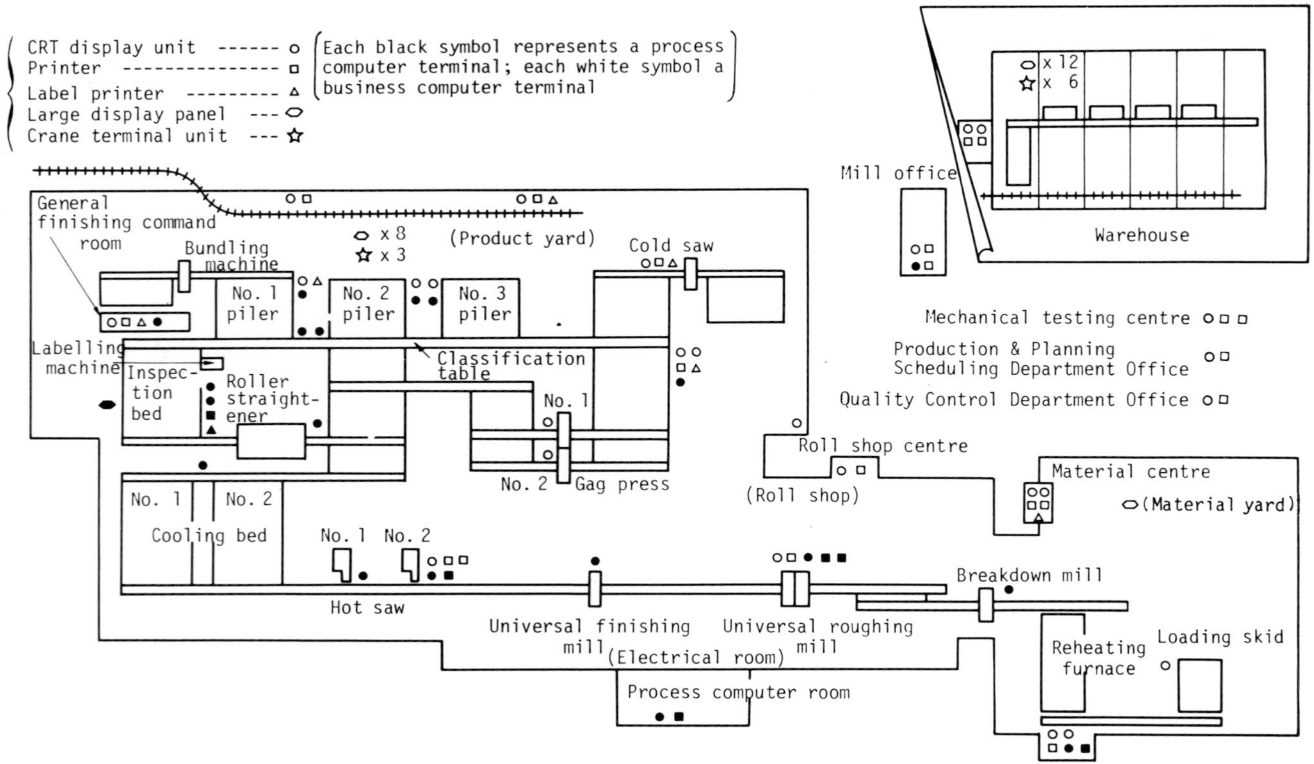

Fig. 22.2 Layout of the large shape mill.

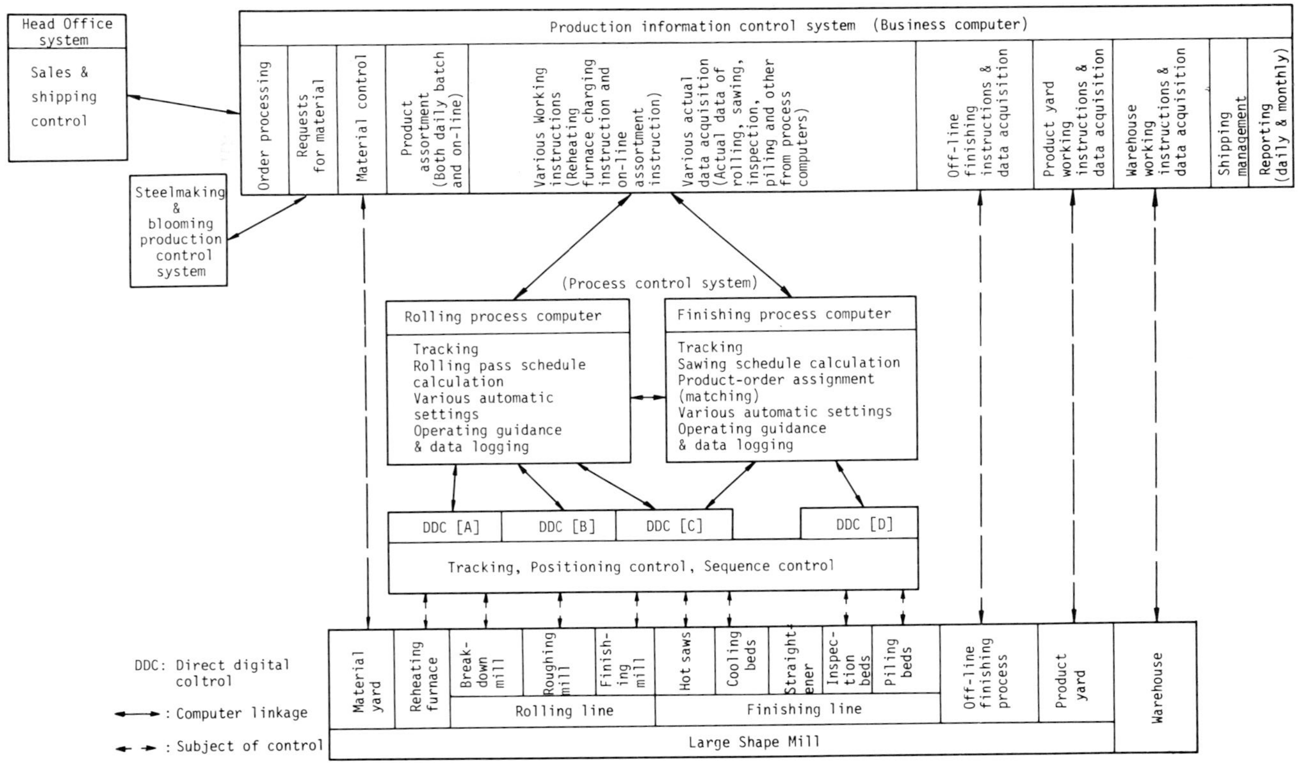

Fig. 22.3 Functional scheme of the system.

DDC computers, there is also a backup computer which is normally used for equipment diagnosis.

22.6 Process Computer Systems

The functions of the process computers are shown in Table 22.2.

Tracking

Automatic tracking is performed throughout the production lines. Products travelling through the finishing line are transported at a speed of 4 m/s and treated at a rate of a little more than 10 seconds, with the maximum number of products tracked being as large as 2000. For such a large number of products travelling at such high speed to be tracked automatically on a piece-by-piece basis, the types and the locations of 250 sensors were carefully selected and a logic capable of recognizing any abnormal movement of products (such as collisions, reverse flows, etc.) was incorporated.

Rolling pass schedule

The mill for groove-type rolling employs a stored schedule system. For rough and finishing rolling mills, an on-line pass scheduling system based on a numerical model has been implemented.

As shown in Fig. 22.4, the rolling pass schedule calculation involves setting up rolling conditions based on material information (such as quality and chemical composition) and production information (such as manufacturing size, rolling temperature, etc.), using draft schedule, rolling temperature and mill setting models.

To optimize the rolling pass schedule, the control system includes functions such as adaptive correction of the rolling load calculation for measured loads, and adaptive zero-point correction of roll openings according to measured product dimensions.

Furthermore, the system is provided with feed-forward functions for the next pass or next mill setting, thus contributing to improvements in product size and shape accuracy and to the prevention of mis-rolling.

Assignment of products to orders

Information concerning the assignment priority of orders based on product specifications, delivery date and so on is received from the business computer. Based on the results of product inspection at the inspection beds, the process computer automatically carries out suitable assignments.

After assignment, product labels are issued by the label printer and automatically affixed to the corresponding products by the labelling machine.

Table 22.2 Process computer functions

Classification	Function	Automation item
Rolling process computer	1. Material tracking (entry into reheating furnace—finishing mill) 2. Rolling pass schedule calculation 3. Mill pacing 4. Various automatic settings 5. Operation instruction display (CRT) 6. Data acquisition and summation, logging 7. Data transmission (to business computer)	(a) All line tables (reheating furnace entrance—finishing mill) (b) Rolling mill (breakdown, rougher and finisher) Automatic setting of roll gaps, side guide, web guide, tilting table, etc. (c) Other equipments Automatic setting of reheating furnace pusher, extractor, walking beam, tongue cut saw, shift transfer and measuring instruments.
Finishing process computer	1. Product tracking (sawing—piling) 2. Product sawing schedule calculation 3. Product-order assignment (matching) 4. Various automatic settings 5. Operation instruction display (CRT) 6. Data acquisition and summation, logging 7. Data transmission (to business computer)	(a) All line tables (sawing—piling) Selection of the range of synchronization; starting and stopping (material positioning) control (b) Sawing sequence Automatic setting of hot saw and stopper gauge (c) Piling equipment Automatic setting of piling mode, piler pusher and elevator (d) Various transfers Automatic setting of the stroke of the cooling bed and inspection bed (e) Labelling machine Issuance of product labels and labelling sequence control

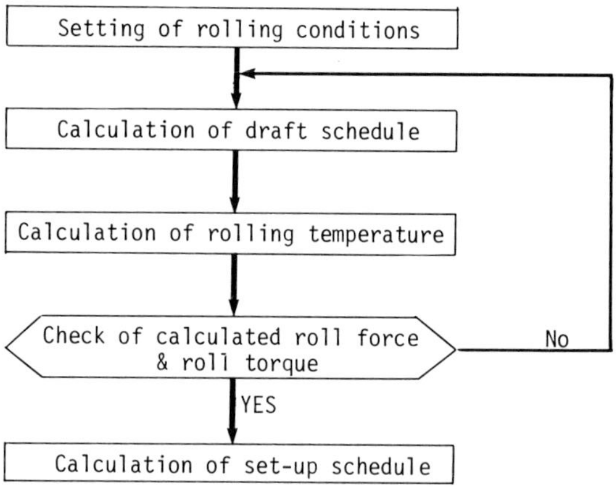

Fig. 22.4 Rolling pass schedule model.

22.7 The Product Assortment System

Product assortment is an operation which calculates the rolled length of material and then computes a combination of product lengths such that the ordered products can be laid out on the material with the highest yield.

In the lot production control system where requests for material are made on the basis of standard size product assortment (which provides a basis for rolling and sawing) plays a very important role.

As shown in Fig. 22.5, the product assortment system actually consists of three systems: the daily batch assortment system, the on-line assortment system and the rolled length control system. The contents and effects of these systems are as follows.

Daily batch assortment system

The daily batch assortment computation is carried out on the day before the rolling date on the basis of the calculated weight to provide the materials to be charged to the reheating furnace. The following processing sequence is involved in this function.

Orders are grouped according to sectional dimensions, standards, and specifications and materials are grouped according to sectional shape, material grade, specifications, and so on.

Then, for each group of orders a material lot, which satisfies such conditions as standards and specifications and which is most cost advantageous, is selected.

In this system, a linear programming method is applied in order to automatically obtain the optimum concatenation.

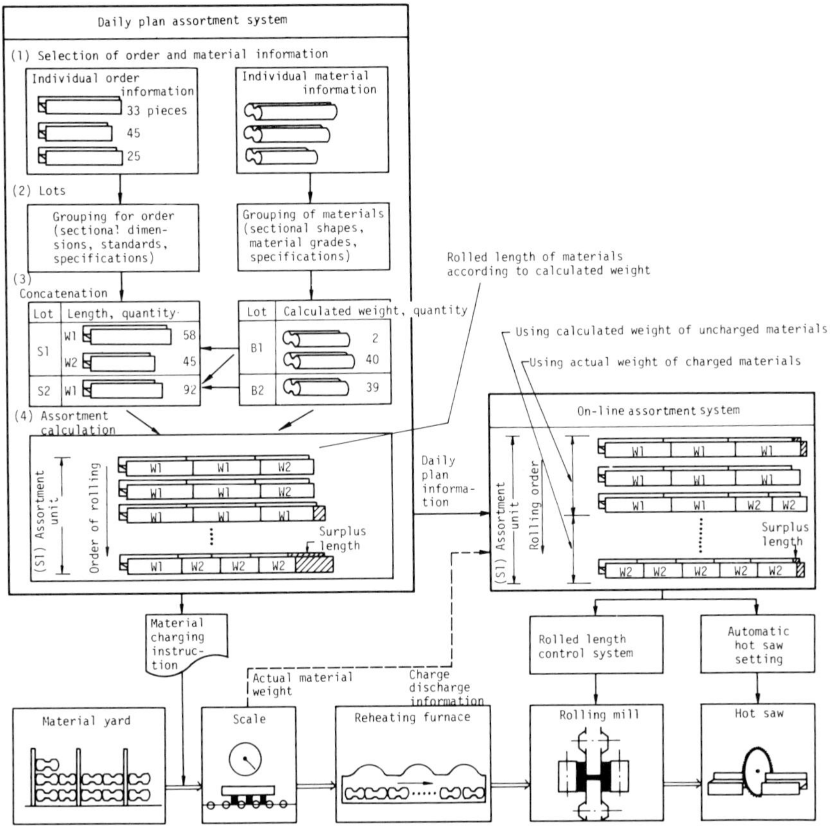

Fig. 22.5 Product assortment system.

Finally, between the order and material lots, an assortment calculation is carried out so as to obtain ordered products with the highest yield taking account of the following conditions.

(a) improvements of sawing efficiency and cooling bed loading efficiency;
(b) improvement of workability of the piling bed.

On-line assortment system
The on-line assortment is carried out every hour based on the actual weight of charged materials, so as to increase the accuracy of the rolling and sawing plan, to improve yield and to preclude surplus products.

The system is unique and, unlike the conventional systems in which on-line assortment is carried out for each material just before sawing, it is designed to optimize the assortment unit as a whole.

Rolled length control
The rolled length of material is controlled by the process computer so that no surplus product will be produced and the yield will be maximal.

22.8 Automatic Equipment

22.8.1 The Labelling Machine

As shown in Fig. 22.6, the label printer issues a product label and the labelling machine affixes it automatically to the corresponding product according to the instruction from the finishing line process computer.

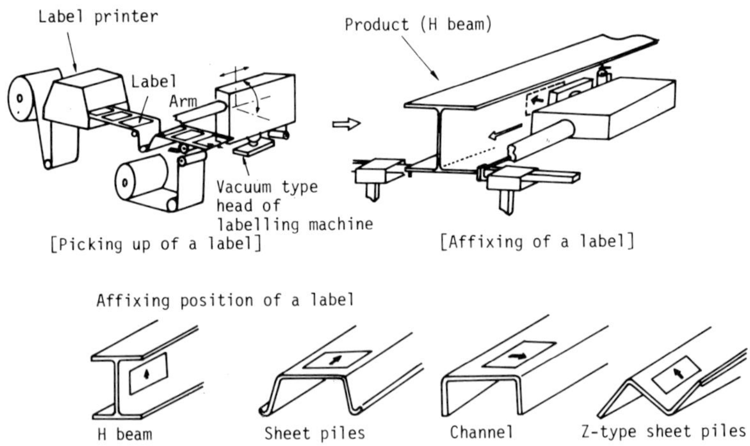

Fig. 22.6 Label printer and labelling machine.

22.8.2 Product yard control system

Figure 22.7 shows the product yard control system. The products automatically piled in the final phase of the finishing line operation are loaded on to trolley-cars and sent to the product warehouse. All working instructions and data acquisition relating to the storing, loading and shipping operations at the product yard are performed using wireless crane terminal units and large display panels. As a result, the crane crew can now work without the aid of wiring hands. Moreover, the issuance of trolley loading records and the management of temporary storage space can be performed without the help of

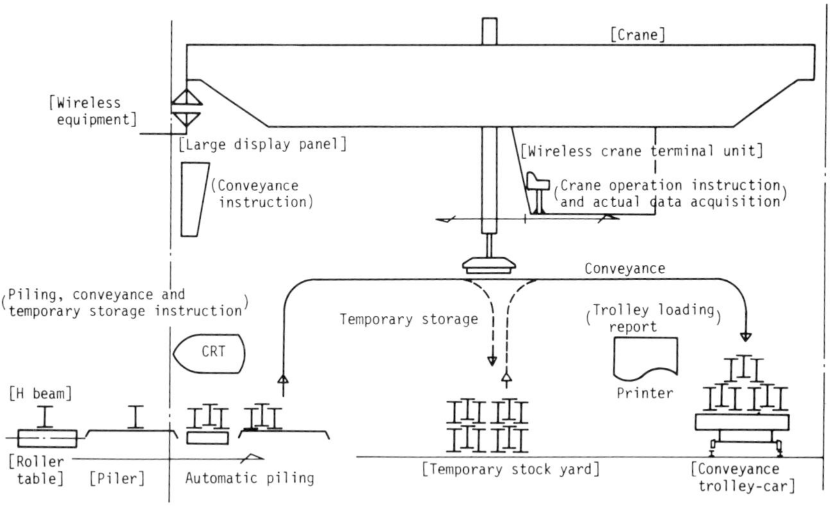

Fig. 22.7 Product yard control system.

administrative personnel. In this manner, considerable savings have been made in manpower.

22.8.3 Automatic Product Warehouse Operation System

As shown in Fig. 22.8, automation of warehousing operations has been accomplished by means of relatively simple equipment (roller tables and transfers) for saving labour.

Spaces for products within the warehouse are selected automatically by the computer according to shipping destination and delivery dates. The products transported from the large shape mill on trolley-cars are accepted at the inlet transfer, from which they are sent to the predetermined shed by means of roller tables and transfer equipment. The operation of all this equipment has been completely automated. The working instructions for crane transportation from the exit transfer to the storage place are given via a large display panel, and data acquisition is performed by means of the wireless crane terminals. These functions have resulted in significant savings in the number of transfer operators, wiring hands and administrative personnel.

22.9 Quality Control System

A systematic quality control which consists of the following functions is conducted.

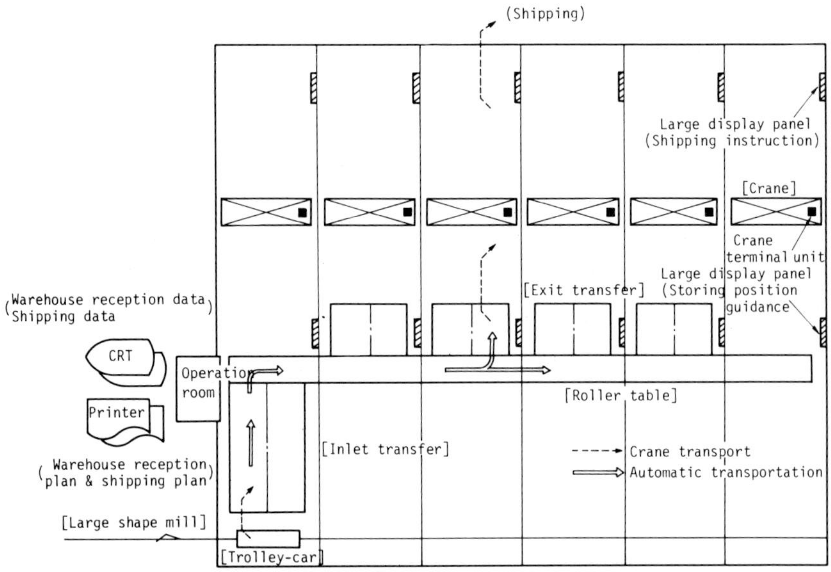

Fig. 22.8 Automatic product warehouse operation system.

Quality design

Quality design as the first step of quality control is to determine the adequate production specifications, which may satisfy the quality specifications requested by customers with minimum cost.

In order to computerize this design procedure, a database of many kinds of technical standards and also customer oriented standards is built up. Provided with this information, the design system automatically determines the optimal production specifications composed of more than a hundred items.

Built-in quality

In the entire production lines, every material and product is tracked piece by piece and its historical data accessed automatically. Based on this information, the following functions for built-in quality are installed in the system.

(a) Reheating furnaces are automatically controlled in order to make materials at a uniform temperature.
(b) Dimensions and temperatures of materials at the rolling line are controlled by an advanced simulation model of rolling based on measurements of temperature at eight points during rolling and consecutive measurements of dimensions at each mill.
(c) According to the results of product inspection, the system assigns a suitable order to it, and issues its product label. The system also has func-

tions such as automatic piling, and working instructions to conditioning processes.

Quality assurance
Automatic quality judgement for quality for each product is conducted based on the measured dimensions and temperatures in the rolling and finishing processes and the results of surface inspection.

Sampling tests instructed by the computer are also performed at the testing centre. If the results are abnormal, the system warns the operator and rejects the corresponding products.

Quality analysis
As with other production control systems in TOPS, the system keeps voluminous production history data to improve the design standards described above. This data acquisition and analysis involves:

(a) Detailed production history of each product from melting to finishing is collected automatically through the production control systems and kept in the database for about a year.
(b) A special hierarchical data structure is introduced for effective entry, reference and storage of data.
(c) Various application software concerning data reference, computation, classification and statistical analysis are provided, and engineers have ready access to them.

22.10 Results

The production control system for the large steel shapes mill has been in operation since April 1975 when the factory began test working. Subsequently some improvements were made in system functions, and the initial aims have been realized.

22.10.1 Operating parameters

As shown in Table 22.3, the system has contributed to considerable improvements such as saving labour, raising yield and shortening delivery times.

22.10.2 Quality

As for dimensions of products (Fig. 22.9), the accuracy has been raised to an

Table 22.3 Benefits of the production control system

Items		Effects
Labour savings		(persons)
Factory		98
Administration		28
Warehouse, etc.		44
	Total	170
Improvement in yield		2.5 (%)
Decrease in surplus ratio		4.1 (%)
Increase in operating ratio		0.5 (%)
Increase in efficiency		5 (T/H)
Reduction of delivering time		9 (days)

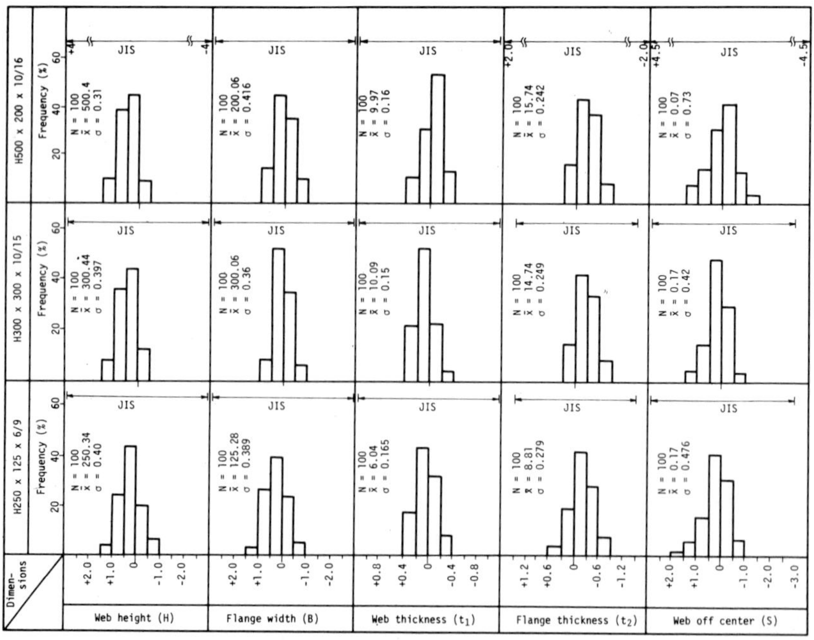

* JIS; Japanese Industrial Standards

Fig. 22.9 Dimensions accuracy of large beam.

extremely high level. Moreover, stable rolling and adequate temperature control have made products of good surface.

There has been no occurrence of product mis-treatment.

22.11 Conclusions

This system is the first instance of on-line production control covering the entire production line and using the management method which is appropriate to a large shape mill. Thus, the mill has achieved the highest yield and operating efficiency in this area of industry as shown in Table 22.4.

Table 22.4 Composition of operating parameters

		Others in Japan			
	Kashima	A	B	C	D
Amount of rolling (t/M)	53 507	50 773	40 913	43 942	69 311
Rolling yield (%)	97.2	95.8	96.2	94.9	96.0
Workers (persons)	206	433	405	305	503
Productivity (T/M/persons)	259.7	117.3	101.0	144.0	137.8

The computer system technology installed uses the so-called CPU one master file system instead of a conventional hierarchical structure. This technology has proved to be a turning point in production control in the iron and steel industries, and has become a guiding principle for new systems.

After the mill started, the system has been able to adapt flexibly to such changes in production conditions as additional products (e.g. sheet pile, channel) and to changes in material supply (ingot to continuous casting slab). Thus, it has contributed to the supply of high quality products to customers at a lower price.

Acknowledgement

In conclusion, the authors would like to express their special thanks to IBM Japan Ltd and to Nippon Electric Co. who co-operated in developing this system.

Bibliography

Toyoda, T., "Production Control System of Large Shape Mill at Kashima Steel Works". The 8th IBM Plant MIS Symposium in Japan (1976), IBM Japan.

Ishihara, K., Saito, K., and Kokubu, T., *et al.*, "Automation of Large Shape Rolling Mill". *Toshiba Review*, 31 (1976), No. 7, Toshiba Electric Co.

Shimamura, N., Nishimura, H., and Saito K., *et al.*, "Fully Computerized Wide Flange Beam Mill and Finishing Line". AISE Annual Convention (1977).

Misaka, Y., Makino, T., and Misawa, T., "Computer Control System of H Shape Universal Mill". The 28th Joint Symposium of Plastic Working in Japan, No. 122 (1977).

Tokuyama, H. and Ueno, N., "The Cutting Stock Problems in the Iron and Steel Industries". *Operations Research '81*, Proceedings of the 9th IFORS International Conference. North-Holland (1981).

23

A Minicomputerized Automatic Layout System for Two-layer Printed Wiring Boards

Ikuo Nishioka, Takuji Kurimoto, and Hisao Nishida, Central Research Laboratory, Sharp Co, Ichimoto, Tenri, Nara, Japan, and Isao Shirakawa and Hiroshi Ozaki, Department of Electronic Engineering, Osaka University, Suita, Osaka, Japan

23.1 Introduction

In order to reduce cost and improve reliability, high density packaging is one of the most important objectives in the design of electronic systems. In the assembly of such systems, two-layer printed wiring boards are still frequently used to provide the necessary connections among functional modules. Hence, the layout scheme to realize high density wiring on such boards is of practical importance.

Generally, the circuit modules to be interconnected by means of a two-layer wiring board are mounted on the first side of the board, each with connector pins inserted through conducting holes, called *plated-through holes*. Each set of the pins to be made electrically common can be connected by con-

CAD/CAM and MIS in Japan
ISBN 0-12-682580-7

ductor paths, or wire routes, composed of horizontal and vertical segments on the first and second sides, respectively, and *vias* used to connect segments on different sides. Thus, the layout process for such a wiring board can be divided into two phases: the first is the placement of circuit modules on the first side; the second is the routing of conductor paths on both sides. A variety of approaches have been proposed and extensively developed (see, for example, excellent literature surveys[1, 2, 3]) to solve these problems. However, in most instances routing algorithms are non-iterative, that is, wires are laid out one at a time in a specified order, and once they are laid out they cannot be modified later. It is essential, therefore, that wires be routed in the "optimum order" subject to some criterion. The ordering scheme is consequently highly critical,[4, 5] but none of the schemes so far proposed is entirely satisfactory. As the wiring density for a board increases, the flexibility of any router based on non-iterative algorithms decreases. To cope with this, new approaches are continually under investigation.[5, 6]. On the other hand, an interactive layout scheme based on a minicomputer[7] is a practical alternative.

This chapter proposes a minicomputerized layout system for *two-layer* printed wiring boards, which operates on a PDP 11/40 computer coupled with a TEKTRONIX 4014 display terminal in conjunction with a 4953 graphics tablet. This system is distinctive mainly in that the routing scheme, composed of line-search and maze-running routines, is automatic in the following sense: a user who has little acquaintance with specific features of the two routines has only to try to modify the current wire pattern so as to make an open space in order that each wire route still lacking can be laid out automatically by applying one of the two routines.

This system has been in use, and is contributing towards reducing the cost and time incurred in manufacturing printed writing boards.

23.2 Hardware Configuration

An outline of the hardware configuration of the system is shown in Fig. 23.1. This system operates on a PDP 11/40 computer with 24 K 16-bit words of core memory and with 1.2 m words of disk storage, coupled with a TETRONIX 4014 display terminal in conjunction with a 4953 graphics tablet. In addition, the system is equipped with a card reader, paper tape reader/punch, 9-track magnetic tape unit, and a console typewriter. A digitizer is used off-line to provide input data concerning the location of each circuit module on the board. A drum-plotter is used to check the drawing of the wire pattern generated by the routing process, and a photo-plotter is for photo realization of the completed wire pattern.

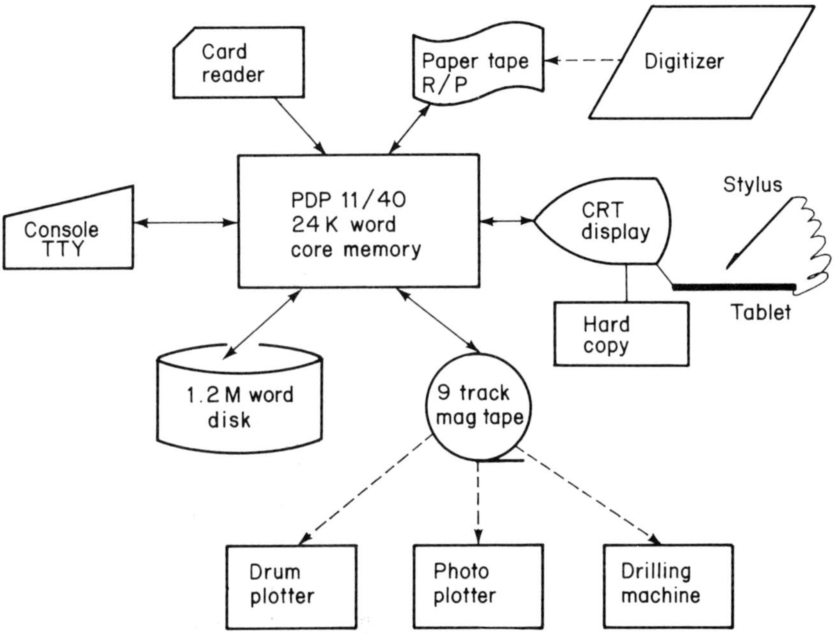

Fig. 23.1 Hardware configuration.

23.3 Software Configuration

The software configuration of the system is outlined as shown in Fig. 23.2, where each routine has the function summarized in Table 23.1.

Note that in this system the design database is accessed by independent routines such as:

(1) Line-search routine (*LINE*),
(2) Maze-running routine (*MAZE*),
(3) Interactive editing routine (*S-GRAPH*),
(4) NC tape generating routine (*NC Tapes Generating A*),
(5) Checking routine (*S-CHECK*),
(6) Replacement routine (*PLACE*).

LINE and *MAZE* constitute the routing scheme of this system, which is outlined later. *S-GRAPH* is for modifying the current wire pattern with the aid of a CRT display so that each wire route still lacking can be laid out automatically by applying any of the two routing routines, as discussed later. *NC Tapes Generating A* is to produce magnetic tapes to be input to a CALCOMP 1036 drum-plotter and for back-up store. *S-CHECK* is for checking auto-

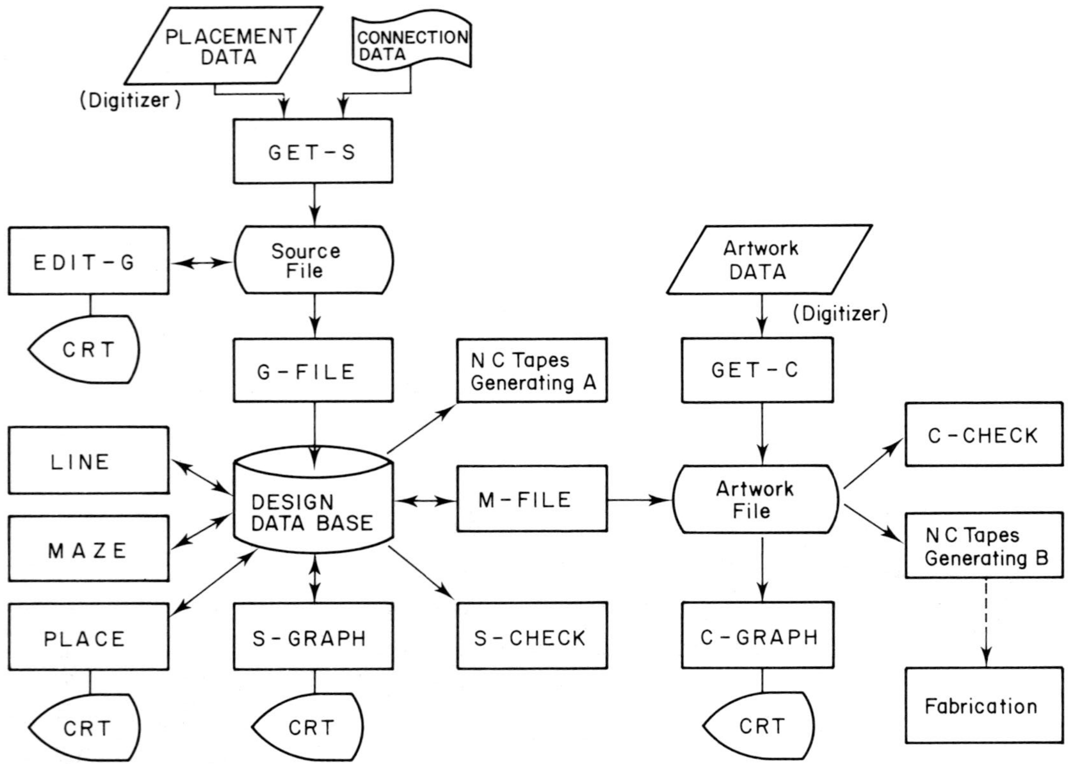

Fig. 23.2 Software configuration.

Table 23.1 Function of each routine

Routine name	function
GET-S	To read input data and to construct the source data file
EDIT-G	To correct errors/which are detected as the input is processed, directly at the graphics display
G-FILE	To build the design database from the source data file
M-FILE	To manage the design database
LINE	Line-search routine
MAZE	Maze-running routine
S-GRAPH	Interactive editing routine for the design database
C-GRAPH	Interactive editing routine for the artwork file
NC Tapes Generating A	To generate NC tapes to be input to a drum-plotter for check drawing
NC Tapes Generating B	To generate NC tapes to be input to a photo-plotter and a drilling machine
S-CHECK	To check shorting and disconnection-errors which may arise in executing S-GRAPH
C-CHECK	Closeness checking for wire pattern in executing C-GRAPH
PLACE	Replacement routine

matically any shortening among wire routes associated with different *signal nets** and the disconnection of wire routes asociated with the same net, which may arise in executing modifications of the wire pattern using the CRT display in *S-GRAPH*. In addition, *PLACE* is optionally provided to modify the initial placement of the modules on the board, using the CRT display.

23.4 Implementation Overview

Normally, a design originates from data describing the pin locations of circuit modules on the board and the net interconnection. As the input is processed, numerous error conditions are detected. These errors may be corrected directly at the graphics display without the delay caused by paper tape or card handling. Once the input has been read, the design database is constructed, and the system awaits the next operation.

As to implementing the routing scheme, in its first run, the process is fully automatic with the use of the net ordering scheme, to be stated later, and begins with the line-search routine for all nets in the specified order. Then, for those nets which are still unconnected, the maze-running routine is executed in the previously specified order.

After executing the maze-running routine, the wire routes still missing in

* A signal net indicates a set of the pins to be made electrically common.

the current pattern are laid out according to the following policy. For each net still unconnected, a user first tries to modify the current wire pattern so as to make an open space such that the wire route between an isolated pin and the connected part of the net can be generated through the open space. He then restarts the routing process, ordinarily the maze-running routine. Graphic updating and automatic routing can be repeated until the user is satisfied with the wire pattern. Thus, in this system, a new wire route is generated in principle by automatic routing rather than manual routing, because the latter is complicated, liable to error, as well as time-consuming, when the wiring density is high.

We must also pay attention to the fact that we are often confronted with various non-standard modules to be mounted on a board. In such a case, it is desirable that a user can execute graphic updating, observing a realistic artwork pattern on the screen. To this end, we provide a powerful tool named C-GRAPH, as will be stated later.

Once a wire pattern layout is completed, facilities are available through this system to obtain a full array of design documentation. This documentation includes artmasters for wiring board fabrication, check drawings, NC tape generation for drilling machines, and back-up store.

23.5 Board Specification

The effective area for wiring use is divided into columns and rows, both with the same width (1.27 mm or 1/20 inch), which is half the interval between two consecutive pins of ordinary DIP (Dual In Line Package) ICs. Let a square region located in column x and row y be designated as a *cell* (x, y). Henceforth, suppose unless otherwise stated, that:

(a) ICs are placed on the board with dual pin-lines oriented vertically;
(b) horizontal wire segments lie in rows on the first side: vertical segments in columns on the second side;
(c) any plated-through hole or via occupies a cell, such that no two may be located pairwise adjacent cells, where cell (x, y) is said to be adjacent to cell (x', y') if and only if $|x - x'| + |y - y'| = 1$.

The wiring area of 512 rows by 512 columns is the maximum size applicable to this system, for which 32 768 words (= 524 288 bits) are provided to form the *bit map* necessary in the line-search routine, with 8192 words (equivalently, 256 rows by 256 columns) in core and excess words on disk.

23.6 Routing Scheme

A line-search algorithm and a maze-running algorithm constitute the routing scheme. One of them may be used for specified signal nets, one at a time, in a predetermined order, or for a single net indicated by a user.

23.6.1 Line-Search Algorithm

In the following, an outline of the line-search algorithm adopted in this system is described, by which usually 80–90% routing can be attained. The line-search method, originally proposed by Mikami and Tabuchi[8] and extensively used, consists essentially in finding a wire route by searching for those candidates which may possibly be chosen as horizontal and vertical line segments of the wire route. The algorithm, based on Yamamura, Shirakawa and Ozaki,[9] is so sophisticated as to be implemented on a minicomputer.

Throughout the routing process, each signal net is partitioned into two; one being a set of the isolated pins and the other a set of pairwise connected pins. For an incomplete net, the routing procedure attempts to create a wire route to connect each isolated pin to the connected wire pattern so far generated for the net.

To describe the line-search algorithm, we need some definitions as follows.

Associated with a pin t_i located at cell (x_i, y_i), let us denote by a segment of level 0, written by $l_0(t_i)$, a possible wire segment candidate in one of the following forms:

(a) a maximal horizontal line segment to be newly drawn on the first side from cell (x_i, y_i),

(b) a vertical segment from (x_i, y_i) to unoccupied $(x_i, y^i \pm 1)$ and a maximal horizontal one to be newly drawn from $(x_i, y_i \pm 1)$, both on the first side.

A segment of level 1, written by $l_i(t_i)$, indicates a possible wire segment candidate in the shape of a maximal vertical line segment to be newly drawn on the second side from a cell located on any $l_0(t_i)$, and a *segment of level 2*, written by $l_2(t_i)$, a possible wire segment candidate in the shape of a maximal horizontal line segment to be newly drawn on the first side from a cell located on $l_1(t_i)$, which does not coincide with any $l_0(t_i)$.

A rough sketch of the line-search algorithm is as follows: Let $T = \{t_1, t_2, \ldots, t_n\}$ be a signal net to which the algorithm is to be applied, and let US and CS denote the *unconnected set*, and the *connected set* of the net, respectively.

The algorithm consists essentially in finding those wire routes to connect each $t_i \in US$ to some of the wire segments connecting pins in CS. The choices are limited to three (or less) horizontal wire segments and two (or less) vertical segments. This limitation on wire pattern generation speeds up the search.

Initially (when $CS = \phi$), the algorithm seeks an interval $[x_1, x_2]$ over which segments $l_0(t_i)$ of level 0, with $t_i \in US$, are most congested. Any of those pins $t_i \in US$, for which $l_0(t_i)$ pass over this interval, are added to the CS and deleted from US.

The algorithm then seeks wire routes composed of segments of level up to 1, as follows: pick up any pin $t_u \in US$, and search for the following segments:

(a) a segment $l_0(t_u)$ of level 0 which intersects any vertical wire segment already in the connected part of the net;
(b) a segment $l_0(t_u)$ of level 0 which meets a segment $l_0(t_c)$ of level 0 for any $t_c \in CS$;
(c) a segment $l_1(t_u)$ of level 1 which intersects a segment $l_0(t_c)$ of level 0 for any $t_c \in CS$.

If such segments are found, then wire routes to connect t_u to the connected part can be defined, based on these segments. Choose the shortest* of these routes, and add it to the connected part, with t_u transferred from US to CS. Repeat this process for the remaining pins in US.

At the next stage, wire routes composed of segments of level up to 2 are generated, as follows: pick up any $t_u \in US$, and seek the following segments,

(1) a segment $l_1(t_u)$ of level 1 which intersects any horizontal wire segment, already in the connected part of the net;
(2) a segment $l_2(t_u)$ of level 2 which intersects a segment $l_1(t_c)$ of level 1 for any $t_c \in CS$.

If such segments are found, wire routes to connect t_u to the connected part can be defined. Choose the shortest of these routes, and add it to the connected part, with t_u transferred from US to CS. Repeat this for the remaining pins in US.

In the computer implementation of this algorithm, the dominant process of finding segments of level 0, 1 and 2 makes extensive use of the bit map. Every time a new wire route for a net is generated, all information about its wire segments is stored in core. The data are only transferred to disk at the termination of the routing process for this net.

This algorithm ordinarily achieves 80–90% of the total wiring; the rest is processed with the maze-running algorithm.

23.6.2 Maze-Running Algorithm

The maze-running method, first proposed by Lee[10] and later modified by

* By the shortest route we mean the one which occupies the least number of cells.

Geyer[11] and Rubin,[12] consists essentially in finding a wire route between two cells by expansion around one until the other is reached. It cannot fail to find a route to connect the cells whenever one exists, but it requires much more processing time than the line-search method.

The algorithms so far proposed aim to seek the shortest route and/or reduce the number of vias. In their application to two-layer wiring boards, it is possible that a fairly long *vertical* wire segment drawn on the *first* side,* can prevent a number of horizontal wire segments from being generated across it on the first side later on, hence lowering the wiring density of the board.

To overcome this, the maze-running algorithm is divided into two phases.

The first phase, provided to avoid the situation outlined above, introduces *directional limitation facilities* such that the vertical (horizontal) portion of any wire route between two consecutive vias does not exceed some specified length on the first (second) side. Although this limitation may result in a wire route, otherwise possible falling may be found, increased routability should result.

The second phase is based on Chiba, Shirakawa and Ozaki,[13] and is such that its application to an incomplete net yields a tree-shaped wire pattern, which connects an isolated pin to the other isolated ones and the connected part of the net, whenever such a tree-shaped wire pattern exists. The following is a brief description of the second phase.

(1) Given a signal net $T = \{t_1, t_2, \ldots, t_n\}$ for which US and CS indicate the unconnected set and the connected set, respectively. Select a pin $t_i \in US$, and let the cells on both sides occupied by this t_i be the *start* cells and the cells occupied by the other pins in US or by the connected part for this net the *target* cells. Mark the start cells and let them be the front cells.

(2) Repeat the following for each front cell while one exists then go to (3):
For any front cell (x, y), examine whether or not there is any target cell in its adjacent cells still unmarked, and let (x, y) be a non-front cell. If a target cell is found, then go to (1), otherwise mark all unoccupied cells adjacent to (x, y), and add them to the front cells.

(3) If such a target cell is the one occupied by a pin in US, then mark it and add it to the front cells. Otherwise (in this case, such a target cell is occupied by the connected part), mark all the cells on the connected part, and add them to the front cells. Then go to (2).

(4) If all the cells occupied by the pins in T are marked, then go to (3), else return to (1).

(5) From each marked target cell, settle the wire route backward to the start

* Note that in this chapter the first side is principally for horizontal wire segments, the second for vertical segments.

cells by tracing marked cells, one by one, in the reverse order of the search, and then halt.

At this stage, all the pins with their corresponding cells marked are connected to the start pin t_i.

In the computer implementation of this algorithm, the wiring area of a given board is paged into blocks of 32 rows by 32 columns, so that the blocks can be processed, page by page, in the core of the minicomputer in such a fashion that once the front line of the search enters into a different page to some depth from the boundary, that page is substituted for the old one in the core so as to be searched next.

23.7 Ordering Scheme

It is possible for a user to input digitized or displayed manual layout into the system. In this case, the signal nets to be ordered are those which remain unconnected after such a pre-assignment.

In this system the input data for such remaining signal nets are divided into two classes. The first is DATA-1 consisting of those nets given by the user, which are to be arranged in order of priority. The second is DATA-2, composed of the remainder, which are left entirely to the ordering scheme.

Traditional ordering schemes tend to attach importance to the length of potential wiring for each net. By contrast, the ordering in this system esteems the relative relationship among the pin locations for each net.

The priority order for the nets depends on the assumption that ordinary DIP ICs are placed with dual pin-lines oriented vertically,* and is given roughly as follows:

(a) nets within a certain horizontal spread;
(b) nets within a certain vertical spread;
(c) nets of high density pin-distribution.

A brief description of the ordering scheme is given as follows.

Let U be a set of the nets in DATA-2, and for each net $T_i = \{t_{i1}, t_{i2}, \ldots, t_{in}\} \in U$, with each pin t_{ik} placed at cell (s_{ik}, y_{ik}), let

$$M_x^i \triangleq \max_k \{x_{ik}\}, m_x^i \triangleq \min_k \{x_{ik}\},$$

$$M_y^i \triangleq \max_k \{Y_{ik}\}, m_y^i \triangleq \min_k \{Y_{ik}\},$$

$$D_x^i \triangleq M_x^i - m_x^i, D_y^i \triangleq M_y^i - m_y^i.$$

* In this case, there are generally far more obstacles for horizontal wire segments than vertical ones.

Then, for specified constants $a = 20$, $b = 12$, and $c = 50$, sort U so as to be put into a list $\Sigma = (\Sigma_1, \Sigma_2, \Sigma_3, \Sigma_4)$ by the following procedure.

For $S_1 \triangleq \{T_i \in U | D_x^i \leqslant a\}$, make a list Σ_1 in ascending order of D_y^i, and put $U \leftarrow U - S_1$.

For $S_2 \triangleq \{T_i \in U | D_y^i \leqslant b\}$, make a list Σ_2 in ascending order of D_x^i, and put $U \leftarrow U - S_2$.

For $S_3 \triangleq \{T_i \in U | D_x^i \leqslant c\}$, make a list Σ_3 in ascending order of D_y^i, and put $U \leftarrow U - S_3$.

For each $T_i \in U$, calculate $d_i \triangleq (D_x^i + D_y^i)/n$, where $n \triangleq |T_i|$, and make a list Σ_4 in ascending order of d_i.

Here, it should be added that the list Σ thus obtained is permitted to be altered in part by the user.

23.8 Interactive Editing Routine

Interactive editing provides two kinds of facilities for graphics updating. The first is S-GRAPH to update the current wire pattern symbolically displayed on the screen; the second is C-GRAPH to complete the final composite artwork pattern to be input to a photo-plotter, based on Artwork File.

In the process of implementing S-GRAPH, once a user attains 100% wiring, he has only to modify the current wire pattern until he is satisfied with it, or until he is convinced that it is free from any trouble latently possible in the subsequent stage. On the other hand, if 100% wiring is still not realized, he may find a better pattern by adding new wire segments and/or modifying the current pattern. However, when the wiring density is high, such manual layout becomes so complicated that it is time-consuming, and furthermore, liable to wiring errors. To cope with this difficulty, the editing routine is organized to supply the user with a variety of sophisticated instructions convenient for modifying wire patterns. Through these he can easily make an open space in the current wire pattern in order that a wire route still missing can later be laid out automatically by applying either the line-search or maze-running routine.

The instructions available for graphics updating are listed as shown in Table 23.2, and are classified as

—Window	—Select	—Add
—Enlarge	—Move	—Grid-on
—Reduce	—Delete	—Scale
—Shift	—Repeat	

Table 23.2 Operating command list for graphics updating

1. *Control mode*	
XEQ	Execute buffered commands
MKDR	Load drawing from master file to working drawing
STOR	Store working drawing to back-up file
EDIT	Enter edit mode
EXIT	Close edit mode
2. *Edit mode*	
A.	*Edit control mode*
XEQE	Execute buffered commands and redraw picture
DBUF	Delete all buffered commands
CTRL	Enter control mode
B.	*Window manipulation*
FIT	Fit drawings into screen
WIND	Enlarge specified part of drawing into window
DWIN	Decrease magnitude by half
SHFU	Move window up by 1/2 window height
SHFD	Move window down by 1/2 window height
SHFL	Move window left by 1/2 window width
SHFR	Move window right by 1/2 window width
C.	*Level manipulation*
EDTL	Make specified levels editable
XEDT	Exclude editable levels from display

User interaction with the editing commands is carried out through the use of a stylus. These available instructions are displayed at the right of the screen, and the user can select one of them with the stylus. At the top of the screen, accepted instructions are also displayed.

Some implementation examples are shown in Figs 23.3 and 23.4.

C-GRAPH is a routine to complete the artwork layout pattern of a board at the final stage, based on the Artwork File. In its implementation, the layout pattern is realistically displayed, such that each wire segment has a specified width and each plated-through hole or via is in the shape of a circle with a specified diameter. To manipulate the layout associated with unstandardized circuit modules, facilities are available to alter the grid size, as desired by the user, with precision up to 0.01 mm.

In addition to the commands listed in Table 23.2, the following are also available in C-GRAPH.

Table 23.2 Operating command list for graphics updating (*continued*)

D.	*Select manipulation*
SELA	Select all components
SELC	Select centre of specified component
SELE	Select end of specified component
SELP	Select specified pins or through-holes
SNET	Select all components of specified signal net
UNSA	Unselect all selected components
UNSC	Unselect all selected centres of specified components
UNSE	Unselect all selected ends of specified components
UNSP	Unselect specified pins or through-holes selected
E.	***Component manipulation***
ADD	Add component on specified point
DELT	Delete selected components
MOVE	Move selected components
SWAP	Change existing level of selected components
CHGS	Change signal net selected
CHGD	Change spindle size for selected pins
CHGR	Change photo aperture for selected pins
CHGL	Change photo aperture for selected lines
NAMS	Display statistics for selected component
LOCK	Switch to prohibit any oblique line
CUTL	Cut specified line into two lines
DNET	Display all components of selected signal net
DTBL	Display details of used parts
F.	***Grids and rulers manipulation***
SCAL	Display x, y axes
GRID	Grid display switch
GDON	Grid on-off switch
DLOC	Display abs. co-ordinates at specified point
DRUL	Display ruler between two specified points
WGRD	Change working grid
REAL	Switch display mode from (to) symbolic to (from) realistic (or composite)
DIVD	Divide the straight line between two specified points into k equally parts, so that k line segments with the same width can pass across the line

23.9 Replacement Routine

This is optionally provided for modifying the placement given by a user, and is applied mainly when the wiring performance through the first run of the automatic routing scheme is 80% or less.

If the user wants to replace modules on the board, he can accomplish it by

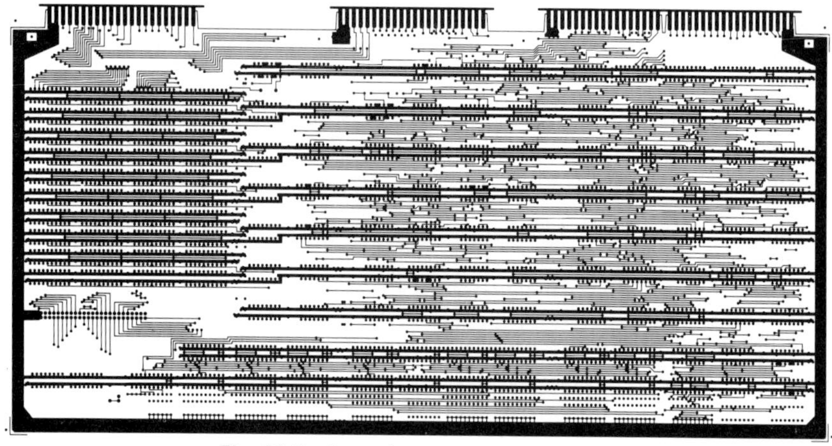

Fig. 23.3 Artwork pattern on the first side.

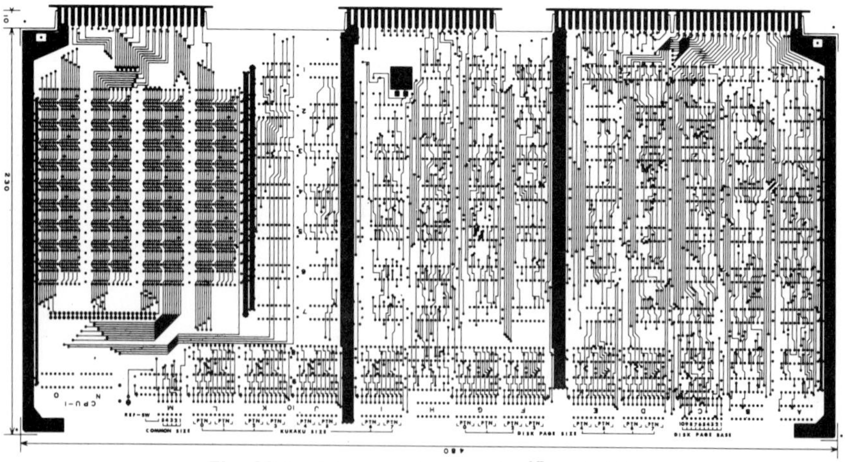

Fig. 23.4 Artwork pattern on the second side.

the use of graphics. In this routine, a display of the modules on the board can be viewed. In order to analyse the placement quality visually, minimum spanning trees for some or all of the nets can be displayed. Through the use of interactive display techniques, the effects of module movement and exchange can be instantaneously realized.

23.10 Implemented Results

This system has been used for the layout and manufacture of more than 300 two-layer boards, all with 100% routing. For equipment to be made on an experimental basis, as opposed to finished products, 100% routing is now always sought.

To see the wirability of the router in this system, we have compared it with APD-W of IBM, Japan, which is a batch program extensively used in Japan.

Statistics of all the five sample boards are as follows.

Size: 200 rows by 204 columns.
Number of modules mounted:

# connectors	22-pin × 4	equivalently
# ICs	16-pin × 53	14-pin × 81 ICs
# LSIs	24-pin × 8	

Module density: 1.26 inch^2/14-pin IC

Implemented results for all these five boards are shown in Table 23.3, where *#connections* indicate the total sum of required connection pairs, that is, the total sum of $|T_i| - 1$ for all nets T_i, *LINE* and *MAZE* denote wiring performances of the line-search and maze-running portions, respectively (associated with this system these two routines are without interaction).

Table 23.4 shows a part of the implemented results by this system, where *Wiring Area* indicates the effective area for the routing process, *Module Density* denotes an area in inch^2/IC equivalent, and *Rerun of MAZE* means the wiring portion performed by the maze-running routine after modification of wire pattern in S-GRAPH.

Figures 23.3 and 23.4 show the artwork patterns on the first and second sides, respectively, of a board output by a photo-plotter.

23.11 Conclusion

We have proposed a minicomputerized automatic layout system for two-layer printed wiring boards, and outlined its functional organization. This system works in practice, and almost all the wiring boards of manufacturing use so far processed by this system have been completed with 100% routing. Thus, this system is contributing much towards reducing the cost and time incurred in laying out wire patterns.

Development is continuing on more sophisticated automatic routing and placement algorithms, as well as on an efficient scheme for the assignment of gates to circuit modules.[14]

Table 23.3 Comparison of this system with APD-W

Datum No.	1		2		3		4		5	
Systems	APD-W	This	APD-W	This	APD-W	This	APD-W	This	APD-W	This
#nets	157		179		249		202		178	
#connections	324		352		420		343		370	
#used-pins/IC	7.9		8.6		10.3		8.7		8.8	
LINE	79.63%	94.75%	83.81%	98.30%	75.71%	90.21%	78.43%	93.40%	79.19%	90.54%
MAZE	16.36	2.78	13.92	1.42	16.43	4.33	16.61	3.70	15.40	5.95
Total	95.99	97.53	97.73	99.72	92.14	94.54	95.04	97.10	94.59	96.49
#isolated-pins	15	9	11	1	35	32	21	15	24	21

(note here that all these boards have been completed with 100% routing by the next run of MAZE in S-GRAPH)

Table 23.4 Implemented results

Board number	Board size (inch)	Wiring area ($inch^2$)	# pins	Module density ($inch^2$/IC)	# nets	# connections	Wiring performance (%) LINE	1st MAZE	Rerun of MAZE	TOTAL
1	9.5 × 9.1	65.4	1092	0.84	137	190	82.11	15.78	2.11	100
2	12.3 × 11.6	130.9	2283	0.80	220	338	74.26	17.46	8.28	100
3	6.1 × 11.4	57.1	1306	0.61	190	335	76.12	18.51	5.37	100
4	9.5 × 10.9	66.9	1196	0.78	123	169	78.70	15.38	5.92	100
5	9.5 × 10.9	98.9	1487	0.93	106	169	78.11	14.79	7.10	100
6	10.3 × 9.0	83.7	1160	1.01	294	530	91.32	8.68	0	100
7	9.5 × 11.0	43.2	456	1.33	68	101	97.03	2.97	0	100
8	12.3 × 12.8	141.6	1768	1.12	260	507	97.44	2.56	0	100
9	12.4 × 11.7	119.9	1496	1.12	237	407	73.71	17.69	8.60	100
10	12.3 × 11.6	112.5	1665	0.95	319	652	90.49	6.90	2.61	100
11	12.5 × 12.5	119.8	1811	0.93	287	517	74.85	18.38	6.38	99.61
12	8.5 × 12.9	93.7	1105	1.19	218	316	87.66	8.86	3.48	100
13	9.5 × 11.5	82.8	1457	0.80	316	604	78.97	16.23	4.80	100
14	12.4 × 12.0	124.1	2270	0.77	263	467	83.08	8.56	7.75	99.36
15	12.4 × 11.6	128.1	1617	1.11	345	635	78.74	14.96	6.30	100

References

1. Breuer, M. A. (ed.), *Design Automation of the Digital Systems*, vol. 1. Englewood Cliffs, N.J.: Prentice-Hall, 1972.
2. Nakahara, H., "Computer-aided interconnection routing: General survey of the stage-of-the-art," *Networks*, **2**, 167–183, 1972.
3. Hightower, D. W., "The interconnection problem – A tutorial," *Proc. Design Automation Workshop*, pp. 1–12, 1973.
4. Abel, L., "On the automated layout of multi-layer planar wiring and a related graph coloring problem," *Coordinated Sci. Lab. Rept.*, No. R-546, Univ. Ill., 1972.
5. Rau, B. R., "A new philosophy for interconnection on multilayer boards," *Proc. Design Automation Conf.*, pp. 225–231, 1976.
6. Allen, J. R., "A topological adaptable cellular router," *Proc. Design Automation Conf.*, pp. 161–167, 1976.
7. Patterson, G. L. and Phillips, B. H., "A proven operational CAD system for p.w.b. design—based on a mini-computer and featuring fully automatic placement and routing," *Proc. Design Automation Conf.*, pp. 259–264, 1976.
8. Mikami, K. and Tabuchi, K., "A computer program for optimal routing of printed circuit conductors," *Proc. IFIP Congress*, pp. 1475–1478, 1968.
9. Yamamura, H., Shirakawa, I., and Ozaki, H., "A line-search method for the route connecting problem on 2-layer printed circuit boards," *IECE Trans.*, vol. 57-A, pp. 671–678, 1974 (in Japanese).
10. Lee, C. Y., "Algorithm for path connection and its applications," *IRE Trans. Electronic Computer*, vol. EC-10, pp. 346–365, 1961.
11. Geyer, J. M., "Connection routing algorithm for printed circuit board," *IEEE Trans. Circuit Theory*, vol. CT-18, pp. 95–100, 1971.
12. Rubin, F., "The Lee connection algorithm," *IEEE Trans. Computer*, vol. C-23, pp. 907–914, 1974.
13. Chiba, T., Shirakawa, I., and Ozaki, H., "A maze-running method for the route-connection on two-layer printed circuit boards," *IECE Trans.*, vol. 59-A, pp. 247–253, 1976 (in Japanese).
14. Nishoka *et al.*, "An approach to gate assignment and module pacement for printed wiring boards," *IEEE Trans. on Computers*, vol. C-29, pp. 681–688, 1980.

The paper was published in 1977, and the system has been continually maintained and polished up as follows.

(1) The system now provides with layout schemes for analog PWBs, hybrid ICs, LSIs, as well as digital PWBs.
(2) Regarding digital PWBs, multi-layer boards with up to 16 signal layers can be treated.
(3) Regarding automatic design scheme for digital PWBs, the system now provides with automatic gate assignment and module placement.
(4) The software is fundamentally hardware independent, but a VAX 11/750 or 780 is most commonly used.
(5) Raster scanning colour graphics are more commonly used rather than storage graphics.
(6) 3-Dimensional mechanical design CAD/CAM system has been developed and added to the application family.

24

An Integrated Information System in an Ironmaking Process

Kawasaki Steel Corporation, Tokyo, Japan

24.1 Introduction

Scientific methods for blast-furnace (BF) operation are becoming established. In ironmaking, the complicated reactions in the blast furnace have recently become clearer owing to the development of various sensors and the use of statistical techniques. After years of in-house development, Kawasaki Steel completed a BF operation stabilizing system (GO-STOP system). Subsequently, the company developed a large-scale on-line database as an integrated ironmaking information system covering ore treating, sintering and blast-furnace plants, because the control of the ore preparation process, which precedes the ironmaking process, is very important in the control of the blast-furnace operation. This system not only performs daily operation and control, but also aids the development of ironmaking techniques and their application to operation by making use of a database and various techniques of analysis and modelling. As a result, there has been remarkable progress in the standardization of blast-furnace operation, an improvement in the level of control, and a great contribution has been made to the maintenance of stable operation. At the same time, in the steel industry great results

CAD/CAM and MIS in Japan
ISBN 0-12-682580-7

have been achieved in resource and energy conservation, which are topical areas in the steel industry.

24.2 Background and Objective of Development

The ironmaking department is responsible for the basic process in an integrated iron and steel works. Stable operation is therefore the first requirement. The energy which the Japanese steel industry consumes is as high as 15% of the total energy consumption in Japan, and the ironmaking department consumes 75% of the total energy consumption of the Japanese steel industry. In recent years, there have been remarkably steep rises in the prices of raw materials and fuel. Hence, energy-saving and resource-saving have become equally important. An improvement in the quality of pig iron has also recently become necessary to reduce the costs in the next process, the steelmaking department. Another of the great problems is how to lengthen furnace life by stable operation of the blast furnace, because it requires enormous investment in plant and equipment.

The problem common to all processes in the ironmaking department is that quantitative analysis of the processes is difficult because of the existence of many complex operating parameters and the fact that there is no information available to express correctly the process conditions. Direct measurement of the process is especially difficult in the blast furnace, and the large size of the equipment produces great inertia in the process. The above-mentioned problems are therefore very serious. Consequently, there is great reliance on the skill and experience of operators for blast-furnace operation and control.

To solve these problems, Kawaski Steel has pushed forward with the development of an integrated information system in the ironmaking process for the following purposes.

(1) Abundant operational data are accumulated in a database through the development of sensors. The development of operating techniques has advanced and scientific operational control has been achieved by encouraging the utilization of these data.
(2) The standardization of operation is carried forward and proper operating instructions are given.
(3) This new system aims at not only optimization within the ironmaking department, but also an optimization of the works as a whole by strengthening the links with other peripheral systems such as the energy supply and demand system of the works.

Figure 24.1 shows the ironmaking process.

24.3 The History of Systematization in Ironmaking

The attempt to systematize the ironmaking process was begun in the period from 1965 to 1970. Development was carried out by an in-house team made up from staffs of the Research Laboratories and departments concerned at the Chiba and Mizushima Works. Figure 24.2 shows the basic concept of the development.

The history of systematization is shown below taking blast furnaces as an example.

Mizushima No. 1 and No. 2 BF system (1969)
Process computers were introduced and blast-furnace process analysis was started for the first time after the construction of the Mizushima No. 2 blast furnace.

Chiba No. 5 BF system (1971)
The GO-STOP system began operation and it is based on blast-furnace process analysis and application of sensor techniques. This system was very effective in preventing furnace irregularities.

Chiba No. 6 BF system (1977)
A large-scale instrumentation and computer system was introduced at the construction of the Chiba No. 6 blast furnace, which may be said to be the results of the accumulation of all the company's blast-furnace techniques. This system marked an important stage in the development of an integrated system by its use of centralized control by process computers, establishment of a medium-scale database, information exchange with adjacent processes and a central computer using networks, etc.

Integrated ironmaking information system (1976–)
The control of preceding processes, such as blending of the raw materials and sintering plant, is very important in the control of blast-furnace operation. Therefore, rather than develop a system for each process, it was decided to systematize the whole ironmaking processes. The development of this information system was started under company-wide support so that it could be used at the Mizushima Works where process computers were then becoming obsolete. The use of this information system for BF-related functions started in 1979 and gradually extended to the sintering plant and raw material yards.

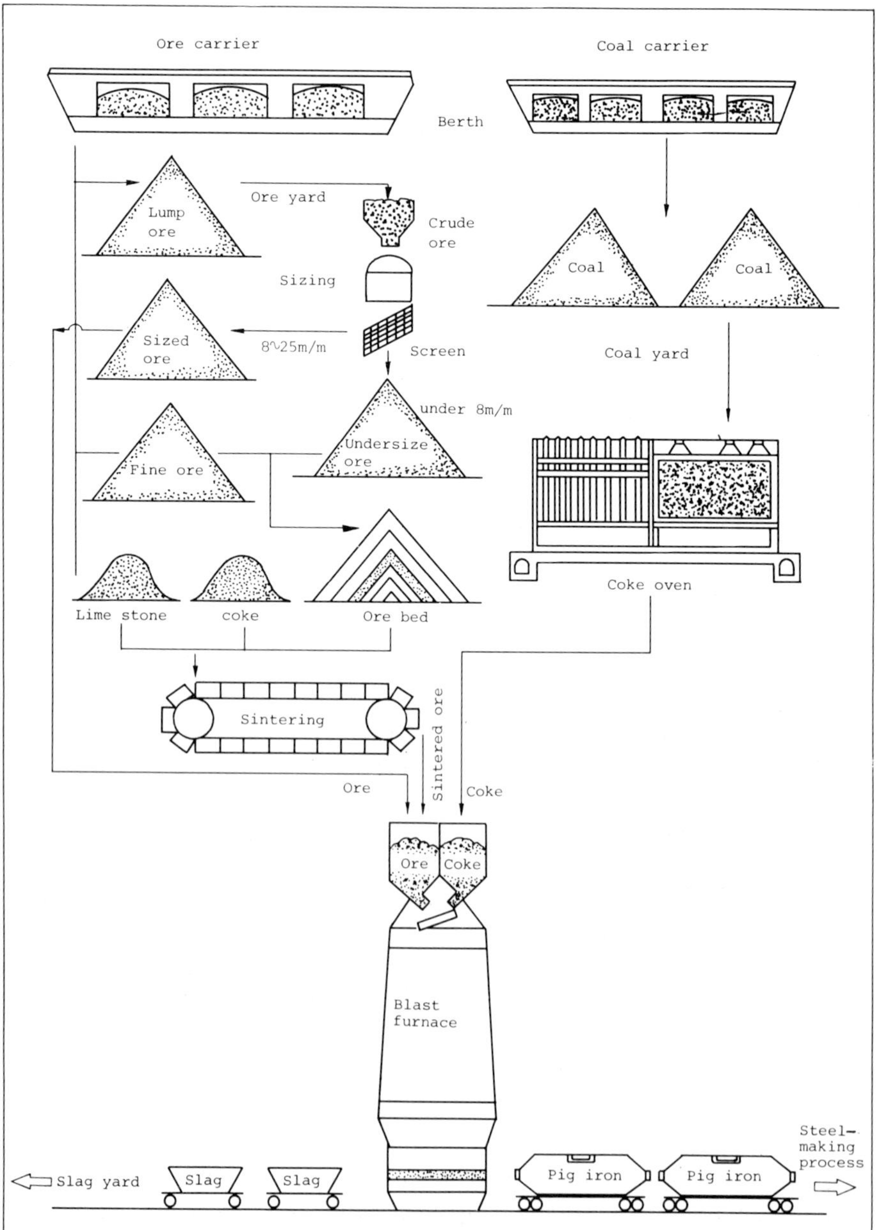

Fig. 24.1 Ironmaking process.

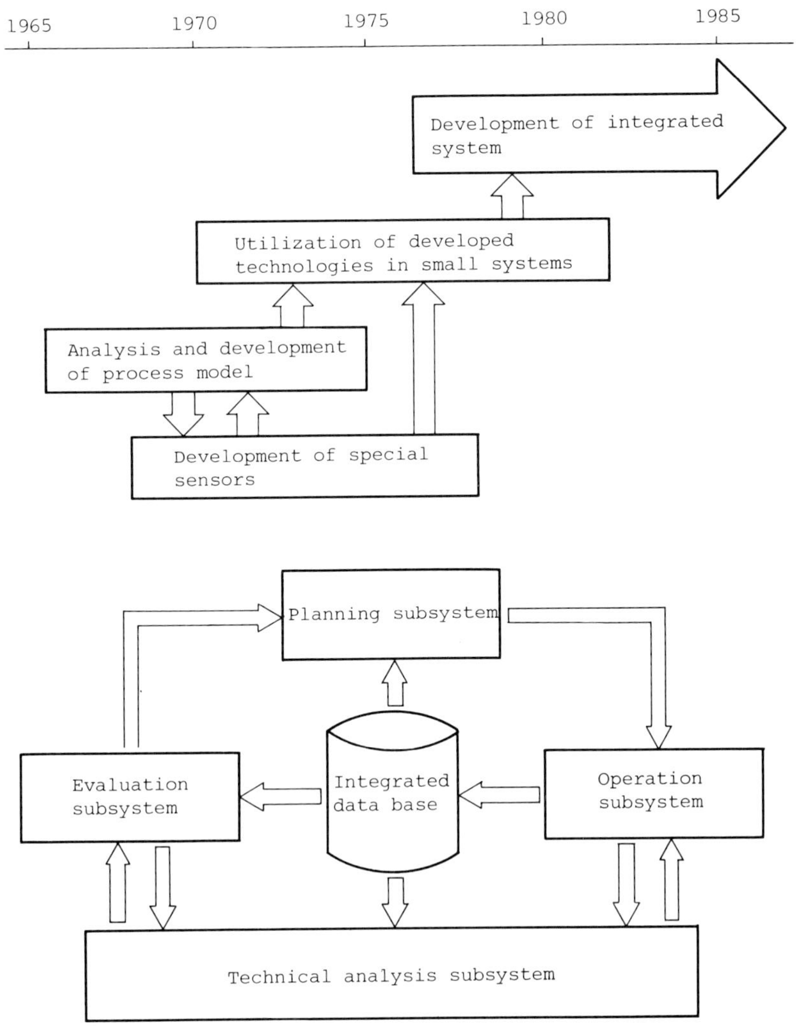

Fig. 24.2 Process of systematization (above) and system functions (below).

24.4 Features of the System

24.4.1 Integrated System

The process in the raw material yards, sintering plant and blast-furnace plant was developed into an integrated system. At the same time, a large-scale database incorporating the operational data and plan data of these processes was built in the central computer to achieve the integrated system.

24.4.2 Clear Division of Functions

The structure of system functions was divided into four—plan, operation, evaluation and technical analysis—to clarifying the system's functions (Figure 24.2).

Furthermore, data gathering was allocated to the process computers including sensors, measuring instruments and control equipment, the remaining functions being undertaken by the central computer, thus clarifying the allocation of functions.

24.4.3 Establishment of Database and Promotion of Its Use

A large-scale database was established in the central computer and operational data was systematically gathered and accumulated. Various kinds of terminal equipment and software were provided so that data analysis could be made in an interactive manner. Thus there was a development of operation techniques. System utilization is as much as 10 to 20% of the total computer processing time of the whole works.

24.4.4 Integration of Process Computers

The raw material yards, sintering plants and blast-furnace plants are each provided with a process computer. They are all hooked-up to a stand-by process computer. All of the four process computers are of the same type. Thus system reliability is improved and system development and maintenance simplified.

24.4.5 Progressive Expansion of the System

This system is designed to be expanded in stages because the scale of development is very large when the process control of the raw material yards, sintering plants and blast-furnace plants is included, especially when the remodelling of various kinds of equipment is carried out at the time of the scheduled relining of a blast furnace.

24.4.6 Efficient System Development

The method adopted is one where typical systems are developed by the system department, while atypical functions are developed by the ironmaking department using the language of the end user. The two departments therefore perform efficient development activities.

In addition, a host preparation system was established in which the central computer performs the development and maintenance of software for the process computers.

24.5 Outline of the System

24.5.1 Planning Subsystem

The planning subsystem was developed by placing emphasis on the raw material blending plans for each process in the raw material yards, sintering plants and blast-furnace plants, and also on blast-furnace operation plans. This planning subsystem allows case studies to be made easily and interactively, and permits optimum human judgement. This system aims at not only the optimization of ironmaking, but also at integration of the whole works through an organic linkage with the energy supply and demand optimization system at the works level.

24.5.2 Operation Subsystem

This subsystem is used for the operation and the operational control of the raw material yards, sintering plants and blast-furnace plants. Various kinds of preset control and tracking are allotted to the measuring instruments, control equipment and microcomputers. Control and optimization is carried out by the process computers. A typical example is the overall assessment of the condition of the blast-furnace by the GO-STOP system. Data gathered by the operation subsystem are fed on-line to the central computer and accumulated in the ironmaking database.

24.5.3 Evaluation Subsystem

In this subsystem, operational data are gathered, accumulated and processed and the results are evaluated by comparing them with projected values. For example, actual monthly raw material distribution is compared with planned distribution. Simultaneously, daily, 10-day and monthly reports (such as daily sintering operation logs, and monthly reports on blast-furnace temperature control) are output. Chinese character terminals, graphic displays, etc. are used so that the operator can easily see and understand the information.

24.5.4 Technical-Analysis Subsystem

The efficient development of techniques such as operation analysis was aimed at by using the overall ironmaking database and the language of the end user. An interactive language was introduced to facilitate the modelling of the process, technical analysis and model development. At the same time, efficient operation analysis was achieved by representing analysis results on graphic displays.

24.6 Functions of the System

Typical functions utilizing developed models are described in the following.

24.6.1 Ore-Bed Quality Estimation System

Quality control of ore beds is important for keeping the chemical composition of the sinter constant.

Raw materials for sinter are stacked in multiple layers on ore beds and are reclaimed by a double-wheeled reclaimer. This system simulates the condition of an ore bed on which more than twenty kinds of sintering raw materials are uniformly mixed, and facilitates suitable action such as the stacking sequence. In addition to analytical values and stacking amounts for each ore, data on the angle of repose and stacking models for the longitudinal and lateral directions of a bed are used. This permits simulation of the chemical composition of a bed more accurately. Stacked ore amounts are measured on-line using a newly developed device for processing signals from stacking position detectors and conveyor scales with the aid of a microcomputer. Figure 24.3 shows an image-plane output of the interactive simulation and variations in the estimated composition of the raw mix on the conveyor when it is reclaimed. Standard deviation of SiO_2 in the sinter was improved from 0.15% to 0.09% by using this system.

24.6.2 Mathematical Model of Sintering

In the sintering process, the improvement of sinter quality control and productivity was achieved by modelling the reactions and heat transfer in a sinter bed, and by simulating both the shape of the combustion zone in the longitudinal direction of a sintering machine and the heat patterns in a sinter bed. An automatic measuring device was developed to measure air volume

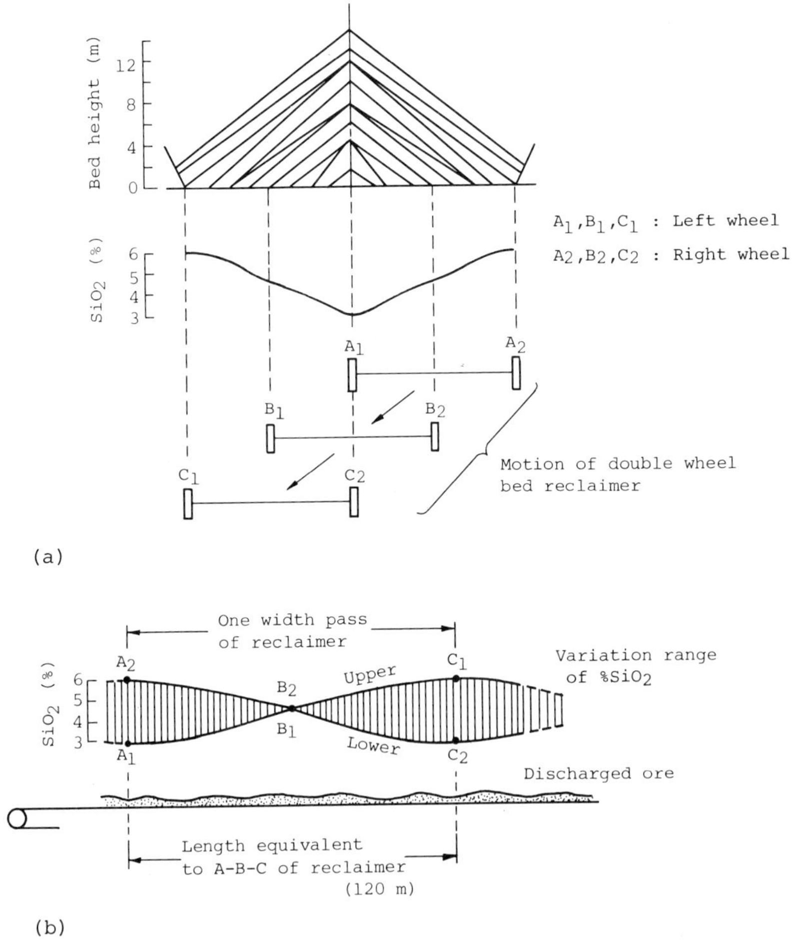

Fig. 24.3 Diagram of ore bed quality model. (a) The % SiO_2 distribution along the bed width direction; (b) the % SiO_2 cyclic variation range along the longitudinal direction of a discharging conveyor.

distribution, which is important input data in this model. Furthermore, a radio transmission type heat pattern measuring device was developed to verify the mathematical model.

This mathematical model is based on equations of heat and material balance in a segment of a sinter bed, and the solid and gas temperature in a bed is calculated by integration. Rapid temperature changes in a sinter bed can be simulated with a high degree of accuracy using this model.

Figure 24.4 shows an example of the results of calculations of the distribution of a combustion zone above 900°C.

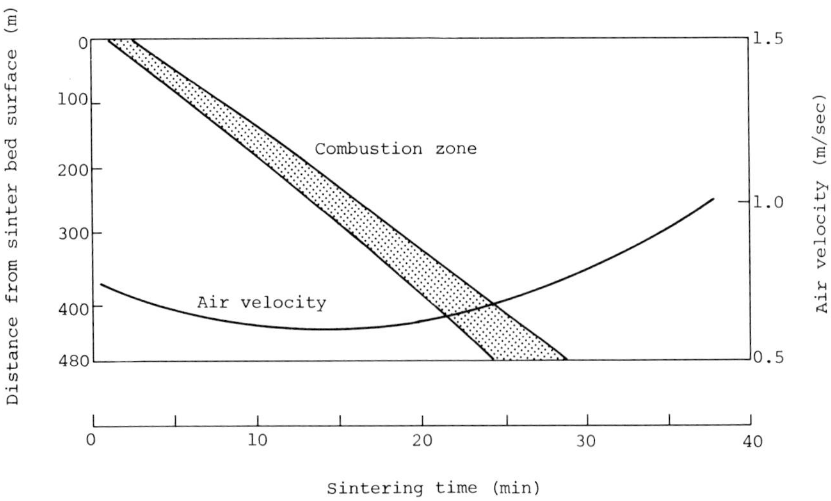

Fig. 24.4 Distribution of combustion zone and air velocity in sintering bed.

Indices representing the sintering process, such as heat retention index and cooling index, were obtained from the results of these calculations and their relations to sinter quality and operating conditions were analysed. Operation has been improved, sinter quality has been stabilized, and the fuel ratio has been reduced (coke consumption: approximately 4%) by using this system.

24.6.3 BF Operation Design System

Improvements have been made over a period of almost twenty years in the development of mathematical models of the ironmaking process. At Kawasaki Steel, the gas utilization rate is estimated with a high degree of accuracy by correlating it with the hydrogen content at the front of tuyere. The company's mathematical model has been put to practical use not only in process analysis, but also in daily BF operation design.

The coke ratio, generated gas volume, etc. are determined by the operating conditions such as blast temperature and blast humidity. The cost of hot metal, including the fuel consumption of the hot stoves, is calculated.

The blast furnace is a source for generating the heat energy consumed at a works. Therefore, control of the volume of generated BF gas is necessary in order to reduce the energy cost of the entire works. In conjunction with the works' energy cost evaluation system, this system is used for examining and determining blast-furnace operating conditions.

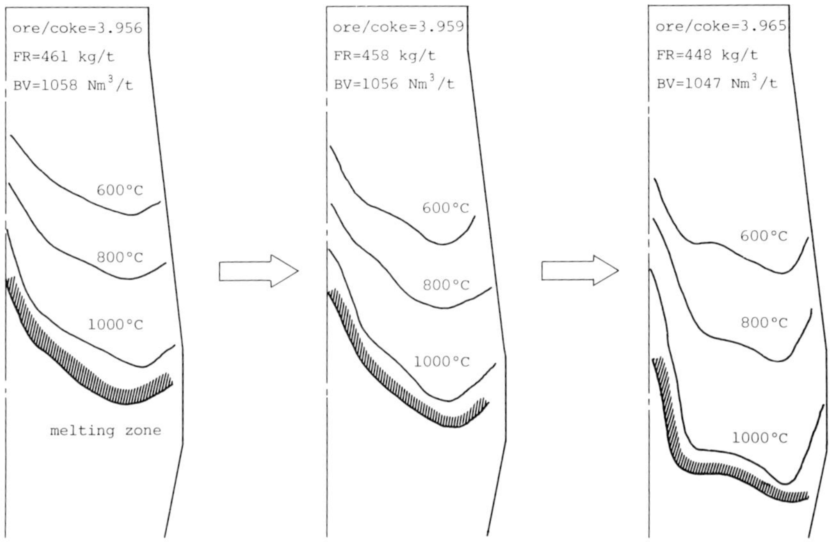

Fig. 24.5 Temperature profile in blast furnace with the change of fuel ratio and blast volume.

24.6.4 Furnace Condition Control by Multi-Cylindrical Model

In a large blast furnace with a large hearth diameter, it is important to ensure a stable furnace condition to control the temperature distribution in the furnace, radial distribution of descending feed, gas flow distribution, etc. The concept of cohesive zones has been developed by Kawasaki Steel into a unique model. Here, the blast furnace is divided into concentric cylinders and the temperature distribution in the furnace is estimated from the material and heat balance in the divided regions. The company has continued to obtain verification using this model.

In addition to basic data on blast-furnace operation, data on radial distribution of temperature and gas composition obtained by furnace top probes and shaft probes are used for calculations adopting this model. Figure 24.5 shows the results of estimated temperature distribution in the furnace using this model.

The correlation between blast-furnace operating conditions and the temperature distribution in the furnace has been examined using this model, and efforts are being made to discover and analyse the shape of cohesive zones when the blast furnace produces hot metal with a low silicon content. Thus proper burden distribution control is performed to maintain a desirable cohesive-zone shape.

The observation of the burden profile at the furnace top is indispensable

Mechanical type	Microwave type	Laser type
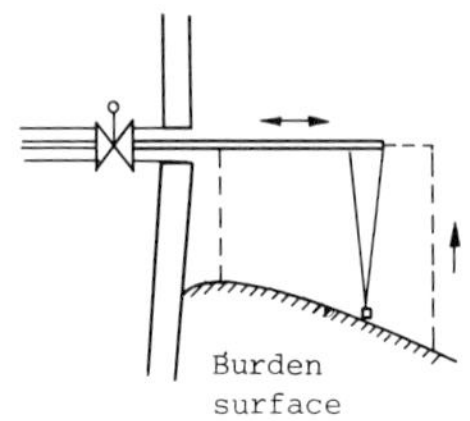	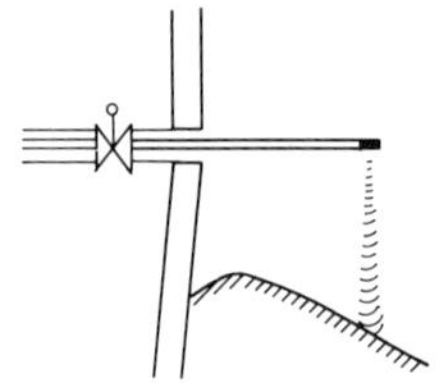	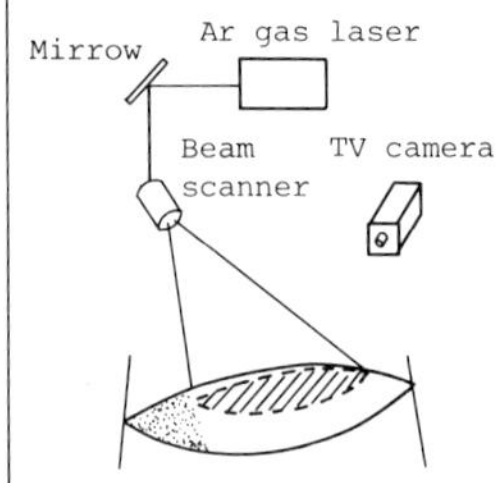
Measuring time 60 sec Accuracy ±50 mm Measuring point 10 (Discontinuous)	Measuring time 120 sec Accuracy ±130 mm Continuous measurement in the radial direction	Measuring time 2 sec Accuracy ±50 mm Continuous measurement of burden surface

Fig. 24.6 Burden surface profile meter of blast furnace.

for proper burden distribution control. For this purpose, mechanical, microwave and laser profile meters were developed (Figure 24.6).

The shape of cohesive zones can also be estimated by observing the gas flow distribution in the furnace. This gas flow distribution is indirectly measured by measuring the temperature and composition distribution of the top gas.

Figure 24.7 shows data on the gas composition distribution in the furnace measured using horizontal probes. Figure 24.7(a) directly represents the radial gas composition distribution in three different periods of operation: A, B and C. There are many overlaps in the three periods and it is difficult to form a judgement. In Figure 24.7(b), this data is reanalysed by the discriminatory function method, and is clearly separated in the discrimination space.

Data detected by sensors are rearranged in patterns using a discriminatory function. Operating instructions are given to operators based on this method to prevent furnace irregularities and maintain an optimum gas flow distribution.

24.6.5 GO-STOP System

One of the most serious furnace irregularities is chilling. Chilling occurs due to the interaction of several factors. The GO-STOP system is an on-line computer based blast-furnace operation method, developed to rapidly detect fur-

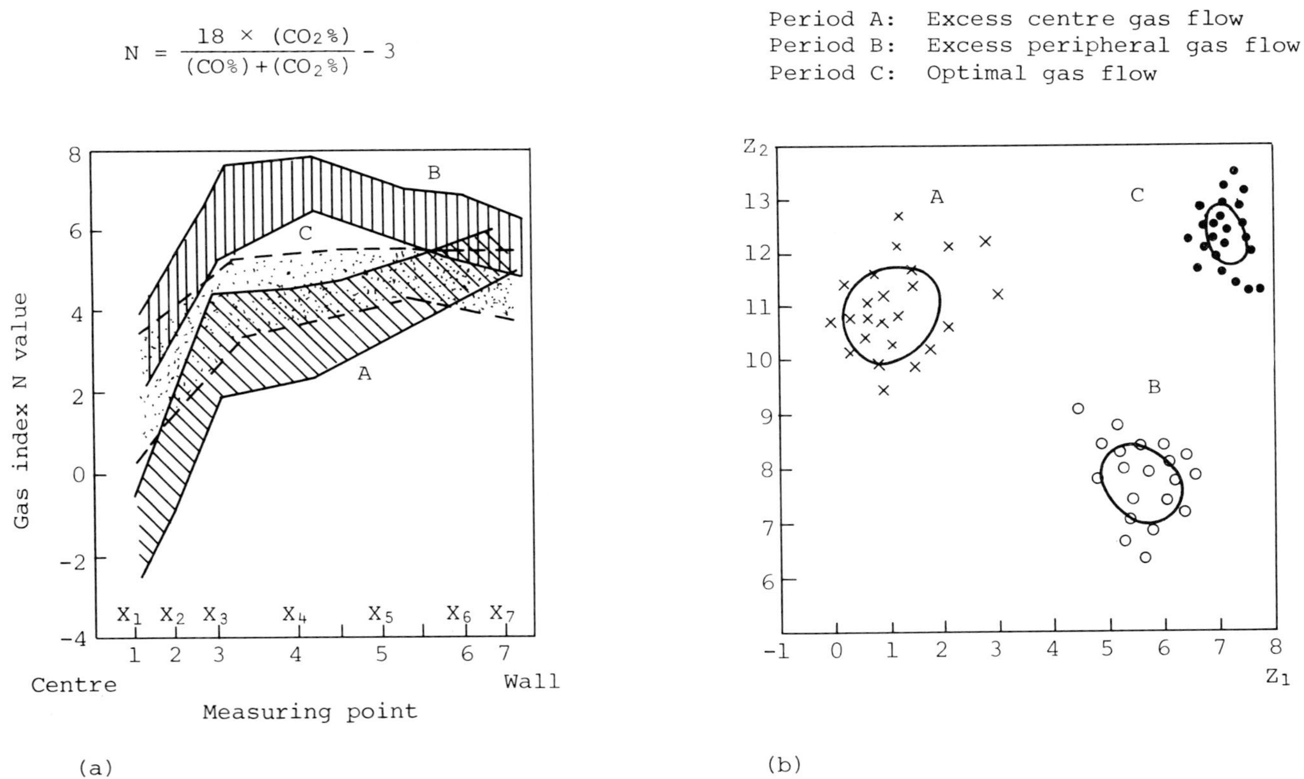

Fig. 24.7 Classification of gas flow pattern in the blast furnace. (a) N volume pattern in the radial direction of the top; (b) discrimant analysis of gas flow pattern.

nace irregularities and maintain a stable furnace condition by taking suitable action.

The system was developed by analysing and examining furnace irregularities such as chilling based on the analysis of past operational data and theories of the ironmaking process. This system is shown in Figure 24.8.

Operational data necessary for judging the furnace condition are calculated as various indices that are closely related to the descending condition of the burden, furnace heat, amount of residual molten slag, etc. The furnace condition is judged from the levels of the eight factors, and variations in the four factors, shown in Figure 24.8. Results are judged by three stages: GO (good), STOP (caution) and BACK (worsening). They are represented on a CRT in the form of a spider's web graph shown in Figure 24.9. In accordance with these instructions, the operator takes the prescribed action (for example, reduction of the blast volume).

This system has been applied to all the company's blast furnaces and since its introduction has contributed greatly to stable blast-furnace operation without causing serious furnace irregularities, in spite of a low fuel-ratio operation.

24.7 Effects of Introducing the System

In the raw material preparation department, the introduction of this system has enabled the properties of blast-furnace raw materials to be improved and their high quality to be retained. Operating techniques in the blast-furnace department have also been remarkably improved. As a result, a stable blast-furnace condition has been maintained. Indices that indicate a stable blast-furnace condition are improved productivity, reduction of the fuel ratio and stable quality. Although an increase in production cannot be expected in the present economic environment, there has been no decrease in production as a result of problems. In spite of the changeover to oil-less operation, the fuel ratio has been positively decreasing, thus contributing greatly to reduction in the cost of hot metal. There has also been an improvement in quality, and the decrease in the silicon content of hot metal has resulted in advantages for the steelmaking process. Both a decrease in the fuel ratio, and a low silicon content of hot metal, involve a risk of chilling in the blast furnace. They can be achieved therefore only when the furnace condition is stable.

A recent decrease in tuyere breakage caused by an unstable furnace condition is also an important advantage.

Kawasaki Steel has achieved many satisfactory deals involving the export of blast-furnace operation techniques—for example, guidance in blast-fur-

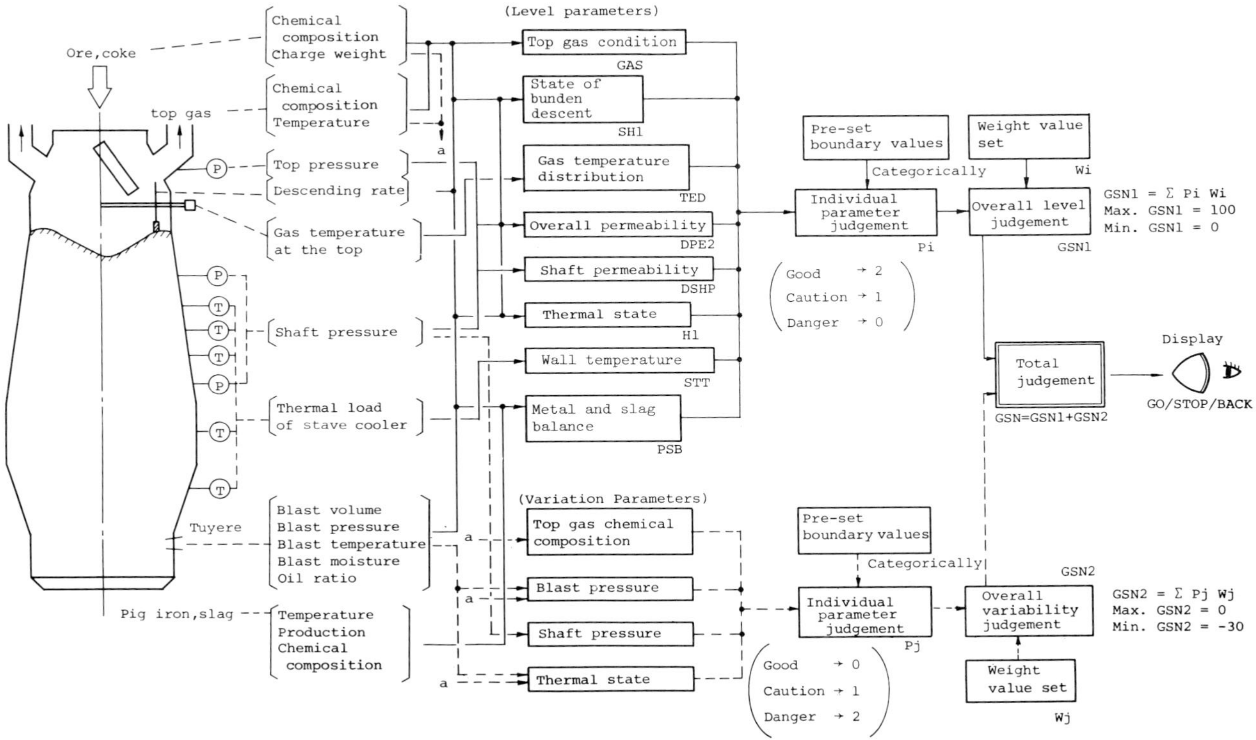

Fig. 24.8 Decision flow of blast furnace condition by GO-STOP system.

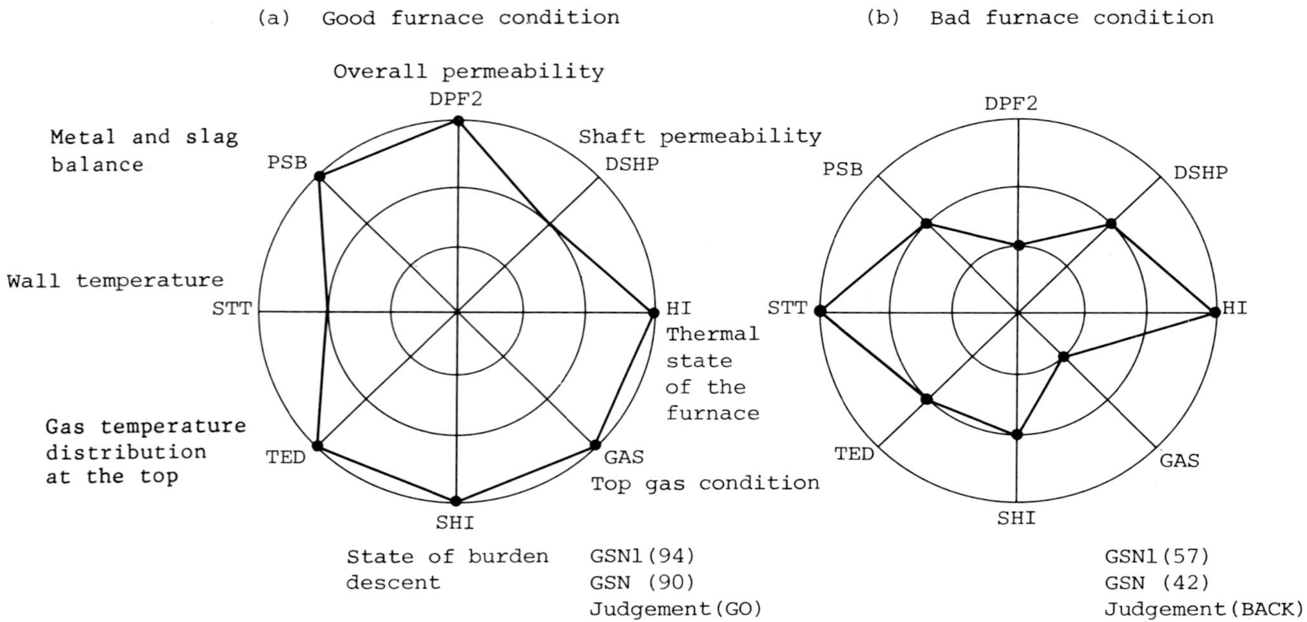

Fig. 24.9 Examples of judgement by GO-STOP system.

nace operation and equipment improvement for USINOR of France, furnishing the GO-STOP system to Thyssen AG of West Germany, and assistance in blast-furnace operation for the Bethlehem Steel Corporation in the USA.

25

The NASD System
NKK Advanced Ship Design System

Y. Hattori, Y. Ikeda and K. Haga, Nippon Kokan KK, Yokohama, Japan

25.1 Background of the System

25.1.1 Characteristics of shipbuilding

In the advanced shipbuilding industry, production process in the assembly stage is similar to the automobile assembly line although production rate differs because of the size of product. However, there is a fundamental difference between both industries. The shipbuilding industry is characterized by contract-oriented production while the automobile industry is based on mass production, which denotes the difference in the framework of production.

The design department in the automobile industry is wholly independent from the production factory, compared to the design department in the shipbuilding industry which cannot be independent. Each design of a ship is different, resulting in a closer tie between design and production in the total production sequence. This means that a wide range of communication usually by means of design drawings is often necessary between the design section and the production section (Figure 25.1)

The role of a drawing is traditionally regarded as an "idea presentation", but must actually be regarded as "data transmission". More than 10 000 drawings are used for each ship, and 70% of them are the working plans drawn by the production design section.

CAD/CAM and MIS in Japan
ISBN 0-12-682580-7

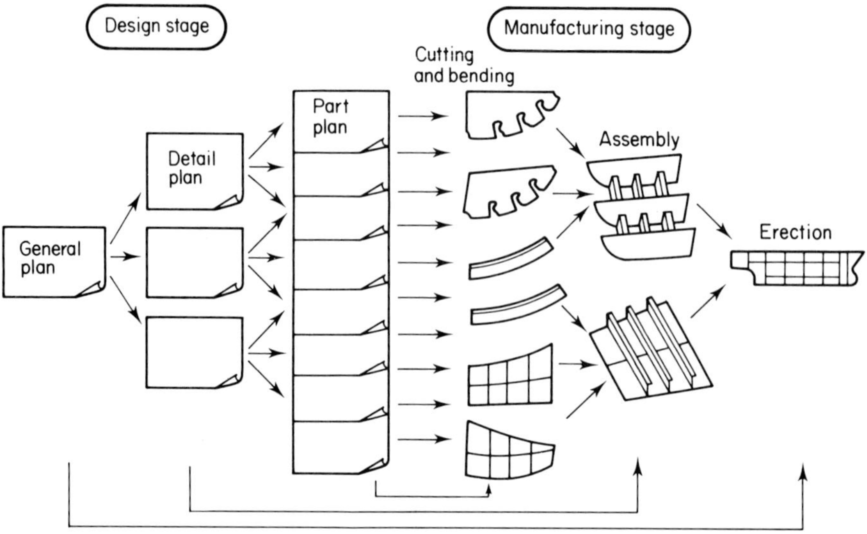

Fig. 25.1 Data transmission in shipbuilding.

The basic idea of the NASD (NKK Advanced Ship Design) system is to handle this data transmission, which is carried out conventionally by drawings.

25.1.2 System Philosophy

Much effort has been spent during the last decade in almost every big Japanese shipyard on implementing a total automated design system. These efforts have not yet given the desired effects. Why is total integration an unrealistic concept? Perhaps the approach has been wrong.

There should be many design and production processes. The fact that so many exist is the basis of shipbuilding technology itself. Therefore to pick only one as the procedure carried out by a program is short sighted.

In the NASD system, neither design procedure nor production process is fixed: only drawing are handled. The important aspect of the NASD system is to feature easy manipulation of drawings. New techniques for creation, modification and deletion of pictures in the computer memory have introduced a new rational method of data transmission fundamentally different from the traditional.

The role of data description has changed. The geometric representation of a drawing has two aspects, pictorial (analogous representation) and dimensions and size (digital representation). In the traditional sense, the main feature of a drawing is an analogous representation, and dimensions are considered as an auxiliary function.

With the NASD system, the digital representation is adopted as the main feature, and the picture in analogous form is regarded as an auxiliary function.

The advantages of this exchange between main function and auxiliary function of data are summarized below.

The first is the improved accuracy of the total shape description (e.g. traditionally, drawings have the accurate dimension at the specified point only, and the remaining points are left uncertain). The second advantage is in quick transmission of data (with manual drawing it is impossible to respond as rapidly to the frequent need for changes in drawing). The third advantage is the preparation of NC (numerically controlled) information for NC machines.

25.2 The NASD System

25.2.1 Fundamentals of the NASD

Configuration of the NASD system is illustrated in Figure 25.2. The system handles two groups of data. The first consists of the basic data which are created in the early stage of design—hull form data, structural location data and scantling data. Once each design step is completed, the design data are stored in the computer memory and referred to in the following processes such as determination of frames, deck lines, butts, seams, longitudinals and inner structures. These data compose the so-called hull-shape file which we call the hull master file (HMF).

The second group of data are the picture data generated by designers, draftsmen and loftsmen. All the data are frequently referred to in the succeeding picture processing.

The typical data in this second group are picture data which are manipulated on a graphic display device by the designer. For this data, we have the picture file called the drawing file (DF). The second group of data also includes several other files which contain pictures for various purposes.

25.2.2 Database

The NASD database consists of several files. The following are brief descriptions.

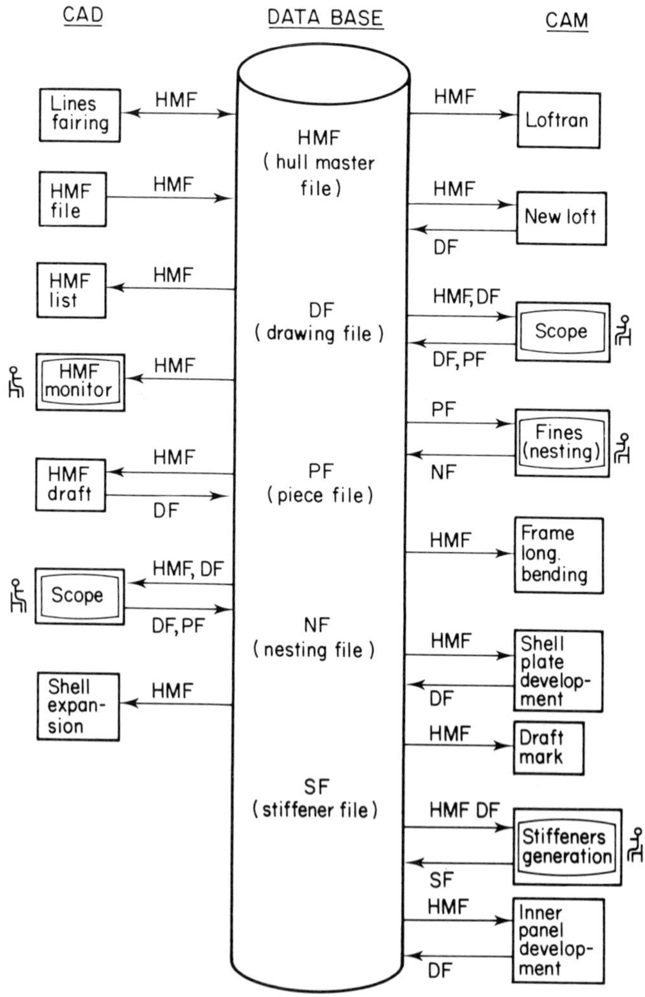

Fig. 25.2 Database and applications of the NASD system.

Hull master file (HMF)
Geometric data of hulls are stored in HMF. Almost every NASD program has access to HMF for storage and/or retrieval. This file contains principal dimensions of ships, frame section lines, seam lines, butt lines, longitudinal frame data, inner structure data, etc.

Drawing file (DF)
This file stores graphic information of design drawings in numerical form. Design drawings can be stored, retrieved and manipulated by SCOPE, which is described later.

Piece file (PF)
Contour data of completed parts are stored here by SCOPE.

Nesting file (NF)
By nesting program FINES, only arrangement information (i.e. co-ordinate values of guide points) is stored in this file.

Stiffener file (SF)
Data for small pieces such as stiffeners and brackets are created by SMART and stored in SF.

25.2.3 Application Programs

The NASD system includes many programs connected with the database. The major programs are as follows.

(1) LINES fairing. This program generates faired hull shapes from rough offsets and stores them in HMF as frame section lines.
(2) HMF file. The geometric hull data other than frame section lines are stored in HMF by this program.
(3) HMF list. The contents of HMF can be printed out in tabulated form by this program.
(4) HMF monitor. The data in HMF can be monitored on the graphic display screen by this program.
(5) HMF draft. The data in HMF can be drafted by this program in batch mode. Figure 25.3 shows a sample of output.
(6) Shell expansion. Shell expansion drawings (Figure 25.4) are generated by this program.
(7) SCOPE. SCOPE is an interactive drafting and part generation program using a graphic display (Figure 25.5). The predecessor of SCOPE was called GLOFT which was used only for part generation. We revised GLOFT by adding some functions and renamed it SCOPE. Figure 25.6 is an example of parts generated by SCOPE.
(8) LOFTRAN and NEWLOFT. LOFTRAN is the part generation language processor. Before GLOFT was developed, LOFTRAN was the only method to generate parts. NEWLOFT is also the part generation language having the same syntax as LOFTRAN, but its output is stored in the Drawing File. We can modify the output of NEWLOFT by SCOPE.
(9) FINES. FINES is the nesting program using a graphic display (Figure

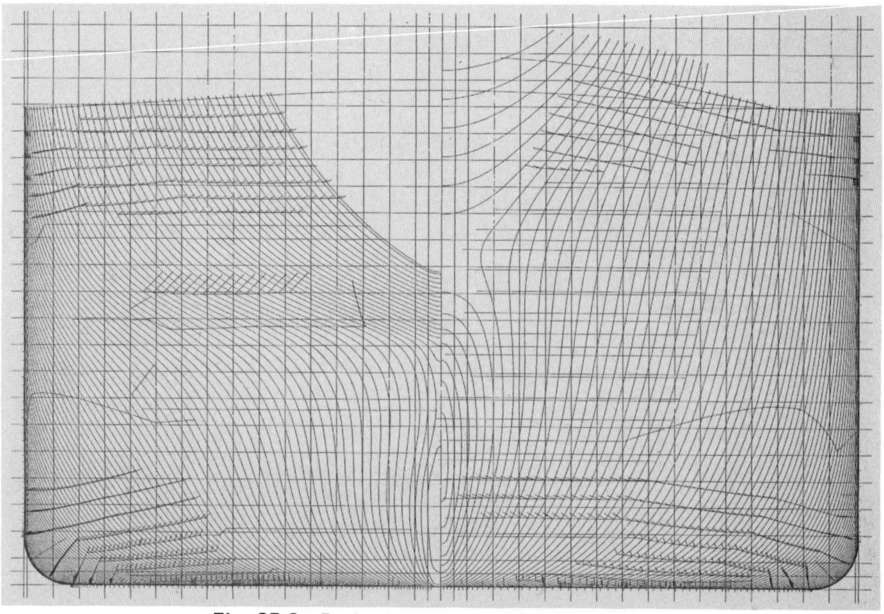

Fig. 25.3 Body plan generated by HMF DRAFT.

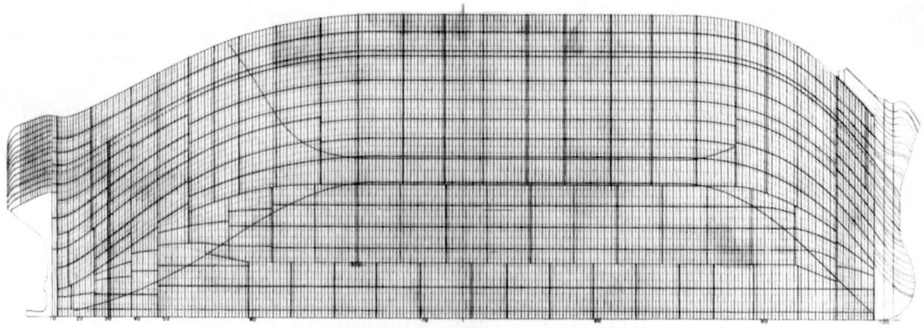

Fig. 25.4 Shell expansion plan.

25.7). The parts data generated and stored in the Piece File by SCOPE or LOFTRAN are nested by FINES.

(10) Shell development. Three kinds of output are generated by shell development programs. These are:

(a) developed shell plate drawings (Figure 25.8);
(b) template drawings;
(c) block offsets.

Fig. 25.5 SCOPE, interactive parts generation system.

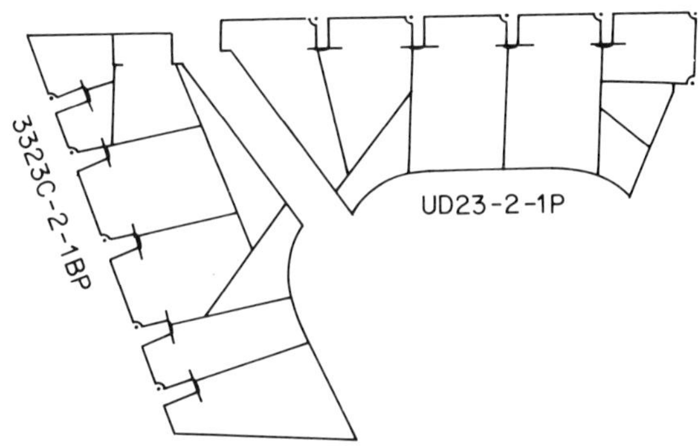

Fig. 25.6 Sample of parts.

(11) Frame bending. Frame bending is performed by the inverted line method. This program generates inverted lines (Figure 25.9) from the information in HMF.

(12) SMART. SMART is the program to calculate the exact shape of stiffeners or brackets which are attached to the main structural members whose pictures are previously defined by SCOPE. The shape of the stiffener can be determined interactively by specifying the position on the main structural member and the end condition. Figure 25.10 is a check

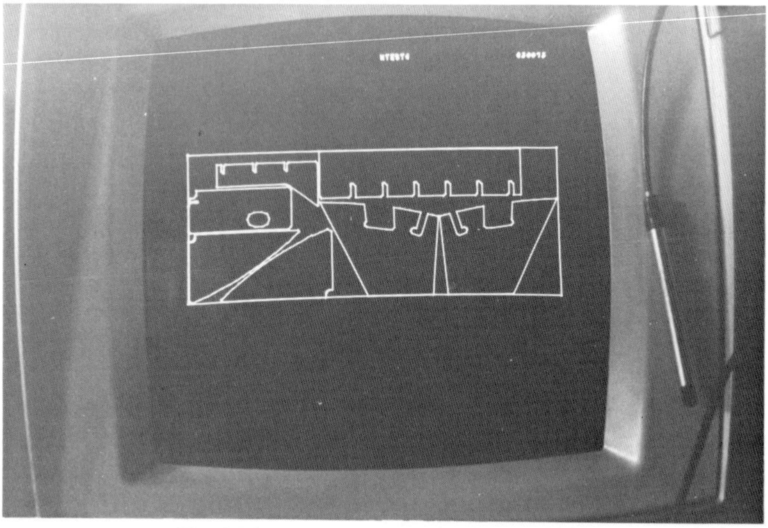

Fig. 25.7 FINES, interactive nesting system.

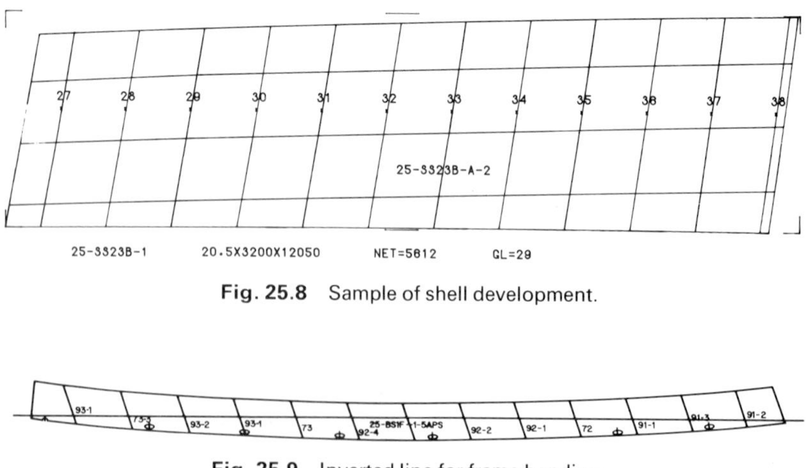

Fig. 25.8 Sample of shell development.

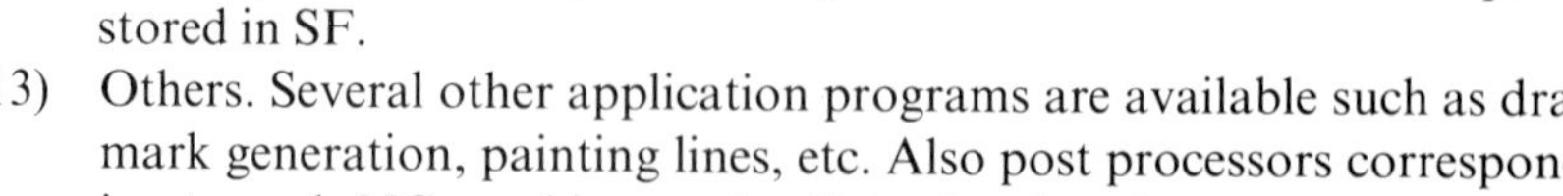

Fig. 25.9 Inverted line for frame bending.

drawing of stiffeners generated by SMART. The resultant output is stored in SF.

(13) Others. Several other application programs are available such as draft mark generation, painting lines, etc. Also post processors corresponding to each NC machine, and utilities for data base management are available.

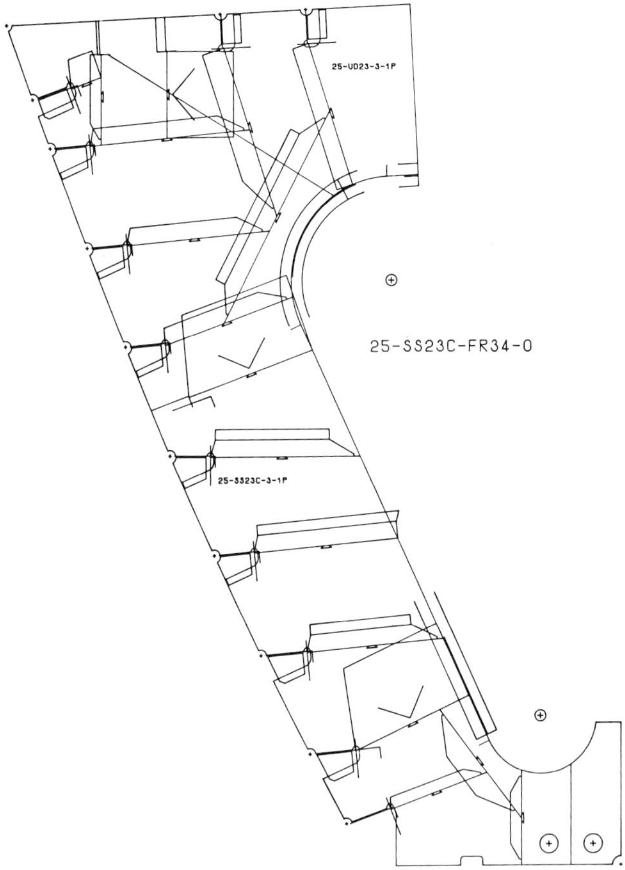

Fig. 25.10 Check drawing of stiffeners.

The NASD system is distinguished from other similar systems by its effective utilization of graphic displays. SCOPE, FINES and HMF Monitor are interactive graphics application programs which are key features of the NASD system.

25.3 Practical Use

25.3.1 Practice in a Design Office

Typical examples of the usage of the NASD system in a design office are categorized into the following two groups.

(1) Input job to HMF.
(2) Preparation of design drawings.

Input job to HMF

For the designer, the first contact with the NASD begins with lines fairing. To obtain faired lines, the designer inputs rough offsets to LINES FAIRING program. The program smoothes these input data, generates faired frame section lines and outputs printed tables and drawings. If the output is not adequate, the input data are modified and computed again. Commonly these processes are repeated several times until the final faired frame section lines are obtained. These lines are stored in the Hull Master File.

After lines fairing, the other hull geometric data such as seams, butts, longitudinals, etc. are stored in HMF by HMF file. The data in HMF should not have errors, because it is basic information used in many application programs. Accordingly, the data should be checked digitally by HMF list, pictorially by HMF draft and interactively by HMF monitor.

Preparation of design drawings

Skeleton drawings can be easily obtained from the data in HMF by the drawing program (HMF draft) and at the same time drawing data are stored in DF. The more complicated drawings (Figure 25.11) can be obtained interactively by SCOPE.

One interesting method which is widely used is to make several similar drawings which have small differences in interactive operation based on a common shape in DF.

The combination of HMF draft and scope is very powerful as mentioned above. The resulting pictures are stored in DF. Thus DF can be used as a filing cabinet for drawings. As the stored data in DF is a kind of soft copy, SCOPE can be used like a copy machine, and many derivatives or secondary drawings can easily be made.

The other advantage of DF is to keep track of unfinished work at any time. When a designer uses a computer program in a batch mode, he must do the whole process at once. However, SCOPE, together with DF, allow him to leave the work at any time and to restart again from the suspended point.

25.3.2 Practice in a Mould Loft

In the production department, the NASD system is used for the following purposes:

(a) part generation and NC tape preparation (data preparation for drafting and cutting);

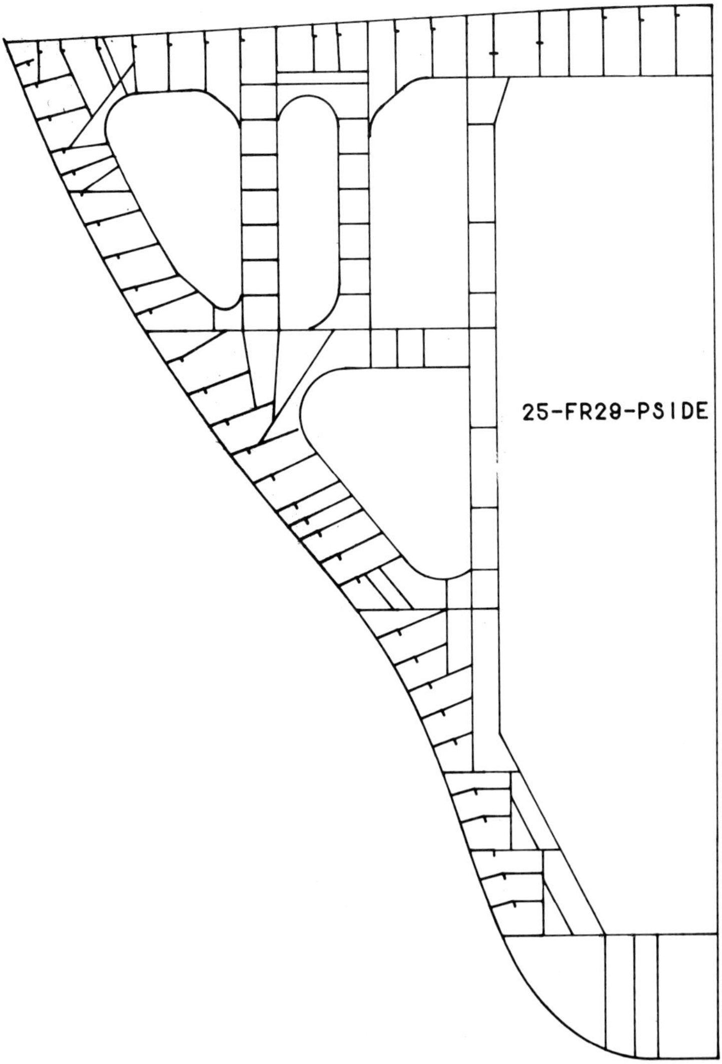

Fig. 25.11 Design drawing by SCOPE.

(b) shell plate development and template offsets table (data preparation for bending or forming);
(c) Frame bending and others

Part generation and NC tape preparation

Part generation is one of the most tedious and complicated operations in shipbuilding. Previously LOFTRAN was used for part generation. Later GLOFT was developed which reduced the lead time for part generation considerably by the use of interactive graphic operation. Experience with both a descriptive language (LOFTRAN) and an interactive system (GLOFT), recognizes the merits and demerits of each technique. NEWLOFT program was developed in order to compensate the limitation of each system. In practice, parts are initially coded in NEWLOFT language, and the results of this program are stored in DF. These pictures are modified and completed by SCOPE operation later.

Three methods for part generation are in current use: LOFTRAN, SCOPE and a combination of NEWLOFT and SCOPE (Figure 25.12). One is chosen to even the labour and graphic terminal load for the part generation job.

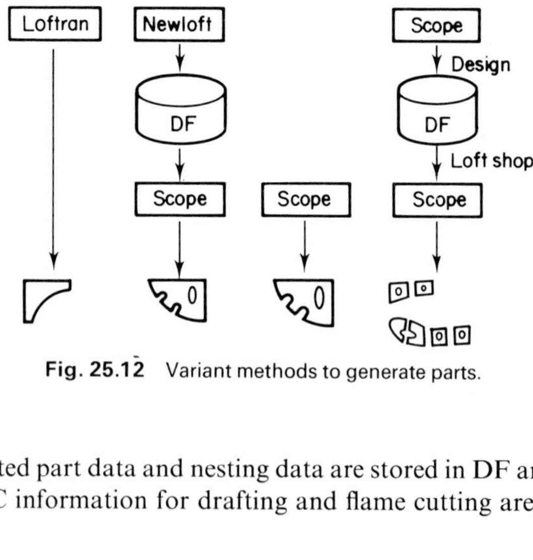

Fig. 25.12 Variant methods to generate parts.

The completed part data and nesting data are stored in DF and NF respectively, and NC information for drafting and flame cutting are created from these files.

Shell plate development and template offsets table
Several programs are prepared for shell plate development. The original data for these programs are fed from HMF.

For bending a shell plate and also for assembling several bended plates into a large panel, the workshop needs templates or supporting needle height tables for checking purposes. There are some programs to calculate this information. The output of shell plate development can be stored into DF for subsequent SCOPE operation or NC preparation.

The shell plate development program is also useful for the optimization of seam landings in the design office.

Frame bending and others
Most longitudinal and transverse frames are bent from shapes by the inverted line method. An inverted line is a curve marked on a straight shape, which becomes straight when a shape is bent correctly. The curve is calculated by the FRAME BENDING program.

Some other programs are prepared for generation of lofting information such as painting lines, draftmarks, etc. Management information such as weight, centre of gravity, cutting and welding length are calculated in connection with NASD programs.

25.4 Concluding Remarks

The data may be stored any time and in any combination regardless of any references which may be made to them. This is the primary benefit of a database system.

However, another even more important advantage is of the database system that it allows the separation of program modifications from the data.

In the NASD system, the database system is independent of the application programs, so it is easy to change and improve the programs and also to add any new application programs to the system without database modification, and vice versa.

Once any computer system is developed, it must be maintained while it is used. Even though the system may perform satisfactorily for the early stages, within a short period it may no longer satisfy the users' needs. Therefore it is important to update the system to meet users' requirements.

The development and the expansion of the application field will be continued.

The following are some conclusions from the discussion meeting which was held recently among members concerning the future of the NASD system.

(1) The NASD system is now aiming towards a more manufacture-oriented basis, whilst always bearing the designer in mind.
(2) The total program of the NASD system is quite large, so further simplification of each program is needed.

Bibliography

Hattori, Y., Ikeda, Y., and Haga, K., NASD-I: NKK Advanced Ships Design System. *Nippon Kokan Technical Report—overseas*, Dec. 1974.

Hattori, Y. and Ikeda, Y., "An application of interactive computer graphics in design and construction system". *Fourth International Symposium on Ship Automation, Genova, Italy, 1974.*

26

Design and Fabrication of Bridges

Shuichi Hasegawa, Director, Yoshihisa Hanamura, Assistant Director, and Kenichi Kyoda, Section Manager, Engineering Division, Yokogawa Bridge Works Ltd

26.1 Introduction

Modern technology has made remarkable progress, particularly in the mass production of individual products. Meanwhile, in other fields of engineering such as bridge fabrication, there are many difficulties in applying similar ideas because these industries deal with large varieties of goods in small quantities which naturally demand multipurpose machines for processing.

The systematic control of these machines is more difficult than those used in mass production. However, the progress in application software and appearance of advanced numerical control (N/C) machines, suggested the possibility of automation and labour saving in those industries that deal with large varieties of goods in small quantities.

In the 1970s, Japanese shipbuilders began to adopt computer and N/C machines in production lines, and this influenced steel bridge fabricators to introduce these tools to automate bridge fabrication. The bridge fabricators are typical of those industries which deal with varieties of parts in small quantities. Because the shapes and sizes of component parts are different for each bridge, an integrated system was developed for plate girder type bridges. After which the development of a general system that would fabricate all kinds of steel structures was examined. However, it was considered imposs-

CAD/CAM and MIS in Japan
ISBN 0-12-682580-7

ible to produce integrated systems for the general types of structures. For these, a "man–machine system" is necessary.

26.2 Features of Total System

Automation of bridge fabrication involves integration of the whole processes from design to fabrication. However, the cost of developing a completely automated system from design to fabrication for all structural types would not be profitable and would not pay for the results.

Steel structures were therefore classified into two types: those suitable for complete automation because the design procedures and treatment are standardized, and those that are not. An integrated system has been developed for the former class of structures, and a man–machine interactive system for the latter.

Figure 26.1 shows a flow-sheet of the total system. The integrated system was developed for plate girder bridges which account for 60% of the total weight of Japanese steel bridge fabrication. This system (as shown in Figure 26.2) flows consecutively through the processes of automated design, drawing and fabrication, and finally yields control information data for the N/C machines in the shop.

On the other hand, the all-purpose system applicable to wide varieties of steel structures is designed as a man–machine system. This system naturally demands a database for efficient data management and conversational language for interactive use and data entry.

A database was therefore developed suitable for bridge design and fabrication together with conversational language ("GIPSY") which deals with pattern processing and data entry of fabrication information. A new-compiler language "NEW DAISY", which is improved ALGOL with functions for pattern processing and database interface, was also developed.

Having produced this basic supporting software, an all-purpose lofting system for the general steel structures was begun in 1976, and was completed in April 1978. Figure 26.3 shows the flow sheet of this system.

Conventional design drawings are used to extract the necessary information to create a structural skeleton. Secondly, in accordance with the skeleton lines, modular processing of the synthetic blocks, component blocks or component pieces of the structures creates information on member shapes to store them on the database.

Because of the consistent structure of resultant output from these processes, the information may be stored on a common database. The system also traps unavoidable input errors and allows updating of the stored data originating from imperfect modular processing of synthetic or component blocks.

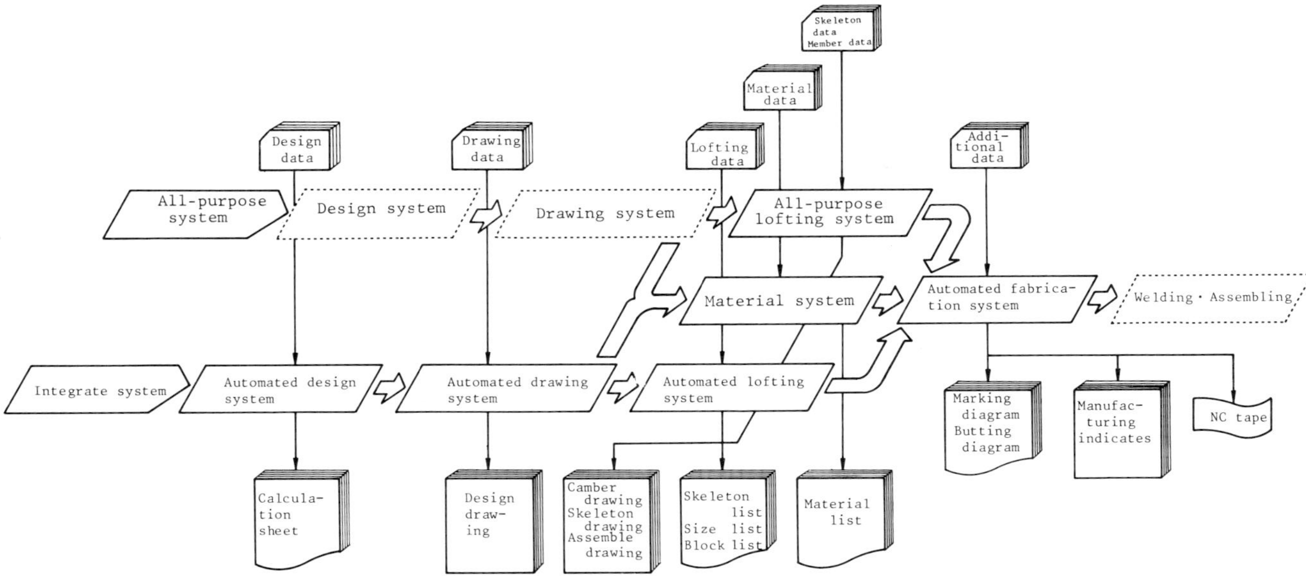

Fig. 26.1 Flow sheet of total system.

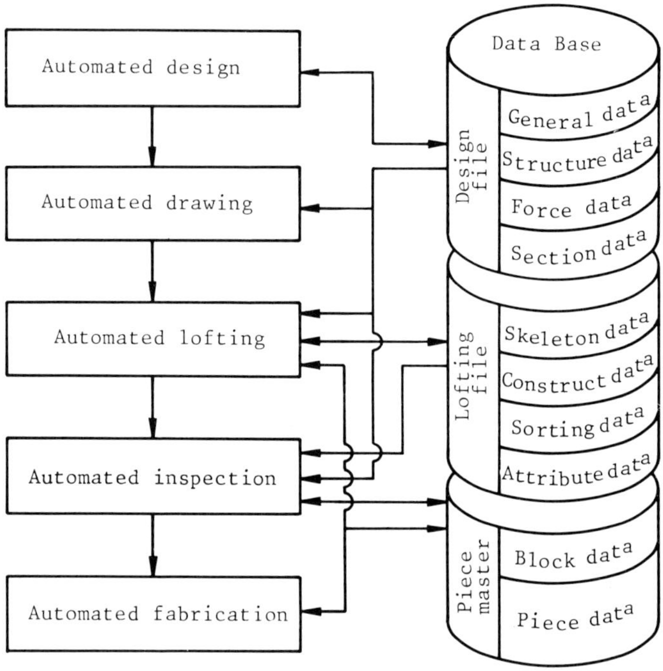

Fig. 26.2 Integrated system.

GIPSY was accessed graphically to correct data using a CRT display. Finally, the complete set of data is transformed into N/C machine control information via the automated fabrication system and transmitted to the shop.

26.3 Administration of the System

26.3.1 History

The integrated system for plate girder bridges was initiated in 1972. Since then, the system has been used for about 300 contracts (500 bridges) involving composite girder and multispan continuous plate girder bridges between 1972 and 1978, whilst the use of automated design and drawing system has made appreciable contribution to labour saving in design work.

Meanwhile, development of the automated fabrication system (which started simultaneously with the automated design system), was finished in 1973. The system has been constantly accumulating experience through the use of N/C machines at the shop.

Later, in 1975, the automated lofting system which interconnects the two

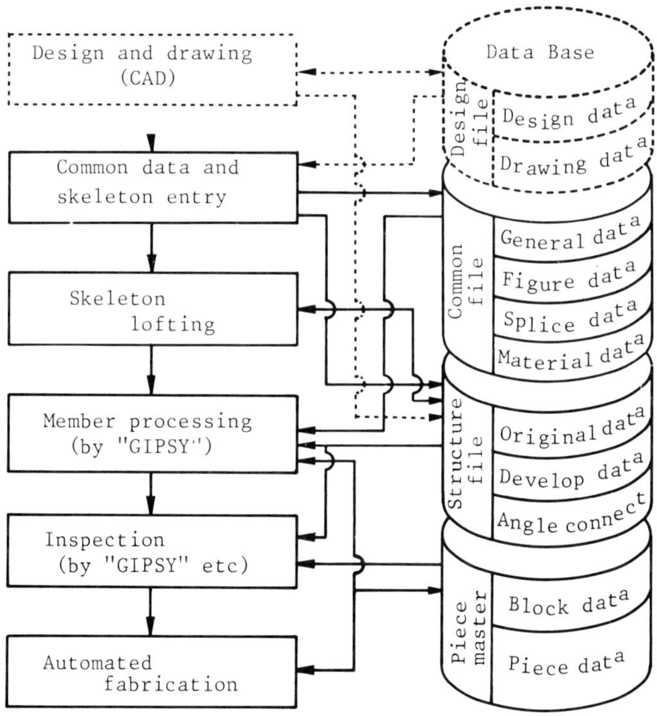

Fig. 26.3 All-purpose system.

systems above was added, thus completing the total system for plate girder bridges. Since then, the integrated system has been employed for all plate girder bridges that were fabricated. The success of the integrated system stimulated the development of the all-purpose lofting system which was completed in April 1978. Although it is too soon to quote results, its application to most of the lofting work is already scheduled and is believed to be effective.

26.3.2 Inspection Performance

For the smooth administration of this kind of large system, thorough inspection of the data at each stage of processing is of key importance. Although every subsystem of the integrated system has its own error-checking functions, it is still difficult to trap all possible errors. We therefore added a visual inspection system which, on demand, prints out or visualizes in graphics the contents of database.

On the other hand, the all-purpose lofting system is not provided with so many error-checking functions as the integrated system. Here, the importance of data inspection is more pronounced because the data are read from the design drawings and are input manually into the program.

Much emphasis has thus been put on the development of a technical supervisory system in which the conversational pattern processing language "GIPSY" plays an important role.

Currently the system submits the contents of the database for inspection and displays them at each stage of the lofting process. The system is thus assisted by supervisory data inspection, and it is only after all shape plans and size lists are checked that the system transfers fabrication information to the shop.

26.4 The Effects of the System

The integrated system for plate girder bridges had been developed for total administration from design to fabrication, with direct saving in labour. Since this was achieved in the design work on a single system, indications were that the effect would be similar with the integrated system. As an example of the dramatic effect of the new system, the design and drawing of a continuous plate girder bridge was completed in 10 man-days compared with 60 man-days required by conventional design procedures.

The automated fabrication system was expected to increase labour saving in the lofting and marking processes. However, it did not comply with all expectations due to the inadequate performance of the N/C machines, and the labour saving rate remained within 60 to 70%. However, the installation of additional N/C machines and a new N/C marking machine may lead to more dramatic savings.

To apply this automated fabricating system the divisions of the shop are lofting, marking and drilling which comprise so-called preliminary processes.

Labour saving is expected to be most marked in the lofting and marking processes. However (as shown in Figure 26.4), statistics reveal that installation of the system has cut down the fabrication cost over 30% compared with the conventional fabrication methods. Since the cost of lofting and marking processes accounts for only 10% of whole processes, this direct effect cannot explain such a result.

In Figure 26.5 the comparison of labour works and the effects of labour saving between the conventional method and the new system is shown. The figure shows close upon 30% labour saving in the processes other than lofting and marking. The result shown in Figure 26.6 may be attributed to the same reason. That is, with the use of N/C machines, improved accuracy and

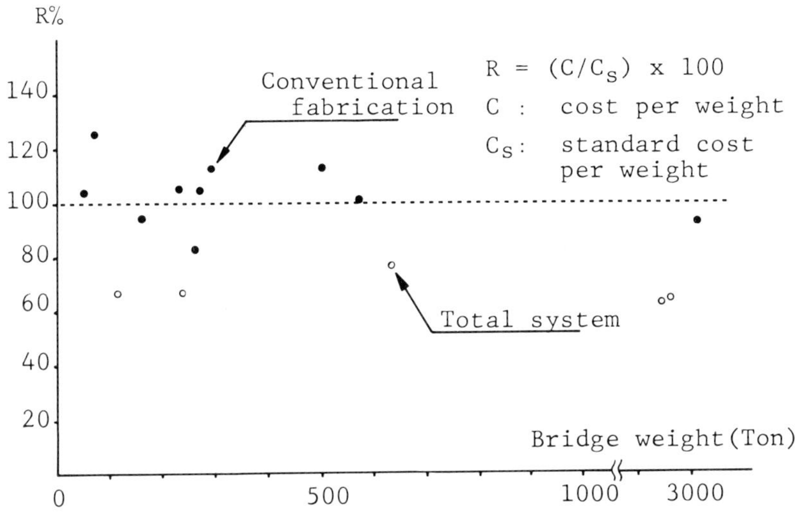

Fig. 26.4 Comparison of fabrication cost.

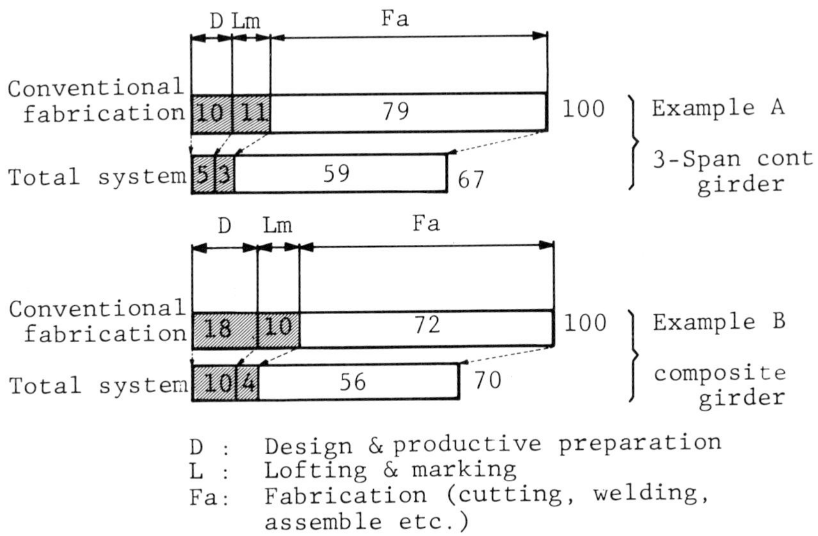

Fig. 26.5 Effect of labour saving D, Design and productive preparation; L, Lofting and marking; Fa, (fabrication, cutting, welding, assembling, etc.)

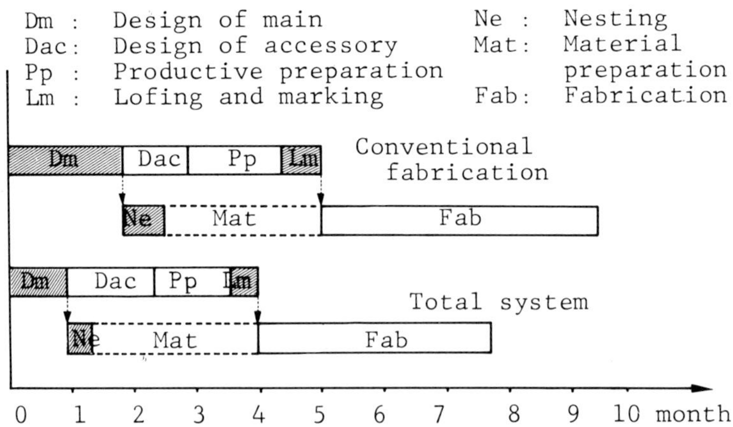

Fig. 26.6 Comparison of fabrication time.

stabilized quality have resulted in less reworking than is the case with conventional fabrication processes.

Shortened fabrication time is another advantage. As shown in Figure 26.6, the design time is shortened by 50% and fabrication time by 20 to 30%.

All these effects have encouraged us to apply the recently completed all-purpose lofting system. However, the efficiency of data creation for N/C machine control information is still in question.

It is therefore planned to restructure the modular processors for synthetic and component blocks to give more efficient data generation.

26.5 Conclusion

The effective employment of electronic digital computers and N/C machines for automation of design and fabrication may be achieved only within a management system. It is recognized that lofting work processed by a computer and the use of N/C machines refine the accuracy of full size development and material size, which contributes greatly to the overall labour saving and shortening of fabrication times.

The total system was initiated by the development of an integrated system for the plate girder bridges, and expanded recently to include an all-purpose lofting system.

Work is currently directed towards the development of an integrated system for box girder bridges, and the automated fabrication of architectural steel frames.

27

MISA–A General-Purpose FEM Program

Takashi Iwaki and Akihiko Maeda, Tamano Laboratory, Mitsui Engineering and Shipbuilding Co Ltd, Okayama, Japan, and Tadashi Ishii, System Headquarters, Mitsui Engineering and Shipbuilding Co Ltd, Okayama, Japan

Nomenclature

SE	super element by which the structure can be specified.
BE	generated basic element.
KN	key node by which the coordinates of SE can be specified.
(ML)SS	(multi-level) substructure.
PPPs	pre- and post-processors.
DOF	degree of freedom.

27.1 Introduction

The finite element method (FEM) is a well established numerical procedure by which problems in solid mechanics can be handled. Extensive applications of this technique to a variety of industrial problems have been made all over

CAD/CAM and MIS in Japan
ISBN 0-12-682580-7

the world. The factors which are usually considered by users in the selection of a particular program are given below.

Capability of analysis
Linear or nonlinear analysis; static or dynamic analysis; element library; load conditions and restraint conditions.

Program availability
Well-organized, well-maintained program; convenient and readable user's manual; adequate computer facilities suitable for the program; reliable and responsible program support.

Computer efficiency
Matrix band width optimization; equation-solving technique; output selection and restart capability.

User's convenience
Data preprocessing features, data checking plots; data generating features; output evaluation and selective processing.

Of the studies on the FEM which have been carried out to date, the majority are concerned with the first of these factors; studies on pre- and post-processors (PPPs) related to features computer efficiency and user convenience are rather inadequate. In cases where the FEM has been used as a design tool, excessive manpower has been spent on data preparation and checking, and evaluation of results.

There is an imbalance in the distribution of studies on the FEM. It is evident that studies have not been directed towards the reduction of analysis costs, as is also pointed out by Herness *et al.*[1]

The various conventional program systems now on the market, such as NASTRAN,[2] ICES/STRUDL,[3] MARC[4] and others, have made it possible to analyse complex mechanical responses, but they usually require excessive computer costs and considerable manual effort to complete a finite element calculation. Mitsui Structural Analysis Program (MISA), was completed in 1974 by the Mitsui Engineering and Shipbuilding Co Ltd. Although the analytical capability of MISA is limited to linear static stress analysis and steady state heat condition analysis, this covers about 80% of the needs of design departments in our company. Special consideration has been given to the saving of manpower and computer time in the development of MISA, the basic specifications of which are briefly described as follows.

Table 27.1 Types of load and restraint condition

	Item no.	Stress Analysis	Item no.	Heat conduction analysis
Load conditions	LS 1	Joint Load	LH 1	Heat Flux
	LS 2	Distributed Load	LH 2	Heat Generation
	LS 3	Parallel Inertia Force	LH 3	Convective Heat Transfer
	LS 4	Thermal Strain	LH 4	Radiant Heat Transfer
	LS 5	Rotational Inertia Force		
	LS 6	Initial Strain		
Restraint conditions	RS 1	Fixed Displacement	RH 1	Specified Temperature
	RS 2	Spring Support		
	RS 3	Specified Displacement		

27.1.1 Analysis Functions

Linear static stress analysis and steady state heat conduction analysis, followed by thermal stress analysis of structures can be dealt with.

27.1.2 Category of Structures to be Analysed

The following kinds of structures can be dealt with: (a) frame structures; (b) shells as an assembly of flat plates; (c) general three-dimensional structures; (d) axisymmetric structures (shells and solids); (e) combinations of the above structures.

27.1.3 Element Library

An adequate element library is needed to idealize various structures into suitable models. The element library therefore consists of twenty-seven kinds of elements for stress analysis and sixteen for heat conduction analysis.

27.1.4 Load and Restraint Conditions

Various types of load and restraint conditions shown in Table 27.1 can be dealt with.

27.1.5 Matrix Solver

The multi-level substructure method which is described in the next section has been adopted, and the LDU decomposition method together with the wave front method are applied to matrix calculation.

27.1.6 Preprocessor

The following capabilities are provided: (a) generation of structure, load and restraint data, using Super Elements as units of input; (b) checking of data at an early stage; (c) renumbering of joint numbers to reduce the band width of the equations; (d) graphic display of the structure, load and restraint data.

27.1.7 Postprocessor

The following capabilities are provided: (a) plot of contour lines or intensity levels of stress components, principal stresses and equivalent stresses; (b) plot of the displacement or deformation of a structure; (c) print-out of the output results suitable for formal reports; (d) combination of the results for load conditions.

Details of the system, especially focusing on PPPs' functions, are given together with their application examples in the following sections.

27.2 General Description of MISA

This section gives an explanation of the multi-level substructure method and the whole program system of MISA. The former is one of the characteristics of the basic specifications for MISA.

27.2.1 The Multi-level Substructure Method

In the case of a large-scale structural analysis of a complicated structure, a structural analysis model can be divided into several substructures (SSs) according to its structural pattern, and input data can be prepared for each SS. A large matrix for the whole structure can be solved according to its hierarchical TREE structure. This procedure is usually called the multi-level substructure (MLSS) method. Matrices are solved sequentially for each SS level (see the next section), from the first level to the last, in order to obtain the various displacements and stresses of all the parts in the structure. Since there are many identical SSs within the structure, the time for data preparation and matrix calculation can be efficiently reduced by use of the COPY function.

The MLSS Method has recently attracted much attention for its many advantages such as high efficiency, easy preparation and checking of data, accurate solutions, etc. The MLSS Method used in MISA has the following characteristics compared with conventional MLSS Methods adopted in other programs.

1. Various types of structures, from one-dimensional structures to three-dimensional ones, including combinations of different types, can be treated.
2. The COPY function is provided for the repetitive SSs of all levels.
3. The COPY function can be applied to repetitive SSs independent of their boundary conditions.
4. The hierarchical TREE structure can be determined automatically by analysing the input parent–child relation of the SSs (see the next section), as can the sequence of forward reduction of the SSs.
5. The DOF of joints for a combined structure can be determined automatically.
6. An analysis model having a different number of division on the common boundary among SSs can be treated by introducing the interpolation function method.

27.2.2 The Program Structure of MISA

The MISA consists of an executive module, three pre- and post-processing modules, six analysis modules (three modules for stress analysis and three for heat conduction analysis). Each module consists of several submodules. The program structure of MISA is shown in Fig. 27.1. In this program system, a dynamic serial structure is introduced, which has the function of bringing the modules needed during execution from the program library into the main storage. Only the necessary modules are called and unnecessary modules are replaced by other required modules at each end of the execution.

The functions of each module are as follows.

Executive module (M01)

(a) Supervising of other modules, decoding of command data for options, and performance of calculations to generate control data.
(b) Decision and performance of the executing sequence of modules including a restart function.
(c) Automatic generation of Job Control Language and management of data files.

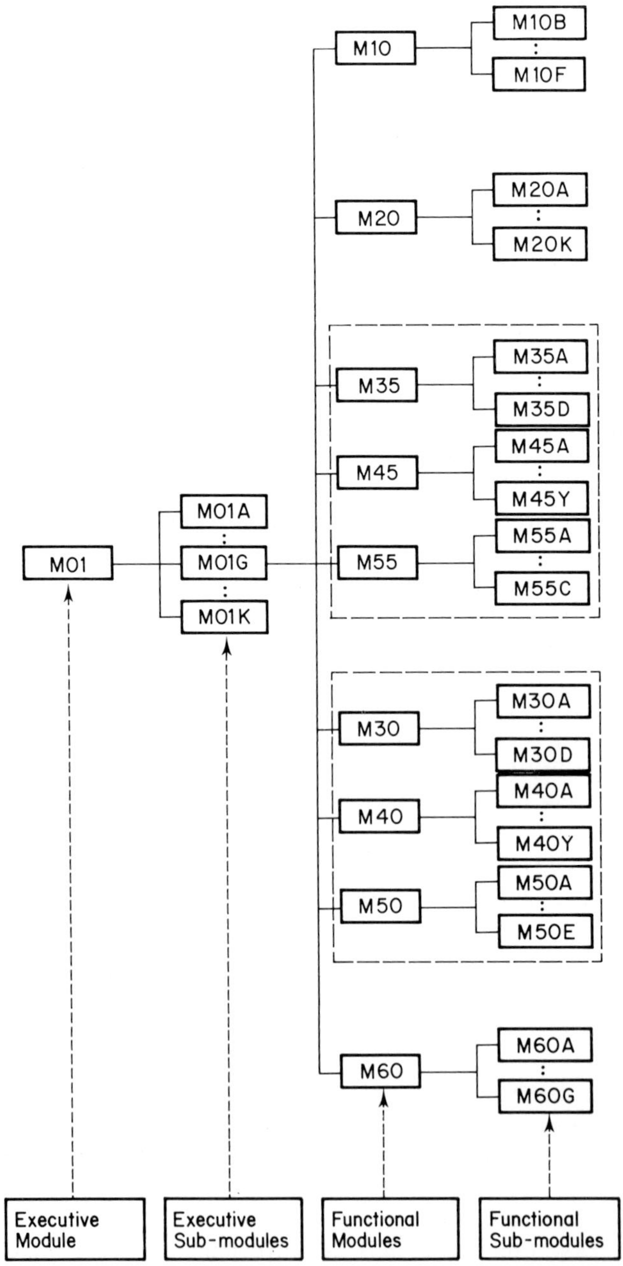

Fig. 27.1 Program structure

Pre- and post-processing modules (M10, M20 and M60).

(i) *Input reader* (M10)

(a) Syntax and logical check of input data and determination of DOF.
(b) Automatic generation of coordinate transformation matrices and high level SS data.
(c) Arrangement of data in each SS and preparation of a table for basic control information for easy reference by other modules.

(ii) *Input generator* (M20)

(a) Automatic mesh division by use of mapping functions and generation of structural, restraint, and load data.
(b) Rearrangement of data in a convenient form and renumbering of joint numbers for solvers and output editors.
(c) Rearrangement of temperature data from the results of heat conduction analysis as load data for thermal stress analysis.

(iii) *Output editor* (M60)

(a) Superposition of load conditions and calculation of principal stresses and equivalent stresses.
(b) Editing and listing of structural, load, restraint, displacement, stress and temperature information for each SS.
(c) Arrangement of lists and plots in a form suitable for formal reports.
(d) Representation of various figures such as perspectives, three-dimensional expressions, contour lines, altered viewpoints and enlarged figures by use of such equipment as dot-printers, drafters, plotters and graphic COMs.

Stress analyser (M30, M40 and M50)

(i) *Matrix generation processor* (M30)

(a) Calculation of element characteristics for stress analysis.
(b) Construction of global stiffness matrices and load vectors of level one SSs.
(c) Forward reduction of level one SSs by use of the wave front method.

(ii) *High level matrix processor* (M40)

Construction of stiffness matrices and load vectors, forward reduction and backward substitution of SSs at level two and above for stress analysis.

(iii) *Stress calculation processor* (M50)

Backward substitution, displacement calculation, balance calculation, and stress and strain calculation of level one SSs.

Heat conduction analyser (M35, M45 and M55)

(i) *Matrix generation processor* (M35)

(a) Calculation of element characteristics for heat conduction analysis.
(b) Construction of global heat resistance matrices and heat load vectors of level one SSs.
(c) Forward reduction of level one SSs by use of the wave front method.

(ii) *High level matrix processor* (M45)
Construction of heat resistance matrices and heat load vectors, forward reduction and backward substitution of SSs at level two and above for heat conduction analysis.

(iii) *Temperature calculation processor* (M55)
Backward substitution, temperature calculation, balance calculation and surface heat flux calculation of level one SSs.

27.3 Preliminary Input Data Processing

M10, a preprocessor, is designed to perform the first step input data processing. In this section, some of its functions are explained in relation to the MLSS method.

27.3.1 Input Data

The organization of input data is best explained by means of an example. The structure model shown in Fig. 27.2(b) can be defined by the simple input model shown in Fig. 27.2(a). Input data for the structure model are shown in Fig. 27.3. The structure consists of six level one SSs (SS01 to SS06) whose TREE structure is shown in Fig. 27.4. Input data mainly include structure data and boundary data, the latter of which are omitted in Fig. 27.3. Input data such as KN, SE and material property data can be prepared independently for each SS. The whole structure is defined by assembling these SSs using global data, which includes TREE data, SS coordinate system data and CONTACT data. These three define the parent–child relationship among SSs, the position of SSs, and the connection of SSs, respectively.

Since SS02, SS03 and SS04 have exactly the same shape as SS01, they can be simply defined as copies of SS01. Only the definition of SS01 is required. Similarly, SS22 can be defined as a copy of SS21 even though they appear only in the TREE structure. The COPY function is useful in reducing the amount of input data and computer time for solving the matrix. Regarding

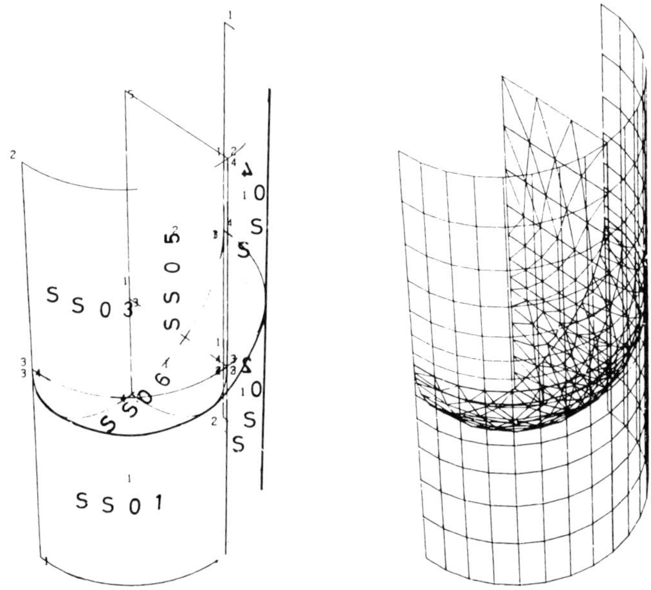

Fig. 27.2 Structural model. (a) Input structure; (b) generated structure.

the coordinate systems in this example, cylindrical and spherical coordinate systems are used for SS01 and SS06, respectively.

27.3.2 Input Data Processing

The first step in data processing is illustrated by using the structure model shown in the previous section.

27.3.2.1 Analysis of TREE structures

The data process in a TREE structure is as follows. Let us call the substructures which define the structure itself, "the level one SSs" and call the one defined by the TREE data, "the high level SSs". The SSs at the upper level are called "Parent SSs", and the lower ones "Child SSs". Defining the level of a Parent SS as equal to the maximum level of the Child SSs plus 1, the whole TREE structure can be determined by the input TREE data as shown in Fig. 27.4. Since a matrix calculation is executed level by level by the MLSS method, the forward reduction procedure can be executed sequentially following the arrows shown in Fig. 27.4 and the backward substitution in the opposite direction. The levels of SSs determined by the above-mentioned method are shown in Table 27.2.

```
                          *STRUCT          Parent SSs      Child SSs
Global  .TREE data        TREE      SS21             SS01SS02
data                      TREE      SS22      4      SS03SS04
                          TREE      SS23             SS05SS06
                          TREE      SS31             SS21SS22
                          TREE      SS41             SS31SS23
        .Inter-SS Coordi- SSCOORD     2       2          90.
         nate System data SSCOORD     3       2          90.       0.              180.     180.
                                    0.          0.       200.
                          SSCOORD     4       2          180.      0.              270.     180.
                                    0.          0.       200.
        .Contact data     CONTACT   SS01      2SS02      1   Indicates identity of the joint at SS01 and at SS02
                          CONTACT   SS05      1SS06      4
                          CONTACT   SS04      2SS03      1SS05     4
                          CONTACT   SS01      4SS03      3SS06     3
                          CONTACT   SS01      3SS02      4SS03     4SS04    3SS05    2SS06    2
                          CONTACT   SS02      3SS04      4SS06     1
SS01    .SS Coordinate    SUBSTRUC  SS01    Cylindrical coordinate system
data     System data      COORD     2001      2          0.
        .Joint data       JOINT       1       2001       50.  r    0.   θ    0.   z
                          JOINT       2       2001       50.       90.       0.
                          JOINT       3       2001       50.       90.       100.
        .Element data     JOINT       4       2001       50.       0.        100.   Division Number for side lines
         (Cylindrical     EQUCMEBE    1     102001     1   2   3   4     8   7   8   7   50.
          Surface)                    4       1  2.                 Joint Number
        .Material Property            2       3  0.5   Weighting factors of side lines
         data             SPCPLAT    10       1   2      5.        7.85            0.3      2.E4
Copied                    SUBSTRUC  SS02      2SS01COPY
SSs                       SUBSTRUC  SS03      3SS01COPY
                          SUBSTRUC  SS04      4SS01COPY
SS05                      SUBSTRUC  SS05              Original SS
data                      JOINT       1                  0.   x    0.   y    50.  z
                          JOINT       2                  0.        50.       100.
                          JOINT       3                  0.        0.        100.
                          JOINT       4                  0.        50.       200.
                          JOINT       5                  0.        0.        200.
                          ETRPMEBE    1    10        1   2   3     200       8   6   6
         (flat plate)                 1     2  0.5       50.        Index for curved side lines
                          EQUPMEBE    2    10        3   2   4     5         6   7   3   7
                                      2     4  2.
                                      5     3  0.5
                          SPCPLAT    10     1   2        5.        7.85            0.3      2.E4
SS06                      SUBSTRUC  SS06    Spherical coordinate system
data                      COORD     3001    2            180.      0.              270.     180.
                                    0.          0.       100.
                          JOINT       1       3001       50.  r    0.   θ    0.   φ
                          JOINT       2       3001       50.       90.       0.
                          JOINT       3       3001       50.       180.      0.
                          JOINT       4       3001       50.       0.        90.
                          ETRCMEBE    1     103001     1   2   4             8   8   8
                                      2     4  2.
         (Spherical                   4     1  0.5
          Surface)        ETRCMEBE    2     103001     2   3   4             8   8   8
                                      3     4  2.
                                      4     2  0.5
                          SPCPLAT    10     1   2        5.        7.85            0.3      2.E4
                          *END
```

Fig. 27.3 Input data.

The table is the basic one representing the relations among SSs, and the sequence of matrix solution.

27.3.2.2 Classification of joints

To completely determine the information concerning the joints is very complicated because they are related to many processings such as classification of internal and external joints, determination of the elimination level, definition of DOF in combined structures, etc. In the MISA system, in order to avoid complexities in describing joint information, joints constituting SE are classified into the following four sets as shown in Fig. 27.5: (a) the joints at the corners; (b) the joints on the side lines except for (a); (c) the joints on the sur-

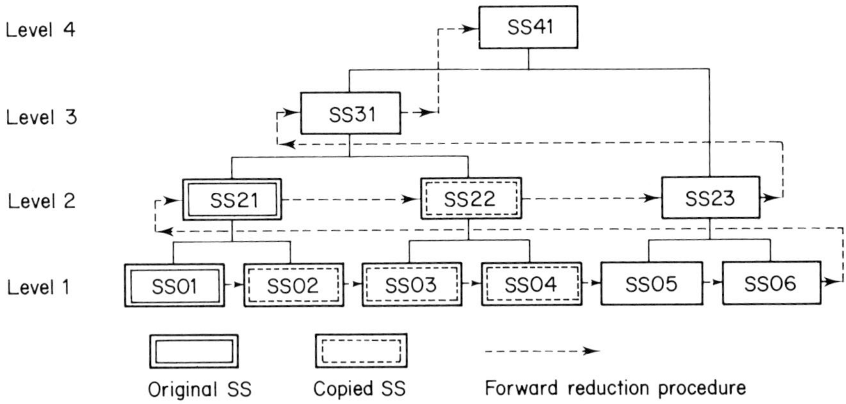

Fig. 27.4 Tree structure and calculation procedure.

Table 27.2 Definition of substructure level

SS NAME	SS41	SS23	SS06	SS05	SS31	SS22	SS04	SS03	SS21	SS02	SS01
Level	4	2	1	1	3	2	1	1	2	1	1
Copy	–	–	–	–	–	SS21	SS01	SS01	0	SS01	0

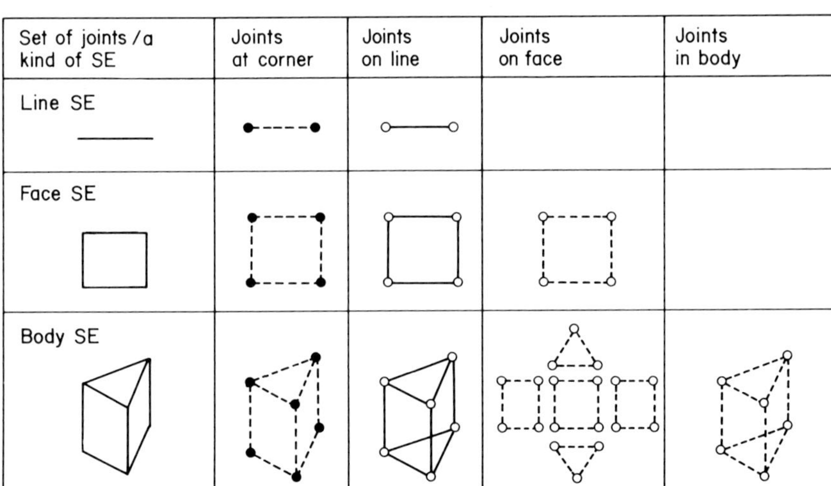

Fig. 27.5 Classification of joints.

face except for (a) and (b); (d) the joints in the body except for (a), (b) and (c). In the figure, the first three of the above-mentioned four sets of joints are shown by these symbols: a dot (·), a solid line (——) and a broken line (– – –), respectively.

This classification provides complete information on the generated joints at the stage before data generation processing, using far less input data than generated data.

27.3.2.3 Determination of degrees of freedom

Some of the joints in a level one SS are sometimes common to some elements which have different co-ordinate systems and DOF. The DOF of such joints can be determined as follows (see Fig. 27.6). Coordinate systems with the minimum DOF required for the definition of the SEs are defined first and then the DOF of the joints are calculated by transferring the coordinate systems to a common one. The DOF of such joints in high level substructures can also be calculated as follows. If the joints in a Parent SS are common to the several child SSs the DOF of these joints can be determined by superposing the DOF of the joints in the child SSs.

27.3.2.4 COPY process

Identical SSs are omittable if and only if the matrices corresponding to the internal joints are identical. Copied SSs are not always connected with other SSs at the identical positions in the original one, and do not always have internal and external joints which are identical with the original one. The classification into internal and external joints is determined by the procedure shown in Section 27.3.2.2. Therefore, in order to apply the COPY function to these identical SSs, internal and external joints must be identical in each SS. As stiffness matrices for SSs with restraints are different from those without restraints, restrained joints must be identified as external ones. The elimination level for a joint whose position is identical in each identical SS is defined as equal to the highest one among the SSs (see Fig. 27.7). The joints with the same joint number in Fig. 27.7(a) are identical (i.e. superposable by coordinate transformation) and their COPY relations are shown in Fig. 27.7(b). First, the elimination levels for the joints are determined by the procedure described in Section 27.3.2.2 without considering the COPY relation (see Fig. 27.7(c)), and then they are modified according to the procedure described in this section. For example, the elimination levels of line 3–4 for SS01, SS02, SS03 and SS04 are determined to be 1, 1, 3 and 3, respectively, as shown in Fig. 27.7(c). Since the maximum elimination level is 3, the line 3–4 for all the

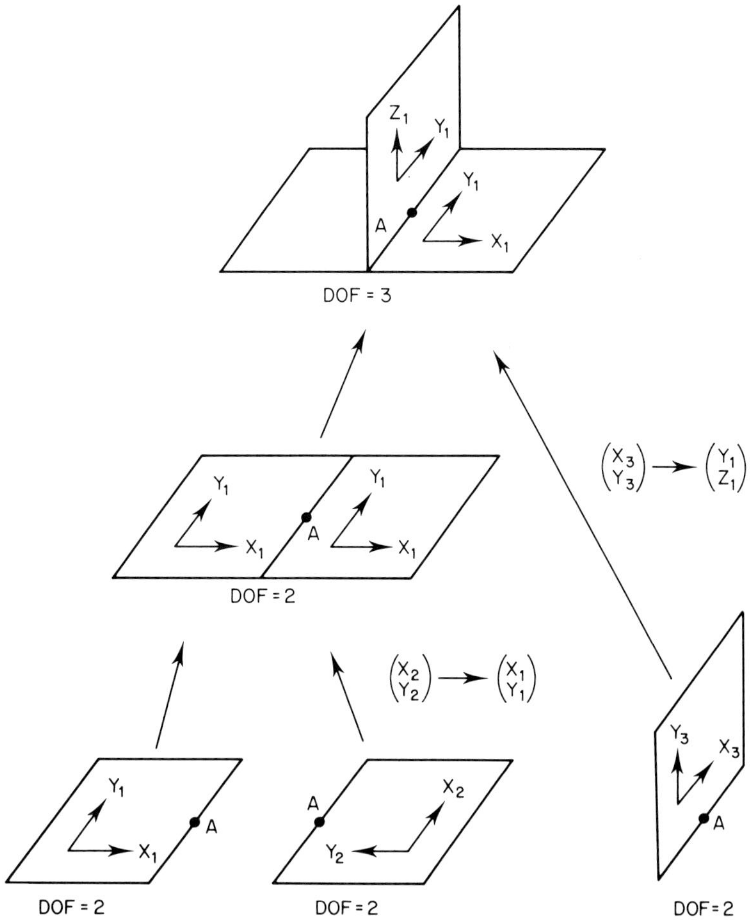

Fig. 27.6 Determination of DOF.

SSs is eliminated at level 3. Similarly, elimination levels of all the joints are determined as shown in Fig. 27.7(d).

27.4 Automatic Data Generation

As stated in Section 27.3, a structure to be analysed can be divided into relatively large structural patterns. They are called "Super Elements (SEs)" and are used to automatically generate a lot of structural, load and restraint data from a minimum amount of input data. The patterns of SEs in this program are shown in Fig. 27.8. A lot of SEs, such as one-dimensional beam elements,

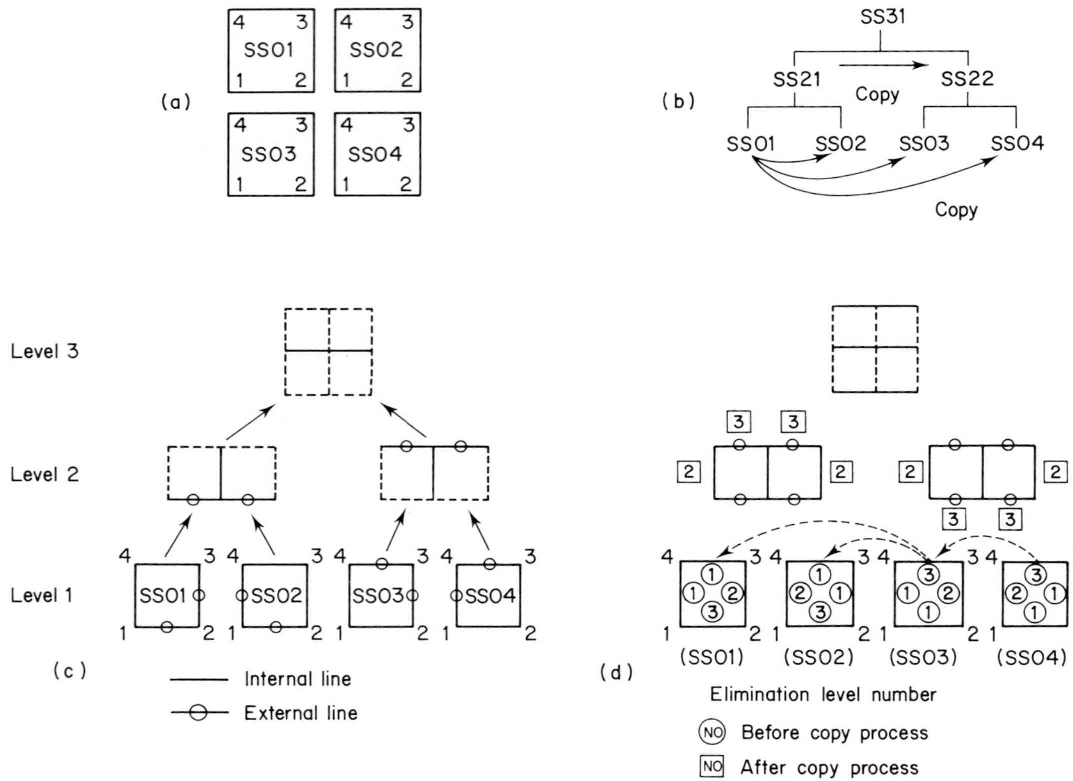

Fig. 27.7 COPY process. (a) Input structure; (b) copy relations; (c) intermediate process; (d) final process.

Type of Super-elements	Pattern of Super-elements	Type of Super-elements	Pattern of Super-elements
1D Element (1) (Beam, Axi-symmetric Shell)	Straight Line Circular Arc Cubic Curve	1D Element (2) (Beam, Axi-symmetric Shell)	1 Mid-side Point 2 Mid-side Point
2D Quadrilateral Element (Membrane, Plate Bending, Axi-symmetric Solid)		2D Triangular Element (Membrane, Plate Bending, Axi-symmetric Solid)	
Quadrilateral Plate on Cylindrical Surface	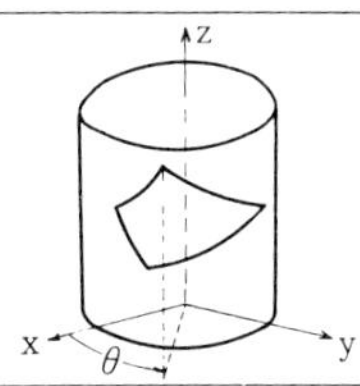	Triangular Plate on Cylindrical Surface	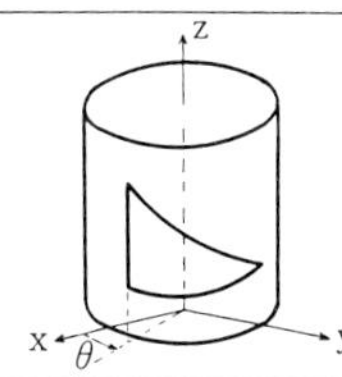
Quadrilateral Plate on Spherical Surface	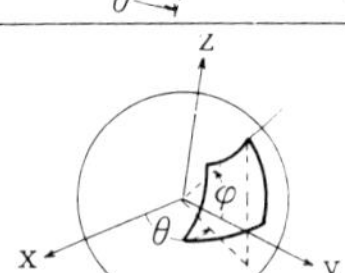	Triangular Plate on Spherical Surface	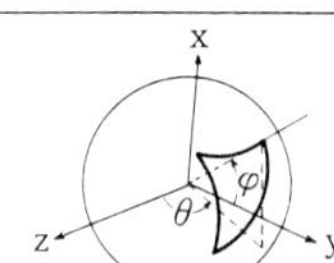
Quadrilateral Plate on Conical Surface	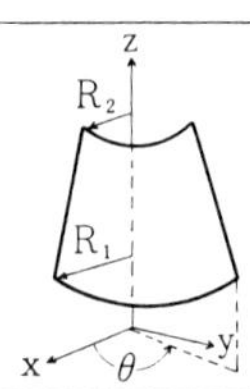	Quadrilateral or Triangular Plate on Generally Curved Surface	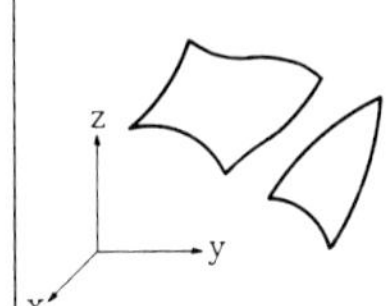
3D Hexahedron Element	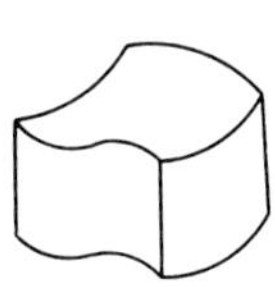	3D Pentahedron Element	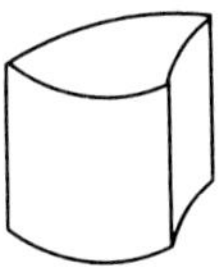

Fig. 27.8 Pattern of super elements.

two-dimensional plane or curved surface elements, or three-dimensional solid elements, have been provided to automatically generate various types of BEs. SEs, which are beam elements, axisymmetric shell elements, axisymmetric solid elements, have been provided with a function to represent the shape of each side in a straight line, circular arc, quadratic curve or cubic curve. A thin curved plate SE with a curvilinear coordinate system has been prepared to make it easy to arrange input data in cases where a curvature (a cylindrical, spherical or conical surface) is treated as an assembly of thin plate elements in practical structural analysis. When sides of a curved surface cannot be represented by a linear relation of components in a curvilinear coordinate system, one or two mid-side points are used to specify the shape of the sides, which makes it easy to represent, for example, a crossing line of two curved surfaces.

27.4.1 Structural Data Generation

In the case of subdivision of the above-mentioned SEs with complicated shapes, the following functions are provided.

(a) An unequal subdivision function which subdivides SEs with unequal division numbers for each side in order to generate fine meshes for parts where either the shape of the structure suddenly changes or singular points occur, and to generate coarse meshes for parts where stress changes very little.

(b) A weighted subdivision function which subdivides SEs with unequal division pitch for each side.

(c) Prevention of subdivision into obtuse triangles.

Each step of the automatic mesh generation procedure is explained below. As an example, a quadrilateral SE is considered here, because a triangular SE can be treated as an assembly of a triangular BE and quadrilateral SE. Input data are the coordinates of KNs, the shape of each curved side, the division numbers and the weighted coefficients for each side (see Section 27.3.1).

As shown in Fig. 27.9(a), consider a square whose side length is two in the "Serendipity" $\xi^\circ - \eta^\circ$ coordinate system and subdivide it without weighting according to the input data. Regarding the general automatic mesh generation scheme, refer to Refs 5 and 6. As shown in Fig. 27.9(b), an arbitrary point (ξ°, η°) can be mapped as (ξ, η) in the $\xi - \eta$ coordinate system by using eqns (27.1) and (27.2) while mid-side points (X) can be transferred from the mid points in $\xi' - \eta'$ to other points in $\xi - \eta$ which are defined by the input weighted coefficients W.

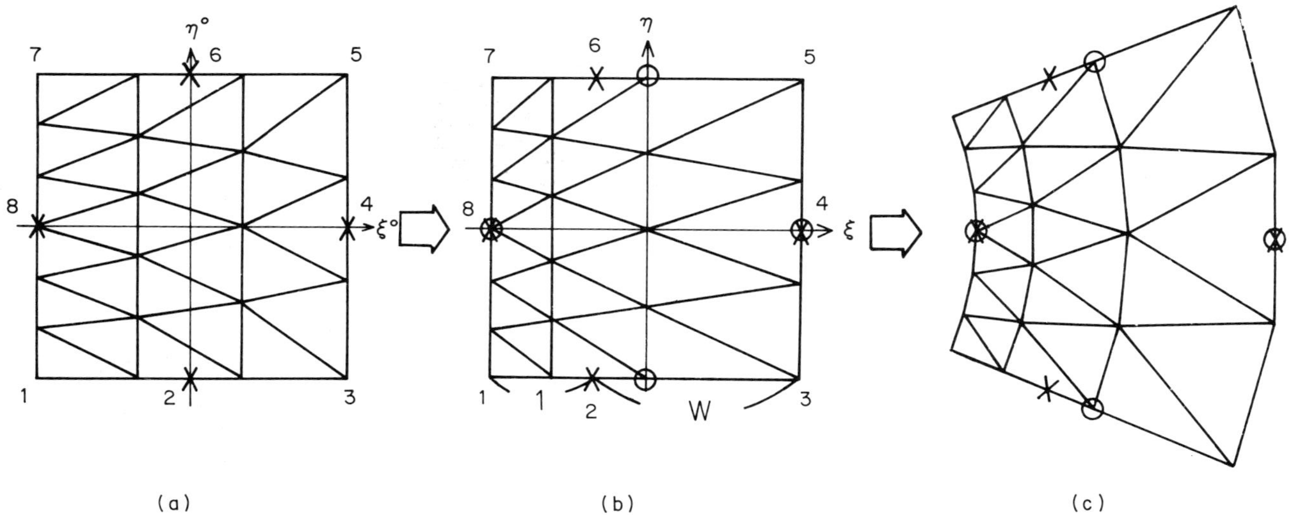

Fig. 27.9 Calculation scheme of joint coordinates. (a) $\xi^\circ - \eta^\circ$ plane; (b) $\xi - \eta$ plane; (c) real plane.

$$\xi = \sum_{k=1}^{8} N_k(\xi^\circ, \eta^\circ)\xi_k \qquad \eta = \sum_{k=1}^{8} N_k(\xi^\circ, \eta^\circ)\eta_k \tag{27.1}$$

where

$$N_1(\xi, \eta) = -\tfrac{1}{4}(1-\xi)(1-\eta)(1+\xi+\eta)$$
$$\vdots$$
$$N_8(\xi, \eta) = \tfrac{1}{2}(1-\xi)(1-\eta^2)$$
$$\xi_1 = -1, \qquad \eta_1 = -1 \tag{27.2}$$
$$\xi_2 = -1 + \frac{2}{1+W}, \qquad \eta_2 = -1, \text{etc.}$$

As shown in Fig. 27.9(c), the coordinates (x, y) of GJs in the real coordinate system x-y can be given by eqns (27.3) which transform the $\xi - \eta$ system into the x-y coordinate system by using the coordinates of KNs and the mid-side points:

$$x = \sum_{k=1}^{8} N_k(\xi, \eta)x_k$$
$$y = \sum_{k=1}^{8} N_k(\xi, \eta)y_k \tag{27.3}$$

In order to precisely represent the real shapes of structures, shape functions with one or two mid-side points (i.e. quadratic or cubic mapping functions) are provided. Using the above calculated GJ coordinates, quadrilaterals can be constructed one by one. The quadrilateral can be divided into two acute triangular BEs by using its shorter diagonal line.

27.4.2 Load and Restraint Data Generation

Load and restraint conditions shown in Table 27.1 can be given by input data prepared for each SE (in the same way as structure data as mentioned above) and are automatically processed by referring to generated joints and elements. Loads on SEs are classified into five load types, which are concentrated loads, distributed loads on a line, distributed loads on a surface, body loads, and others, depending on the kind of SEs and pattern of load distribution. The load data for the above-mentioned load types are generated for each GJ and BE by using mapping functions except for equally distributed loads. Restraints are also classified into categories similar to those for loads. They are restrained points, lines and surfaces. The restraint data, such as spring constants and specified displacements, are generated for each GJ in the same way as load data. For example, let us consider the load generation procedure, using a quadratic mapping function, of a distributed load specified by the eight data on the surface as shown in Fig. 27.10.

(a) Prepare input load data P_k ($k = 1 \sim 8$) at KNs and mid-side points (see Fig. 27.10(a)).
(b) Calculate the load at an arbitrary GJ on the surface by eqn (27.4)

$$P = \sum_{k=1}^{8} N_k(\xi, \eta) P_k \tag{27.4}$$

where $N_k(\xi, \eta)$ are the same mapping functions as eqn (27.2), and (ξ, η) are the weighted joint coordinates in the square as shown in Fig. 27.10(b) (which corresponds to Fig. 27.9(b)).

As the mapping functions, quadratic and cubic mapping functions are provided, automatic generation of input data for arbitrarily distributed loads and restraints, and independent preparation of input data for a structure and its loads and restraints, have become possible because double mapping, which is the mapping of the shape and weighting, is provided in the automatic generation of the structure data. Furthermore, data for loads, such as commonly used pressure loads, parallel inertia forces and rotational inertia forces, can be automatically generated by simple input data given for each SE. Oblique restraint data can also be specified easily by selecting the convenient coordinate system from among the Cartesian, cylindrical, and spherical coordinate systems provided in the MISA.

27.5 Summary of Pre- and Post-Processors' Functions in MISA

In general, the main purposes of PPPs are the standardization and automatic generation of input and output procedures for many kinds of analysis programs in order to achieve labour saving, high calculation efficiency and technical improvements. PPPs for general purpose programs are required to meet the above conditions for all the types of structures shown in Table 27.3.

From the above point of view, PPPs in MISA are classified into three functions which are preprocessing, postprocessing and general processing functions. These are shown in Tables 27.4, 27.5 and 27.6, respectively.

27.6 Application Examples

Two numerical examples are shown illustrating some of the PPPs' features of

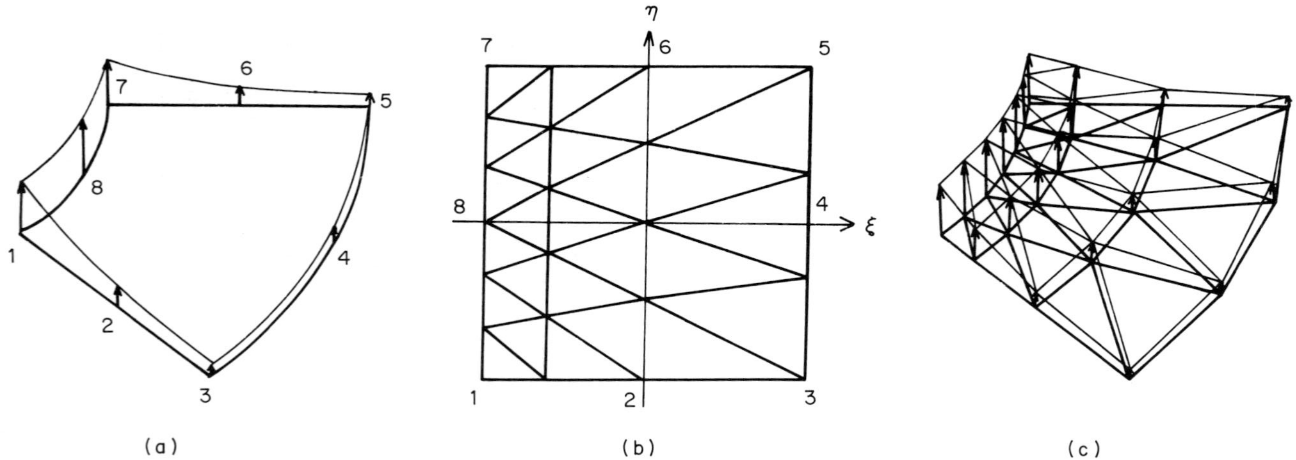

Fig. 27.10 Load generation scheme. (a) Input loads; (b) $\xi - \eta$; (c) generated loads.

Table 27.3 Classification of structures

	Type of Structures	
	2-D	3-D
Element pattern 1-D	(a) Planar truss Planar beam Axisymmetric shell	(b) Space truss Space beam
Element pattern 2-D	(c) Membrane Bending Axisymmetric solid	(d) Space plate Shell as an assembly of flat elements
Element pattern 3-D		(e) Solid

MISA. These examples are taken from a rather limited range of problems in structural design; however, the practicability of MISA in the actual design stage has been demonstrated through more than 300 applications to various kinds of products.

Example 1

The pressure vessel shown in Fig. 27.2 was analysed for stresses produced by a horizontal inertia force applied on the whole structure. It consisted of a cylinder, a hemisphere, a partitioning plate and a skirt. Membrane and bending elements were used for the idealization of the structure. The structure data were automatically generated by using three kinds of functions: the COPY function, the weighted subdivision function, and the various kinds of coordinate system functions as mentioned in Section 27.3.1. The load and restraint data were automatically generated in the same way. The modelling items and the number of input data cards used for this example are shown in Tables 27.7 and 27.8, respectively. The efficiencies of data generation, e_1 for joints and e_2 for elements, are also shown in Table 27.7. A wide variety of drawing functions are prepared for users' convenience. The two of them which were used here, deformation and principal stress distribution on the outer surface, are shown in Figs 27.11 and 27.12, respectively.

Example 2

The pressure vessel shown in Fig. 27.13 was analysed for stresses produced by an internal pressure. It consisted of a cylinder and a head with holes. One twenty-fourth of it was treated by considering its symmetric structural ar-

Table 27.4 Preprocessor's functions in MISA

	Functions	(a)	(b)	(c)	(d)	(e)
		Applicable structures (see Table 27.3)				
	Structure data					
1	MLSS concept	A	A	A	A	A
2	Data generation for high level SSs	A	A	A	A	A
3	Copy	A	A	A	A	A
4	Determination of DOF	A	A	A	A	A
5	Interpolation	A	A	A	A	AL
6	SE library	A	A	A	A	A
7	Input coordinate systems	A	A	A	A	A
8	Unequal subdivision	NA	NA	A	A	A
9	Weighted subdivision	A	A	A	A	A
10	Prevention of obtuse triangles	NA	NA	A	A	A
11	Coordinate calculations for GJs	A	A	A	A	A
12	Ordering of joint number for BEs	A	A	A	A	A
13	Determination of the elimination levels for GJs	A	A	A	A	A
14	Plot of input structure	A	A	A	A	A
15	Plot of generated structure	A	A	A	A	A
	Load data					
1	Data generation by mapping function	A	A	A	A	A
2	Easy input of pressure loads	A	NA	A	A	A
3	Easy input of parallel inerita forces	A	A	A	A	A
4	Easy input of rotational inertia forces	A	A	A	A	A
5	Plot of input loads	PA	PA	PA	PA	PA
6	Plot of generated loads	PA	PA	PA	PA	PA
	Restraint data					
1	Data generation by mapping function	A	A	A	A	A
2	Input coordinate systems	A	A	A	A	A
3	Determination of elimination levels	A	A	A	A	A
4	Plot of input restraints	A	A	A	A	A
5	Plot of generated restraints	A	A	A	A	A
	Miscellany					
1	Renumbering					
2	Updating of input data					
3	Topological map of matrices					

A – Available, PA – Partially available (under development), AL = Available with some limitations, NA – Not available.

Table 27.5 Postprocessor's functions in MISA

	Functions	Applicable structure (see Table 27.3)				
		(a)	(b)	(c)	(d)	(e)
	Displacement					
1	Plot of deformations	A	A	A	A	A
2	Plot of displacement components	A	A	A	A	A
	Stress					
1	Plot of stress components	PA	PA	A	A	A
2	Plot of principal stresses	PA	PA	A	A	A
3	Plot of equivalent stresses	PA	PA	A	A	PA
4	Plot of surface stresses (for bending problems)	PA	NA	A	A	NA
5	Plot of contour lines	NA	NA	A	A	NA
6	Plot of colour intensity representation	NA	NA	A	A	NA
7	Plot of line intensity representation	NA	NA	A	A	NA
8	Selection of maximum stresses	A	A	A	A	A
	Temperature					
1	Plot of contour lines	NA	NA	A	NA	NA
2	Plot of values	A	A	A	A	A
	Miscellany					
1	Balance check lists	A	A	A	A	A

A – Available, PA – Partially available (under development), AL – Available with some limitations, NA – Not available.

Table 27.6 General processing functions in MISA

	Functions
1	Restart (at each stage of calculation)
2	Error check and error treatment
3	Enlargement and rotation of figures
4	Selective plots and lists
5	Load combinations
6	Control of calculation flows
7	Selection of calculation units
8	Simplified JCL
9	Arbitrary choice of output devices
10	Superposition of figures
11	Role of PPPs for other programs
12	Standardized data bank

Table 27.7 Modelling items

	KN	SE	GJ	BE	e_1 = KN/GJ	e_2 = SE/BE
Example 1	12	5	381	475	0.032	0.011
Example 2	24	5	660	420	0.036	0.012

Table 27.8 Number of input data cards

	Structure	Load	Restraint	Others	Total
Example 1	57	18	31	8	114
Example 2	67	6	12	9	94

rangement and load condition. Solid elements were used for the idealization of the structure. All of the structural, load and restraint data were also automatically generated from the simple input data shown in Fig. 27.13 as follows. The structure consisting of one SS was divided into five SEs as shown in Fig. 27.13. The sizes of the generated elements can be gradually changed by using the three-dimensional weighted subdivision function as shown in Fig. 27.14. The load and restraint data were automatically generated in the same way. The modelling items and the number of input data cards used for this example are shown in Tables 27.7 and 27.8, respectively. The efficiencies of data generation, e_1 for joints and e_2 for elements, are also shown in Table 27.7. Regarding representation of the calculation results, figures for three-dimensional structures are usually too complicated to understand visually. To avoid this difficulty, a selective drawing function is provided. Using this function, the principal stress distribution on the inner surface of the vessel is shown in Fig. 27.14. The partial representation enables us to study the state of stress concentration around the hole and the corner of the vessel.

27.7 Future Extension of MISA

As mentioned before, MISA has been developed as a highly user-oriented program system including two analysis functions. The main feature involved in future projections of MISA is the further extension of its PPPs functions.

In general, PPPs are required to have the following functions: (a) conver-

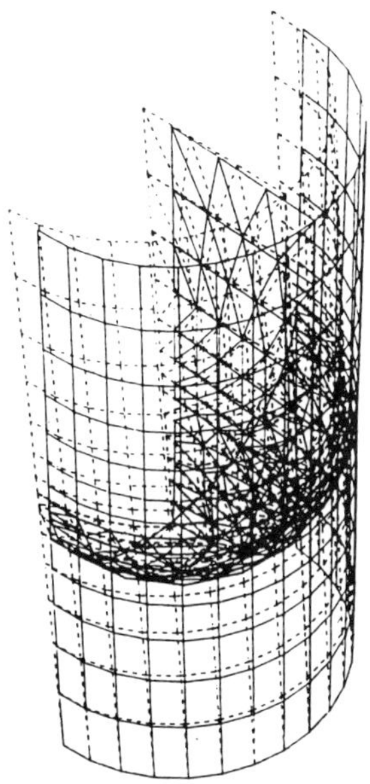

Fig. 27.11 Deformation.

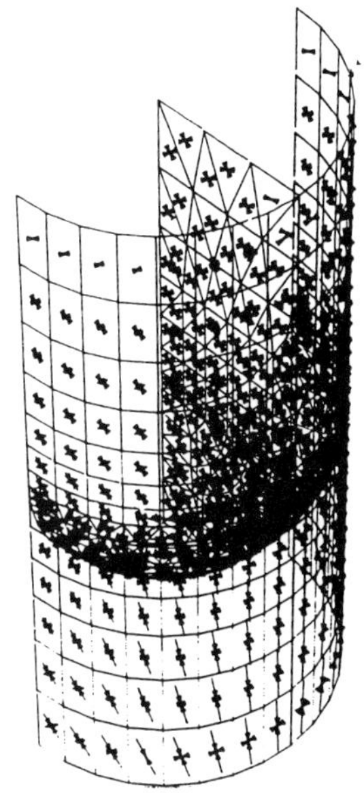

Fig. 27.12 Principal stresses on the outer surface.

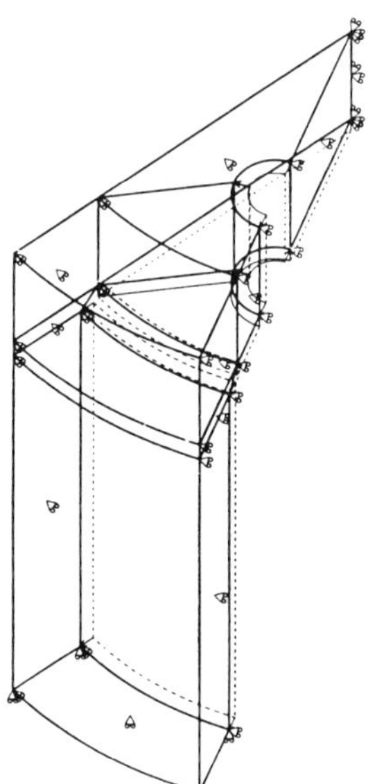

Fig. 27.13 Input structure, load and restraint.

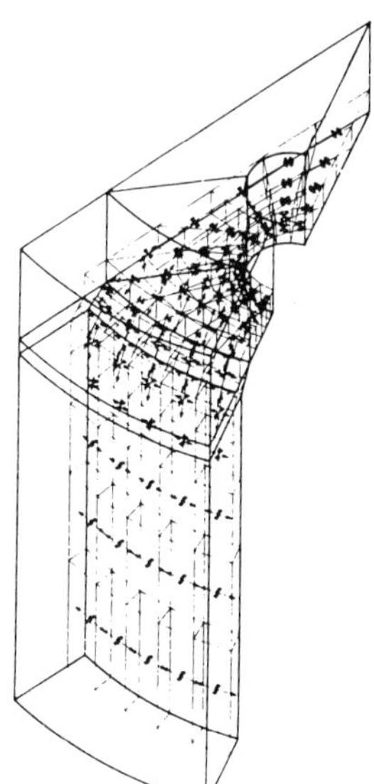

Fig. 27.14 Principal stresses on the inner surface.

sational mode; (b) data generation; (c) graphic supports; and (d) support of other programs.

Concerning item (a), the remote batch processing system has been adopted for MISA. However, in order to improve the efficiency of the actual structural design, the preparation of input data and the display of calculation results by interactive modes, including free format features and graphic CRTs are being investigated for use in MISA.

Item (b) is one of the most highly developed features of MISA compared with other general purpose programs. But its continuous improvement will be required in order to satisfy the various users' needs, such as modification to fit suitable programs, and the introduction of other elements or load conditions.

Concerning item (c), MISA can support five kinds of output devices which are line printers, graphic COMs, dot printers, drum type plotters and flat bed type plotters. In connection with item (a), MISA is now being modified to be able to support CRT graphic displays.

Concerning item (d), at present the PPPs of MISA can support the following special and general purpose programs by means of the standardized MISA data bank system in which generated input data and calculation results are stored:

(i) HEAT-RING. Used for unsteady state heat conduction analysis of axisymmetric structures.
(ii) POST-HEAT. Used for the discrimination and selection of the case of the severest temperature distribution against thermal stresses from the results obtained by the HEAT-RING.
(iii) POST-1592. Used for strength evaluation according to ASME Code case 1592 of three-dimensional or axisymmetric pressure vessels used in a creep temperature range.
(iv) POST-STRESS. Used for strength evaluation according to ASME Code Section III and Section VIII Div. 2 of three-dimensional or axisymmetric pressure vessels.
(v) EP-RING. Used for elasto-plastic analysis of axisymmetric structures.
(vi) ICES/STRUDL (McAuto Version).

In order to execute these programs, special appropriate data must be added to the data generated by MISA. These programs must, therefore, be run as separate jobs from MISA, but the independence of these programs is maintained.

The number of programs currently supported is limited, but will be widened step by step in accordance with users' needs. As a result, users will no longer be restricted to particular structural programs whose input data they are familiar with. Users can prepare input data for a designated analysis program by using one common language for input data generation, the MISA language.

Acknowledgements

The authors wish to express their gratitude to Dr N. Tanaka, Messrs T. Mii, M. Itoh and T. Yoshiki for their supervision in the developing stages. The authors are especially grateful to Messrs S. Nishihara, D. Nomura, H. Hara and Y. Kodera for their work described in this paper.

References

1. Herness, E. D. and Tocher, J. L., *Design of Pre- and Post-Processors, Structural Mechanics Computer Programs*, pp. 887–898. University Press of Virginia (1974).
2. *The NASTRAN User's Manual* (Level 15.5). The MacNeal-Schwendler Co. (1974).
3. *ICES STRUDL-II The Structural Design Language Engineering User's Manual*. MIT Dept of Civil Engineering.
4. *MARC-CDC User Information Manual*, Vol. II. MARC Analysis Co. (1975).
5. Imafuku, I. *et al.* "A generalized automatic mesh generation scheme for the finite element method", *Int. J. Num. Meth. Engng.* **15**, 713–731 (1980).
6. Tanaka, N., *et al.*, "The Mitsui Structural Analysis Program System (MISA) by the Finite Element Method", *Japan Shipbuilding & Marine Engng* **10**(2), 5–22 (1976).

Index